UNIVERSITY LIBRAR
UW-STEVENS POINT
W9-BIY-555

ADVANCES IN CHEMICAL PHYSICS

VOLUME 118

Advances in CHEMICAL PHYSICS

Edited by

I. PRIGOGINE

Center for Studies in Statistical Mechanics and Complex Systems
The University of Texas
Austin, Texas
and
International Solvay Institutes
Université Libre de Bruxelles
Brussels, Belgium

and

STUART A. RICE

Department of Chemistry
and
The James Franck Institute
The University of Chicago
Chicago, Illinois

VOLUME 118

AN INTERSCIENCE® PUBLICATION
JOHN WILEY & SONS, INC.
NEW YORK • CHICHESTER • WEINHEIM • BRISBANE • SINGAPORE • TORONTO

This book is printed on acid-free paper. ♾

An Interscience® Publication

Published simultaneously in Canada.

For ordering and customer service, call 1-800-CALL WILEY

Library of Congress Catalog Number: 58-9935

ISBN 0-471-43816-2

Printed in the United States of America.

10 9 8 7 6 5 4 3 2 1

CONTRIBUTORS TO VOLUME 118

IGOR V. KHMELINSKII, Universidade do Algarve, Faro, Portugal

VLADIMIR I. MAKAROV, Institute of Chemical Kinetics and Combustion, Novosibirsk, Russia; Department of Chemistry, University of Puerto Rico, San Juan, Puerto Rico

SUSUMU OKAZAKI, Department of Electronic Chemistry, Tokyo Institute of Technology, Yokohama, Japan

ALEXEI A. STUCHEBRUKHOV, Department of Chemistry, University of California, Davis, CA

WOLFGANG WERNSDORFER, Laboratoire Louis Néel, Grenoble, France

INTRODUCTION

Few of us can any longer keep up with the flood of scientific literature, even in specialized subfields. Any attempt to do more and be broadly educated with respect to a large domain of science has the appearance of tilting at windmills. Yet the synthesis of ideas drawn from different subjects into new, powerful, general concepts is as valuable as ever, and the desire to remain educated persists in all scientists. This series, *Advances in Chemical Physics*, is devoted to helping the reader obtain general information about a wide variety of topics in chemical physics, a field that we interpret very broadly. Our intent is to have experts present comprehensive analyses of subjects of interest and to encourage the expression of individual points of view. We hope that this approach to the presentation of an overview of a subject will both stimulate new research and serve as a personalized learning text for beginners in a field.

I. Prigogine
Stuart A. Rice

CONTENTS

TOWARD AB INITIO THEORY OF LONG-DISTANCE ELECTRON TUNNELING IN PROTEINS: TUNNELING CURRENTS APPROACH

ALEXEI A. STUCHEBRUKHOV

Department of Chemistry, University of California, Davis, CA

CONTENTS

Advances in Chemical Physics, Volume 118, Edited by I. Prigogine and Stuart A. Rice.
ISBN 0-471-43816-2

I. INTRODUCTION

The studies of long-range electron tunneling (ET) in proteins [1–7] have added a new dimension to the diverse subject of nonadiabatic electron transfer reactions [8, 9]. In addition to traditional questions about reorganization of the medium, driving force, quantum and classical effects, and the magnitude of the transfer matrix element, which have to be discussed now in the context of a specific biological application, conceptually new questions arise about (1) the nature of long-distance electronic coupling, (2) tunneling pathways, and (3) the role of protein dynamics in electronic interactions of donor and acceptor complexes. This chapter presents a review of one theoretical approach that allows one to gain some insights into these new questions. More comprehensive reviews of both theoretical and experimental work in this area can be found in two previous volumes of the *Advances in Chemical Physics* dedicated to electron transfer [10].

The theoretical model that will be discussed consists of a disordered organic medium, such as a protein, with embedded donor and acceptor redox complexes, which are separated by 10–20 Å or more, so that there is no direct contact between them. The redox complexes will typically contain transition metal ions which change their oxidation state upon electron transfer. By electron transfer we generally mean that the total charge of the redox complexes changes by one unit upon reaction. The interpretation of the reaction in terms of electron or hole transfer will be discussed later in this chapter.

We will assume that we know positions of all atoms in the system; the approximate equilibrium configuration is known from crystallographic data for protein structure, and the thermal motion of the protein around equilibrium can be predicted by molecular dynamics simulation. Generally, the system will include not only the protein itself but also the surrounding and internal solvent water molecules.

A prototypical system of this sort is cytochrome c oxidase [11], the terminal enzyme of the electron transport chain of the cell respiratory system. In this protein complex, three metal redox centers are separated by the distance of the order 15–20 Å, and the crystal structure is known to 2.2 Å resolution. Electrons are transferred in this enzyme sequentially from one redox site to another, down the free energy gradient, toward molecular oxygen, the final acceptor of electrons. Another example is Ru-modified metalloproteins, studied by Gray and co-workers. Their system consists of a native protein with a metal ion (Fe(II/III) (cytochrome) or Cu(I/II) (azurin) buried inside the protein body, along with

Ru(III/II)(Bpy2)Im complex attached to a surface His residue. Electron transfer occurs between a reduced native metal center and the oxidized Ru-complex [4–7]. Depending on the position of the Ru-complex on the surface of the protein, the distance between donor and acceptor varies in the range of 14–25 Å. Recently, the crystal structures of some of these modified proteins have become available, and the reaction has been studied both in solution and in the crystal, under well-controlled conditions. Yet another example is much discussed photosynthetic reaction center [2, 3].

We will assume that the donor complex, D, and acceptor complex, A, can be in two oxidation states—reduced and oxidized—and there are three thermodynamically stable states: (1) both D and A are reduced, (2) both D and A are oxidized, and (3) D is oxidized and A is reduced.

The charge transfer reaction is initiated by preparing a nonequilibrium state, by quickly introducing an extra electron on the donor site in a completely oxidized system, or quickly removing one electron from the acceptor site in totally reduced system. The ET reaction is a thermal transition of the system to the most thermodynamically stable state, where acceptor is reduced and donor is oxidized. CcO and Ru-modified proteins are examples of the first and the second types of initiation of the reaction, respectively.

The major mechanism of charge transfer in proteins (e.g., in the examples cited above and many others) is thermally assisted electron (hole) tunneling [8, 9]. A similar mechanism is responsible for electron transport in doped semiconductors at low temperatures, where it is known as variable range hopping [12]. An electron hop between a pair of impurity atoms in a semiconductor is equivalent to charge transfer between a pair of redox sites in proteins.

In semiconductors, at sufficiently low temperatures the excitation of electrons/holes into the conduction band is negligible, and electron transport can only occur via direct transfer of electrons between localized states of the impurity atoms. These localized and weakly coupled states form a narrow transport band which is located in the gap between the valence and the conduction band of a semiconductor.

In proteins, which electronically resemble disordered semiconductors, there also exists an energy gap of several electron volts between the occupied states and the excited states. The energy levels of donor and acceptor are located in this gap, and the energies of the system are such that thermally accessible states are only those specified above: D is reduced and A is oxidized, or D is oxidized and A is reduced. All other electronic states, such as both donor and acceptor reduced at the expense of ionization of a protein medium, or, both donor and acceptor oxidized with one extra electron in the protein medium, are of much higher energy in natural redox proteins.

Other, hopping mechanisms are, of course, also possible, in particular in photoinduced reactions or in “molecular wires” [10, 13–16]. For example, charge transport in DNA is believed to occur via such a mechanism [17]. In general, such reactions involve a significant input of energy, via a photoexcitation or a voltage bias, so that an electron from the donor complex (or a hole from the acceptor) can be directly injected into the "conduction band" of unoccupied orbitals of the system.

A thermal excitation of the transferring electron to the conduction band of the protein is typically very small, and therefore the hopping transfer contribution to the overall rate can be neglected, compared with that of the tunneling transfer. It should be noted, however, that the rate of tunneling transfer falls exponentially with distance between donor and acceptor, while the rate of the hopping transport is less sensitive to distance. Therefore, formally, the latter will always prevail at asymptotically large distances [14].

From a biological perspective, the functioning of redox proteins in electron transport chains via the tunneling mechanism makes perfect sense. Indeed, such a mechanism provides stability of redox complexes (no electrons are lost due to thermal auto-ionization of redox centers), along with maximum efficiency of passing electrons to a specific partner in the chain. (The hopping mechanism—that is, diffusion of electrons between redox centers—would be less efficient in this sense.) Interestingly, the idea that electron tunneling is involved in charge transfer in proteins was proposed by DeVault and Chance in the early 1960s [18], when they observed that at very low temperatures the ET rate does not vanish and becomes independent of temperature. However, a direct experimental evidence of this has been obtained only recently (see, for example, Refs. 3 and 6).

In this chapter, we will be concerned with the tunneling process—that is, the case when only two electronic states of the system are involved in the transition. The key role in the coupling of donor and acceptor electronic states is played by the protein medium in which the reaction occurs. In vacuum, such a long-distance coupling would not be possible because the tunneling barrier would be too high and the overlap of electronic orbitals of the redox centers would be too small. In the protein medium, however, there always exists a number of sequentially coupled intermediate *virtual* states, which result in superexchange coupling of the remote donor and acceptor sites (see reviews [19(a–c)] and references therein). These virtual states effectively lower the potential barrier providing the "tunneling bridge" along which electrons can tunnel with appreciable rates, sufficient for the biological function of redox proteins. The idea of superexchange has emerged in the early work of McConnel [20], Larsson [21], Marcus [8], Newton [19], and others (see review [19a] for a detailed account).

For a number of years, a major question has been whether specific electron tunneling pathways [22–24] exist. This question is still debated in the literature [2, 5], because the pathways themselves are not observed directly, and the interpretation of experimental results on ET rates involves ambiguities. The extremely small tunneling interactions are difficult to calculate accurately. Current theoretical calculations indicate that the specific paths do exist in static protein structures. However, protein motions can result in significant averaging of spatial patterns, and it is not clear how accurately subtle quantum interference effects are described by the present theories. The key to resolving these issues is to perform accurate, first principles calculations of electron tunneling that include dynamics of the protein.

The theoretical challenges here are:

- To develop dependable first principles electronic structure methods that capture the many-electron nature of the problem correctly, and compute the extremely small electronic couplings in proteins.
- To include protein dynamics in electron tunneling calculations and evaluate the dynamic effects.
- To perform calculations on specific protein systems and provide quantitative interpretation and predictions of experimental results.

In this chapter, I briefly describe the method of tunneling currents [25–34] which we developed for the description of electron tunneling in proteins. This method provides a rigorous theoretical framework within which one can address the above challenges.

The method also provides a rigorous solution to the problem pioneered by Beratan and Onuchic: how to find electron tunneling pathways in proteins [22]. Their first pathway model (BO model) offered a simple path-finding strategy, by counting the number of jumps that an electron will make through covalent bonds, hydrogen bonds, or van der Waals contacts, along a tunneling path, and assigning specific weights to each of such jumps. This simple model has been extremely useful for a qualitative analysis of tunneling paths [23, 24]. However, its essentially empirical character imposes natural limitations both on the accuracy and the extent to which such a model can be improved. For example, this model does not include quantum interferences, or the properties of the wave functions of donor and acceptor complexes. Also, it leaves unresolved the issue of the absolute value of the coupling matrix element, the quantity which controls the rate of electron transfer and which can be directly related to experimental data. In an attempt to address these issues, therefore, different and more rigorous quantum approaches have been developed recently by several authors, including the authors of the BO pathway model themselves [23, 24].

The method of tunneling currents is one such approach. It offers a completely different perspective on the problem. Both the tunneling pathways and electronic coupling can be rigorously described now in one formalism, and the method can be implemented using one- or many-electron formulations of the problem. The latter is very important, because accuracy can now be improved without limitation, using more sophisticated levels of electronic structure treatment, while retaining the same tunneling formalism.

We will begin with a review of one-electron methods for the evaluation of electronic tunneling matrix element of very large systems.

II. ONE-ELECTRON THEORY OF LONG-DISTANCE TUNNELING

A. Evaluation of the Tunneling Matrix Element for Very Large Systems

It is difficult to do accurate first principles calculations on whole proteins—they are too big for such treatment. Although some progress has recently been made with linear-scaling electronic structure methods [35] (for standard ground state calculations), and it is possible that one will be able to apply such methods in electronic tunneling problems in the future, the analysis shows that in fact such calculations are unnecessary. The tunneling nature of electronic communication between distant redox sites makes the size of the protein region involved in propagating the tunneling electron relatively small. Therefore, a general approach, which has been commonly used, is to perform the analysis of the proteins in two stages: first to examine the whole protein with one-electron methods, to identify the most important parts of the protein, and to simplify the system—the procedure that we call protein pruning—and then to proceed with more accurate, many-electron *ab initio* treatments. The number of amino acids actually involved in propagating the tunneling electrons turns out not to be very large. The problem then is to identify these amino acids. This goal can be achieved with approximate one-electron methods.

The first stage of the analysis deals with the whole protein. Here, the problem is to devise approximate one-electron Hamiltonians and to develop methods for the effective computation of the tunneling matrix elements of extremely large systems. The main strategy is to do calculations without diagonalization or even inversion of large matrices. A number of techniques have been proposed for the solution of this problem.

1. In a series of studies [36–38] a technique has been developed for computation of the perturbation theory expression for the transfer matrix element in the donor–bridge–acceptor systems [39–42].

$$T_{DA}^{(0)} = \sum_{ij} V_{ai} G_{ij}^{B} V_{jd} \tag{2.1}$$

where V represents the coupling of donor and acceptor complexes to the protein, G^B is the Green function of the tunneling electron in the bridge (i.e., protein), and E is the tunneling energy. Effects of nonorthogonality of atomic orbitals can be included in V and G [23, 37]. A simple method based on sparse matrix techniques that avoids both diagonalization and inversion of large matrices has been developed for computation of the above expression. The computations are reduced to solving a system of sparse linear equations. The solution of the latter problem can rely upon a variety of numerical iterative techniques. The simplest version of the conjugated gradient method, for example, is utilized in the code developed by this author and co-workers. There are other attractive numerical alternatives, such as GMRes, a method developed in fluid dynamics specifically for dealing with extremely large (up to the order of $10^6 * 10^6$) sparse linear systems [37]. With such methods there is practically no limitation on the size of the proteins that can be treated with one-electron Hamiltonians.

2. The above expression for the tunneling matrix element is only the lowest-order approximation in coupling V of the series that has the form

$$T = V + VGV + VGVGV + VGVGVGV + \cdots \tag{2.2}$$

The coupling V, of course, is not small in real systems, and the neglect of the higher terms is always questionable. We have recently shown [43] that the summation of *all* terms of the perturbation series results in the following expression for the transfer matrix element:

$$T_{DA} = \frac{T_{DA}^{(0)}}{\sqrt{(1 - \Sigma'_{aa})(1 - \Sigma'_{dd})}} \tag{2.3}$$

where $T_{DA}^{(0)} = \Sigma_{da}$ is the lowest-order expression given above, and Σ_{aa} and Σ_{dd} are self-energies of the Green's functions of the tunneling electron in donor and acceptor states, and prime denotes the derivative in energy. The corrections account for the delocalization of *the (diabatic) donar and acceptor states* in the protein medium. The expressions for self-energies Σ_{aa} and Σ_{dd} have the same form as $T_{DA}^{(0)} = \Sigma_{da}$ and can be computed using the same technique of sparse matrices as for $T_{DA}^{(0)}$ discussed above. Calculations tested on Ru-modified proteins [4, 5] showed that the new expression gives results which are nearly identical to those obtained with exact diagonalization of the Hamiltonian matrix, even when intermediate resonances are present in the medium (in contrast to $T_{DA}^{(0)}$, which diverges for such cases). The advantage of the former method, of course, is that it can be applied to systems that may be too large for a direct diagonalization.

3. The electronic coupling T_{DA} contains contributions of both electron and hole transfers [19],

$$T_{DA} = T^e_{DA} + T^h_{DA} \tag{2.4}$$

The relative contribution of hole and electron transfers is an important issue [19], which provides insight into the nature of the coupling. In Ref. 36 a method has been developed which allows one to calculate separate contributions of hole and electron transfer to the total amplitude T_{DA} without explicit calculation of molecular orbitals of the protein, and without diagonalization of the Hamiltonian matrix. The method treats the above expression for complex tunneling energies, and it uses analytic properties of the tunneling amplitude in the complex energy plane.

The methods described above effectively allow one to (a) accurately compute electronic tunneling matrix elements for an arbitrary large protein system and (b) find separate electron and hole contributions without recourse to diagonalization or even inversion of large matrices. These methods are suitable for any effective one-electron Hamiltonian description of the proteins. (The development of such Hamiltonians is a separate problem [23, 44], which will be discussed later in this chapter.) The computations of the electronic coupling can be done very effectively either for a given configuration of the protein or for a series of dynamic configurations along its dynamic trajectory [24, 45].

B. Protein Pruning

In the analysis of long-distance electronic coupling in proteins, the first step, which perhaps is the most important one from an experimental point of view, is to identify the amino acids that are involved in propagation of the tunneling electron. Several groups have developed different strategies to achieve this goal [23, 41]. Our method is called protein pruning. The idea is to probe sensitivity of the electronic coupling to computer-induced changes in the protein. The calculations at this stage of the analysis are performed at the semiempirical one-electron level using the methods described above. There is no limitation on the size of the protein that can be analyzed in this way. Although electronic coupling calculated at this stage is not very accurate, due to approximations in electronic structure treatment, the method can identify amino acids that are *not* important for the coupling. These amino acids can be deleted from the protein to simplify the model. Thus, the essence of the method is to eliminate unimportant amino acids (i.e., to prune the protein) and to identify the important ones. The details of such calculations have been discussed in several publications (e.g., Refs. 38 and 46–50).

The pruning procedure naturally leaves donor and acceptor complexes intact, and identifies a number of amino acids that make up the tunneling bridge which connects the redox sites. Typically the pruned molecule contains 10–20 amino acids and redox complexes. The method has been tested in calculations of several Ru-modified proteins [38], ferredoxin, and has been utilized in our recent studies of DNA photorepair enzyme photolyase [49, 51], and cytochrome c oxidase [50].

C. Interatomic Tunneling Currents and Tunneling Paths

The procedure described above can identify the amino acids that are involved in electronic coupling of distant redox complexes. More detailed information about the coupling and the associate tunneling process can be obtained by using the method of atomic tunneling currents [26–29, 34]. The idea is to calculate the relative probabilities that the tunneling electron will pass through various atoms of the protein during the transition from donor to acceptor. This problem has an exact quantum mechanical solution both in one-electron [26] and many-electron formulations [28, 29]. As other pathway models [22–24], the method of tunneling currents aims to obtain information about the tunneling process at the atomic level of description of the protein.

The general idea of the approach is to examine dynamics of charge redistribution in the system "clamped" at the transition state. The tunneling dynamics is described by

$$|\Psi(t)\rangle = \cos(T_{DA}t/\hbar)|D\rangle - i\sin(T_{DA}t/\hbar)|A\rangle \tag{2.5}$$

where $|D\rangle$ and $|A\rangle$ are diabatic states corresponding to localization of the tunneling electron on the donor and acceptor complexes respectively, and T_{DA} is the transfer matrix element. (Notice that both one- and many-electron formulations of the problem will have the same form of the above state.) The periodic change of the wave function from the donor to acceptor state (wave function "perestroika") results in periodic variations of charge distribution in the system. The redistribution of charge is associated with current. It is this tunneling current that we are interested in tunneling calculations.

The redistribution of charge in the system during the tunneling transition can be described in terms of current density $\vec{j}(r,t)$ and its spatial distribution $\vec{J}(r)$ [28],

$$\vec{j}(\vec{r},t) = -\vec{J}(\vec{r})\sin\frac{2T_{DA}t}{\hbar}, \qquad \vec{J}(\vec{r}) = -i\langle A|\hat{\vec{j}}(\vec{r})|D\rangle \tag{2.6}$$

or, in terms of *interatomic* currents J_{ab}, which are introduced as [25, 29]

$$\frac{dP_a}{dt} = \sum_b j_{ab}, \qquad j_{ab} = -J_{ab}\sin\frac{2T_{DA}t}{\hbar} \tag{2.7}$$

where P_a are atomic charges. The total current through an atom is proportional to the probability that the tunneling electron will pass through this atom during the tunneling jump from donor to acceptor [25]. The streamlines of the current $\vec{J}(r)$ represent the whole manifold of Bohminan trajectories [52]. Both interatomic currents J_{ab} and current density $\vec{J}(\vec{r})$ provide full information about the tunneling process and, in particular, about the distribution of the tunneling current in space—that is, about the tunneling pathways (see Figs. 2, 4, and 5 below).

In addition, remarkably, both $\vec{J}(\vec{r})$ and J_{ab} are related to the tunneling matrix element [25–37]:

$$T_{DA} = -\hbar \sum_{a \in \Omega_D, b \notin \Omega_D} J_{ab} = -\hbar \int_{\partial\Omega_D} (d\vec{s} \cdot \vec{J}) \tag{2.8}$$

In the above formula, Ω_D is the volume of space that comprises the donor complex, and $\partial\Omega_D$ is its surface.

In the one-electron approximation, the expressions for tunneling currents have the following form:

$$\vec{J}(\vec{r}) = \frac{\hbar^2}{2m}(\psi_D \nabla \psi_A - \psi_A \nabla \psi_D) \tag{2.9}$$

where ψ_D and ψ_A are donor and acceptor diabatic states, and

$$J_{ai,bj} = \frac{1}{\hbar}(H_{ai,bj} - E_0 S_{ai,bj})(C^D_{ai} C^A_{bj} - C^A_{ai} C^D_{bj}) \tag{2.10}$$

where ai and bj are indices of two atomic orbitals on atoms a and b, H and S are the Hamiltonian and overlap matrices, and C^D and C^A are the coefficients of expansion of states $|D\rangle$ and $|A\rangle$ in the atomic basis set of the system. The total interatomic current between two atoms, J_{ab}, is a sum of $J_{ai,bj}$ over orbitals of these atoms.

The sign of the current is related to its direction; thus, a positive J_{ab} corresponds to current from atom b to atom a.

The total current through an atom is proportional to the probability that the tunneling electron will pass through this atom during the tunneling jump from donor to acceptor. The total atomic current is given by the sum

$$J^+_a = \sum_b{}' J_{ab} \tag{2.11}$$

where the summation is limited only to positive contributions J_{ab}, which describe tunneling current from various atoms b to atom a.

Both the interatomic currents J_{ab} and the total atomic currents J_a^+ can be utilized for visualization of the tunneling process and tunneling pathways. For example, the magnitude of J_a^+ can be taken as an indicator that the atom is involved in the tunneling process (see, for example, Refs. 26, 49, and 50 and; also see Fig. 2 below).

The information about all tunneling paths and their interferences is contained in the matrix J_{ab}, which describes the total tunneling flow in an atomic representation. The analysis of the tunneling flow gives an accurate and rigorous description of where electronic paths are localized in space. For example, if a specific atomic path exists, one can find it using the method of steepest descent; that is, begin from a donor atom, d, and find an atom $b1$ to which the current $J_{b1,d}$ is maximum, then go to atom $b1$, repeat the procedure and find atom $b2$, and so on, until the acceptor atom a is reached. The sequence of atoms d, $b1$, $b2, \ldots, bn$, a is the tunneling path. Of course, this procedure will work only if a single atomic path exists. Usually, the structure of the tunneling flow is more complicated, and many interfering paths exist simultaneously. A more careful analysis of J_{ab} is required in this case.

In addition to pathways analysis, the interatomic currents allow one to calculate, accurately, very small electronic couplings without using perturbation theory. Later in the chapter, we will consider an example of *ab initio* calculation of electronic coupling as small as $10^{-4}\,\mathrm{cm}^{-1}$ using the above methods [34] (see Figs. 4, 6, and 7 below). The accurate evaluation of such small interaction energies from *ab initio* calculations is the main theoretical challenge. Indeed, an alternative straightforward approach could be based on a well-known relation [19]

$$T_{DA} = H_{DA} - E_0 S_{DA} \tag{2.12}$$

where $\hat{H}$ is the Hamiltonian of the system at the transition state configuration, $\hat{H}(Q^\dagger)$, $S_{DA} = \langle D|A\rangle$ is the overlap of the two states, and $E_0 = \langle D|\hat{H}|D\rangle = \langle A|\hat{H}|A\rangle$ is their common resonance energy. The application of the above formula to large systems is impeded, however, by numerical difficulties. The problem is that the two terms in Eq. (4.7) by themselves are large numbers, of the order of $10^4\,\mathrm{cm}^{-1}$, and they almost completely cancel each other since T_{DA} is typically of the order of 10^{-2}–$10^{-4}\,\mathrm{cm}^{-1}$. (In particular, the resonant energy of two states is difficult to calculate accurately.) Therefore, the straightforward method of Eq. (4.7) is expected to fail when the coupling matrix element is small.

Two other *ab initio*-based methods—the Green's function-based method and the GMH method—have been recently developed [23, 44, 53, 54]. Both methods can be used in a semiempirical context as well. Direct methods [based on Eq. (4.7)] have been applied in the past to rather small systems (e.g., Refs. 19, 55, and 56), where the coupling is not difficult to calculate. The tunneling currents approach provides yet another perspective on the problem and introduces new computational techniques for studies of electron tunneling in molecular systems.

Several key points of the formalism of tunneling currents in many-electron formulation are described in the following section.

III. ELECTRON TUNNELING IN MANY-ELECTRON FORMULATION

A. Current Density Operator

The main idea of this approach is the same in one- and many-electron formulations: We would like to describe the time evolution of charge distribution in the system during the tunneling transition. One approach is to trace the time evolution of the local charge density. The changes of charge distribution are described in terms of currents. The corresponding formalism is developed as follows.

The current density of N classical particles with coordinates $\vec{x}_i(t)$ moving with velocities $\vec{v}_i(t) = d\vec{x}_i(t)/dt$ is given by

$$\vec{j}(\vec{x}) = \sum_{i=1}^{N} \delta(\vec{x} - \vec{x}_i)\vec{v}_i \tag{3.1}$$

The quantum generalization of the above equation results in the following expression for the *operator* of current density:

$$\hat{\vec{j}}(\vec{x}) = \frac{1}{2}\sum_{i=1}^{N} \left[\delta(\vec{x} - \vec{x}_i)\frac{\hat{\vec{p}}_i}{m} + \frac{\hat{\vec{p}}_i^{\,+}}{m}\delta(\vec{x} - \vec{x}_i)\right] \tag{3.2}$$

where m is the mass of particles (electrons in our case) and $\hat{p}_i$ is the momentum operator of the ith electron, $-i\hbar\partial/\partial x_i$. The symmetrization of the classical expression makes the current density operator Hermitian. The Hermitian conjugated operator $\hat{p}_i^+$ is assumed to be acting on the left. The total coordinate representation of the above expression thus has the following form:

$$\hat{\vec{j}}(\vec{x}) = \frac{\hbar}{2mi}\sum_{i=1}^{N} \left[\delta(\vec{x} - \vec{x}_i)\frac{\partial}{\partial \vec{x}_i} - \frac{\partial^+}{\partial \vec{x}_i}\delta(\vec{x} - \vec{x}_i)\right] \tag{3.3}$$

Thus, the current density has the form of a one-electron operator. In the same representation, the electron density operator reads

$$\hat{\rho}(\vec{x}) = \sum_{i=1}^{N} \delta(\vec{x} - \vec{x}_i) \tag{3.4}$$

which is a one-electron operator as well. We can now apply the standard technique to calculate matrix elements of these operators for many-electron wave functions.

An alternative approach to the above formalism is to use the second quantization technique, as in Ref. 28. Using the field operators $\hat{\psi}(\vec{x})$ and $\hat{\psi}^{+}(\vec{x})$ (see, for example, Refs. 57 and 58) the above operators can be written as

$$\hat{\rho}(\vec{x}) = \hat{\psi}_{\sigma}^{+}(\vec{x})\hat{\psi}_{\sigma}(\vec{x}) \tag{3.5}$$

and

$$\hat{\vec{j}}(\vec{x}) = \frac{\hbar}{2mi}(\hat{\psi}_{\sigma}^{+}(\vec{x})\nabla\hat{\psi}_{\sigma}(\vec{x}) - \hat{\psi}_{\sigma}^{+}(\vec{x})\nabla^{+}\hat{\psi}_{\sigma}(\vec{x})) \tag{3.6}$$

where summation is assumed over the repeating spin index σ. The formalism of second quantization for treatment of the above expressions (we are basically interested in calculation of the matrix element of the above operators) is somewhat different from that based on the coordinate representation (see Ref. 28). Both, of course, lead to the same results. The coordinate representation is more convenient when nonorthogonal basis sets are used. Therefore, in what follows, we use a more conventional coordinate technique, adopted in the quantum chemistry literature.

B. Tunneling Dynamics in Terms of $\vec{J}(\vec{x})$

The idea of the method is to examine spatial distribution of the current density in a tunneling transition. Suppose two resonant diabatic electronic states $|D\rangle$ and $|A\rangle$ corresponding to localization of the tunneling electron on the donor and acceptor complexes respectively, are coupled by the transfer matrix element T_{DA} [19]. Then, if the tunneling electron is localized initially in the donor state, $|D\rangle$, later in time the total electronic wave function will evolve into a linear combination of states $|D\rangle$ and $|A\rangle$ as follows:

$$|\Psi(t)\rangle = \cos(\theta)|D\rangle - i\sin(\theta)|A\rangle \tag{3.7}$$

where

$$\theta = \frac{T_{DA}}{\hbar}t \tag{3.8}$$

We wish to examine current density and electron density in such a state. For local density we find

$$\rho(\vec{x}, t) = \langle\Psi(t)|\hat{\rho}(\vec{x})|\Psi(t)\rangle = \cos^2(\theta)\langle D|\hat{\rho}(\vec{x})|D\rangle + \sin^2(\theta)\langle A|\hat{\rho}(\vec{x})|A\rangle \tag{3.9}$$

The local density changes in time as follows:

$$\frac{\partial\rho(\vec{x},t)}{\partial t} = \frac{T_{DA}}{\hbar}\left(\langle A|\hat{\rho}(\vec{x})|A\rangle - \langle D|\hat{\rho}(\vec{x})|D\rangle\right)\sin 2\theta \tag{3.10}$$

On the other hand, the local current density in the same state is

$$\vec{j}(\vec{x},t) = \langle\Psi(t)|\hat{\vec{j}}(\vec{x})|\Psi(t)\rangle = i\langle A|\hat{\vec{j}}(\vec{x})|D\rangle\sin 2\theta \tag{3.11}$$

As expected, both density and local current are changing in time periodically with the same frequency, $2T_{DA}/\hbar$.

In both expressions, the time dependence is defined by the factor $\sin 2\theta$. The rest of the expression gives the amplitude of oscillations, which as a function of coordinates gives a spatial distribution of the current/density in the whole system. The spatial part of the current, $\vec{J}(\vec{x})$, is introduced as follows:

$$\vec{j}(\vec{x},t) = -\vec{J}(\vec{x})\sin 2\theta \tag{3.12}$$

Given this definition, the spatial distribution is

$$\vec{J}(\vec{x}) = -i\langle A|\hat{\vec{j}}(\vec{x})|D\rangle \tag{3.13}$$

Using the conservation equation for current,

$$\frac{\partial\hat{\rho}(\vec{x})}{\partial t} = -\text{div}\,\hat{\vec{j}}(\vec{x}) \tag{3.14}$$

and also using the expressions for $\rho(\vec{x},t)$ and $\vec{j}(\vec{x},t)$, we find a relation between local density and current density in a tunneling system:

$$\frac{T_{DA}}{\hbar}\left(\langle A|\hat{\rho}(\vec{x})|A\rangle - \langle D|\hat{\rho}(\vec{x})|D\rangle\right) = \text{div}\,\vec{J}(\vec{x}) \tag{3.15}$$

Finally, surrounding the donor complex by some closed surface, $S_D = \partial\Omega_D$, that will run sufficiently far from it so as to include most of the charge density corresponding to a tunneling electron on the donor site, along with integrating the Eq. (3.15) over the volume comprised by S_D, Ω_D, one finds the key relation between the tunneling matrix element and current density:

$$T_{DA} = -\hbar\int_{S_D} d\vec{s}\,\vec{J}(\vec{x}) \tag{3.16}$$

Equations (3.13) and (3.16) are the most important relations of the method that are used in the calculations described in the following sections. Notice the same form of the above relation both in one- and many-electron formulations.

C. Calculation of $\vec{J}(\vec{x})$. Hartree–Fock Approximation

Suppose states $|D\rangle$ and $|A\rangle$ are one-determinant many-electron functions, which are written in terms of (real) molecular orbitals $\varphi^D_{i\sigma}$ and $\varphi^A_{i\sigma}$ with corresponding spin orbitals $\chi^D_{i\sigma}$ and $\chi^A_{i\sigma}$, where σ is the spin index, $\sigma = \alpha, \beta$. These orbitals are the optimized orbitals obtained from Hartree–Fock calculations of states D and A. The D and A states then have the following form:

$$|D\rangle = |\chi^D_{1\alpha} \cdots \chi^D_{p\alpha} \chi^D_{1\beta} \cdots \chi^D_{q\beta}\rangle \tag{3.17}$$

$$|A\rangle = |\chi^A_{1\alpha} \cdots \chi^A_{p\alpha} \chi^A_{1\beta} \cdots \chi^A_{q\beta}\rangle \tag{3.18}$$

Using standard rules for matrix elements of one-electron operators [58], we have

$$\left\langle A \middle| \sum_i \hat{O}(i) \middle| D \right\rangle = \det(\mathbf{S}_{AD}) \sum_{ij,\sigma} (\mathbf{S}^{-1}_{AD})_{i\sigma,j\sigma} \langle \varphi^A_{j\sigma} | \hat{O} | \varphi^D_{i\sigma} \rangle \tag{3.19}$$

for any operator $\hat{O}$ of the form of Eq. (3.3) or Eq. (3.4). The overlap matrix $\mathbf{S}_{AD}$ in the above equation is given by

$$(\mathbf{S}_{AD})_{i\sigma,j\lambda} = \langle \chi^A_{i\sigma} | \chi^D_{j\lambda} \rangle = \delta_{\sigma\lambda} \langle \varphi^A_{i\lambda} | \varphi^D_{j\lambda} \rangle \tag{3.20}$$

The determinant of this matrix is the overlap integral for states A and D:

$$S_{AD} = \langle A | D \rangle = \det(\mathbf{S}_{AD}) \tag{3.21}$$

Both computational simplification and physical insight are gained by making the molecular orbitals $\varphi^A_{i\sigma}$ and $\varphi^D_{i\sigma}$ in Eq. (3.19) biorthogonal [59–61]. In this case the overlap matrix of states A and D is diagonal,

$$\langle \varphi^A_{i\sigma} | \varphi^D_{j\sigma} \rangle = \delta_{ij} s^\sigma_i \tag{3.22}$$

and for states with p orbitals in α-spin and q orbitals in β-spin

$$\det(\mathbf{S}_{AD}) = \prod_i^p s_i^\alpha \prod_j^q s_j^\beta \tag{3.23}$$

and

$$(\mathbf{S}_{AD}^{-1})_{i\sigma, j\sigma} = \delta_{ij}(s_i^\sigma)^{-1}, \qquad \sigma = \alpha, \beta \tag{3.24}$$

Using the above results and also substituting the current operator Eq. (3.3) for $\hat{O}$ in Eq. (3.19), we find the following explicit form for the spatial part of the current $\vec{J}(\vec{x})$ defined in Eq. (3.13):

$$\vec{J}(\vec{x}) = -\frac{\hbar}{2m}\det(\mathbf{S}_{AD})\sum_{ij,\sigma}\frac{1}{s_i^\sigma}(\varphi_{i\sigma}^A(\vec{x})\nabla\varphi_{i\sigma}^D(\vec{x}) - \varphi_{i\sigma}^D(\vec{x})\nabla\varphi_{i\sigma}^A(\vec{x})) \tag{3.25}$$

This expression gives the total current in the system as a sum of contributions from corresponding orbitals of donor and acceptor states.

This expression is an obvious generalization of the one-electron picture. Now, different pairs of corresponding (overlapping) orbitals of donor and acceptor states contribute to the current density. The smaller the overlap between corresponding orbitals in donor and acceptor wave functions (i.e., the greater the change of an orbital between the D and A states), the greater the contribution of a given pair to the current.

As we will see later, the major contribution to current is due to one pair of external orbitals, which describe the tunneling electron. The orbitals of other ("core") electrons just shift slightly, due to polarization effects. Their contribution enters as an electronic Franck–Condon factor in the expression for the tunneling electron current. This factor is the overlap of the wave function of the core electrons in donor and acceptor states—that is, product of overlaps of individual core orbitals.

The idea that the major contribution to the tunneling process can be reduced to a Franck–Condon dressed one-electron picture is not new. Interestingly, much earlier, Newton and co-workers presented similar arguments and arrived at this picture from a completely different perspective [19, 62, 63].

The shift of the core orbitals in donor and acceptor states does not appear to be significant; that is, their Franck–Condon factor is of the order of unity. This is a surprising result because the *canonical* orbitals of the core electrons change significantly in donor and acceptor states [19, 63], which is in line with a significant redistribution of the charge in donor and acceptor complexes.

The total current in the above expression is a sum of the current due to tunneling charge per se and to polarization effects. In the Hartree–Fock approach, all electrons in the systems are treated on an equal footing, and therefore the expressions for both tunneling and polarization currents are similar in structure. Their physical meanings, however, are quite different; therefore, polarization and tunneling contributions should in fact be treated differently [64], see Section V. A, below.

For practical purposes, it is convenient to express the currents in terms of the atomic basis functions. Each of the molecular orbitals are assumed to be found in a Hartree–Fock calculation as a linear combination of atomic basis set functions ϕ_μ:

$$\varphi_{i\sigma}^{D} = \sum_{\mu}^{K} D_{\mu i}^{\sigma} \phi_\mu \tag{3.26}$$

$$\varphi_{j\sigma}^{A} = \sum_{n}^{K} A_{\nu j}^{\sigma} \phi_\nu \tag{3.27}$$

where K is the total number of atomic orbitals in the basis set of the system.

Then, in terms of atomic orbitals, the expression for currents takes the following form:

$$\vec{J}(\vec{x}) = -\frac{\hbar}{2m} \det(\mathbf{S}_{AD}) \sum_{\mu\nu,\sigma} B_{\mu\nu}^{\sigma} (\phi_\mu(\vec{x}) \nabla \phi_\nu(\vec{x}) - \phi_\nu(\vec{x}) \nabla \phi_\mu(\vec{x})) \tag{3.28}$$

where the density matrix $B_{\mu\nu}^{\sigma}$ is defined as follows:

$$B_{\mu\nu}^{\sigma} = \sum_{i=1}^{(p,q)} A_{\mu i}^{\sigma} \frac{1}{s_i^{\sigma}} D_{\nu i}^{\sigma} \tag{3.29}$$

where summations are limited by p for α spin and by q for β spin.

Using antisymmetry of the above expression (3.28) in indices μ and ν, one arrives at the final expression for currents:

$$\vec{J}(\vec{x}) = \vec{J}_\alpha(\vec{x}) + \vec{J}_\beta(\vec{x}) \tag{3.30}$$

$$\vec{J}_\alpha(\vec{x}) = -\frac{\hbar}{2m} \det(\mathbf{S}_{DA}) \sum_{\mu,\nu=1}^{K} (B_{\mu\nu}^{\alpha} - B_{\nu\mu}^{\alpha}) \phi_\mu(\vec{x}) \nabla \phi_\nu(\vec{x}) \tag{3.31}$$

$$\vec{J}_\beta(\vec{x}) = -\frac{\hbar}{2m} \det(\mathbf{S}_{DA}) \sum_{\mu,\nu=1}^{K} (B_{\mu\nu}^{\beta} - B_{\nu\mu}^{\beta}) \phi_\mu(\vec{x}) \nabla \phi_\nu(\vec{x}) \tag{3.32}$$

where $\det(\mathbf{S}_{DA})$ is given by Eq. (3.23) and $B^{\sigma}_{\mu\nu}$ is defined by Eq. (3.29). Similarly, the overlap matrices can be expressed in terms of the overlap integrals of the atomic orbitals. These expressions have a very simple structure and are directly suitable for programming.

D. Interatomic Tunneling Currents

Tracing the changes in local charge density is one approach to the description of dynamics in a many-electron system. Another approach is to consider dynamics of atomic populations. This approach, of course, is less rigorous, because atoms in the molecule cannot be unambiguously separated due to atomic orbital overlaps [65]. It is quite reasonable, however, to neglect such subtleties in our problem, for the sake of convenience. Indeed, it is most natural to describe the protein medium and tunneling paths in terms of the protein atoms, however unambiguous this description is in the rigorous mathematical sense.

The atomic formalism is developed in a similar way to that of current density. The idea is to derive equations that would describe the charge redistribution in the system during the tunneling transition. These equations can be written in the following way:

$$\frac{d\bar{P}_a}{dt} = \sum_b j_{a,b} \tag{3.33}$$

where $\bar{P}_a$ is the population of the ath atom in the system, and $j_{a,b}$ are exchange currents between atoms a and b. Since the time evolution during the transition between resonant donor and acceptor states is a simple coherent oscillation (we consider the system "clamped" at the transition state), in which the system goes back and forth between two resonant states, the currents $j_{a,b}$ will have a simple oscillating time dependence of the type $j_{a,b} = J_{a,b} \sin \omega t$. The spatial part of the currents $J_{a,b}$ will then describe the distribution of the tunneling flow in the system.

The major steps in the formalism leading to the above equations are the following. First of all, we need an operator of total atomic population that would correspond to P_a.

1. Mulliken Population Operators

Let a set of real functions $\phi_\nu(x)$, $\nu = 1, \ldots, K$ be any particular atomic (Gaussian) basis set that is chosen for an electronic structure calculation of our system. From these functions, in the usual way, we form a set of atomic spin orbitals, $\chi_{\nu\sigma}(x,\sigma') = \phi_\nu(x)\delta_{\sigma\sigma'}$, where σ, $\sigma' = \alpha,\beta$ are spin indices. These functions are utilized in the construction of the molecular orbitals of our system. In the corresponding Hilbert space, the following notation will be used: $|\nu\rangle$, $|\mu\rangle$ …, and so on, for atomic spatial orbitals, and $|\nu\sigma\rangle$, $|\mu\sigma'\rangle$ ..., and so on, for atomic spin orbitals.

The states with the same spin are not, in general, orthogonal. Their overlap is defined by a symmetric overlap matrix $S_{\nu\mu} = \langle\nu|\mu\rangle$, such that for atomic spin orbitals we have

$$\langle\nu\sigma|\mu\sigma'\rangle = \delta_{\sigma\sigma'} S_{\nu\mu} \tag{3.34}$$

In the Hilbert space of spin orbitals, the identity expansion has the following form:

$$\hat{I} = \sum_{\sigma=\alpha,\beta} \sum_{\nu,\mu=1}^{K} |\nu\sigma\rangle S_{\nu\mu}^{-1} \langle\mu\sigma| \tag{3.35}$$

The Mulliken population operator $\hat{p}_{\nu\sigma}$, for state $\nu\sigma$ is defined, then, as follows:

$$\hat{p}_{\nu\sigma} = \frac{1}{2} \sum_{\mu=1}^{K} (|\nu\sigma\rangle S_{\nu\mu}^{-1} \langle\mu\sigma| + |\mu\sigma\rangle S_{\mu\nu}^{-1} \langle\nu\sigma|) \tag{3.36}$$

If any arbitrary state $|\psi\rangle$ is given in terms of the expansion

$$|\psi\rangle = \sum_{\mu\sigma'} C_{\mu}^{\sigma'} |\mu\sigma'\rangle \tag{3.37}$$

then the average value of the operator $\hat{p}_{\nu\sigma}$—that is, the population of the state $|\nu\sigma\rangle$ according to the definition given above—is [58]

$$\langle\psi|\hat{p}_{\nu\sigma}|\psi\rangle = \frac{1}{2} \sum_{\mu=1}^{K} (C_{\nu}^{\sigma *} S_{\nu\mu} C_{\mu}^{\sigma} + C_{\mu}^{\sigma *} S_{\mu\nu} C_{\nu}^{\sigma}) \tag{3.38}$$

If there are N electrons in the system, each of the electrons can occupy the states that we just described. Then, each electron will be associated with its own Hilbert space spanned by functions $\chi_{\nu\sigma}(\xi_a)$, where $a = 1, \ldots, N$ is the index for different electrons. The states belonging to different Hilbert spaces (i.e., states of different electrons) will be distinguished by an additional index a as follows: $|\nu\sigma(a)\rangle$. In terms of these states, we can describe operators acting in different spaces—that is, operators acting on different electrons. Obviously, the operators acting in different spaces do not act on each other; that is, they commute.

Thus, for the ath electron the population operator is written as follows:

$$\hat{p}_{\nu\sigma}(a) = \frac{1}{2}\sum_{\mu=1}^{K}(|\nu\sigma(a)\rangle S^{-1}_{\nu\mu}\langle\mu\sigma(a)| + |\mu\sigma(a)\rangle S^{-1}_{\mu\nu}\langle\nu\sigma(a)|) \tag{3.39}$$

And the operator of the *total* population of the state $|\nu\sigma\rangle$, $\hat{P}_{\nu\sigma}$, is the sum of the above operators over all electrons in the system:

$$\hat{P}_{\nu\sigma} = \sum_{a=1}^{N}\hat{p}_{\nu\sigma}(a) \tag{3.40}$$

The atomic populations are found by summing contributions of all orbitals of a given atom.

2. *Population Dynamics*

The time evolution of the system is described by Eq. (3.7). The average value of the population of any atomic state $|\nu\sigma\rangle$ at time t during the tunneling transition is then given by

$$\bar{P}_{\nu\sigma}(t) = \langle\Psi(t)|\hat{P}_{\nu\sigma}|\Psi(t)\rangle = \frac{T_{DA}}{\hbar}(\langle D|\hat{P}_{\nu\sigma}|D\rangle\cos^2\theta + \langle A|\hat{P}_{\nu\sigma}|A\rangle\sin^2\theta) \tag{3.41}$$

and the rate of change of the population is

$$\frac{d\bar{P}_{\nu\sigma}(t)}{dt} = -\frac{T_{DA}}{\hbar}(\langle D|\hat{P}_{\nu\sigma}|D\rangle - \langle A|\hat{P}_{\nu\sigma}|A\rangle)\sin 2\theta \tag{3.42}$$

We can also evaluate $d\bar{P}_{\nu\sigma}/dt$ from the above equation in a different way. For this purpose, we introduce the *velocity operator* $\dot{\hat{P}}_{\nu\sigma}$,

$$\dot{\hat{P}}_{\nu\sigma} = \frac{i}{\hbar}[\hat{H}, \hat{P}_{\nu\sigma}] \tag{3.43}$$

Calculating the average of the right-hand side of the above equation over the state $|\Psi(t)\rangle$ should give the same result as $d\bar{P}_{\nu\sigma}/dt$ given by Eq. (3.42).

One can easily check that for real states $|D\rangle$ and $|A\rangle$ and for a real and Hermitian operator P, we have

$$\langle D|[\hat{H},\hat{P}]|D\rangle = \langle A|[\hat{H},\hat{P}]|A\rangle = 0 \tag{3.44}$$

and

$$\langle D|[\hat{H},\hat{P}]|A\rangle = -\langle A|[\hat{H},\hat{P}]|D\rangle \tag{3.45}$$

Then, we find

$$\langle \Psi(t)|\hat{\dot{P}}_{\nu\sigma}|\Psi(t)\rangle = \frac{d\bar{P}_{\nu\sigma}(t)}{dt} = -\frac{1}{\hbar}\langle A|[\hat{H},\hat{P}_{\nu\sigma}]|D\rangle \sin 2\theta \tag{3.46}$$

We notice the same time-dependent factor in both Eq. (3.42) and Eq. (3.46).

The goal of the rest of the calculation is to evaluate all matrix elements in Eqs. (3.42) and (3.46) in terms of the molecular orbitals of the donor and acceptor states and to express $d\bar{P}_{\nu\sigma}(t)/dt$ in the form

$$\frac{d\bar{P}_{\nu\sigma}}{dt} = \sum_{\mu} j_{\nu\sigma,\mu\sigma} \tag{3.47}$$

The kinetic coefficients $j_{\nu\sigma,\mu\sigma}$ introduced above clearly have the meaning of exchange fluxes (i.e., currents) of probability per unit time from orbitals $|\mu\sigma\rangle$ to a given orbital $|\nu\sigma\rangle$. Summation over the orbitals of a given atom results in the kinetic equation of the form (3.33).

The calculations of all matrix elements in the above equations are described in Ref. 27, where the explicit expressions for interatomic currents J_{ab} in terms of the molecular orbitals of the diabatic states $|D\rangle$ and $|A\rangle$ are found. The $|D\rangle$ and $|A\rangle$ are obtained by one of the possible quantum chemistry methods. We emphasize that it can be a one-determinant representation (such as in Hartree–Fock [19, 21, 32–34] or DFT calculations [33]), or the many-determinant representation, of some form of the CI calculation. Although the resulting expressions for interatomic currents are somewhat complicated, a significant simplification can be achieved if the single tunneling orbital approximation (STOA; see below) is used.

Thus, the interatomic current J_{ab}, can be obtained from first principles calculations. When the matrix J_{ab} is known, the analysis of the tunneling pathways, and the calculation of the transfer matrix element T_{DA} are carried out as described in the previous sections.

IV. CALCULATION OF TUNNELING CURRENTS

A. Electron Transfer in Ru-Modified Blue Copper Proteins

Gray and co-workers synthesized and studied electron transfer in Ru-modified azurin, in which site-directed mutagenesis is used to introduce a His residue on the surface of the protein that is coordinated to a -$Ru^{3+}bpy_2Im$ complex. In Figure 1 the crystallographic structure of one such system is shown.

Electron transfer occurs between the reduced Cu(I) center and oxidized Ru(III) complex. The distance between Ru and Cu atoms in the system shown in Fig. 1 is about 27 Å. The reaction is initiated by a quick laser-induced oxidation of the Ru(II) complex in the fully reduced (Cu(I)...Ru(II)) system. The electron transfer reaction is monitored in real time by observing the change in the absorption of reduced/oxidized forms of both Cu and Ru complexes. The Ru complex could be attached at different positions, and therefore the distance dependence of the reaction rate could be studied in these experiments. This distance dependence is believed to be controlled primarily by the electronic coupling between Ru and Cu centers.

The analysis begins from one-electron calculations and protein pruning. The calculations were performed as described in Section II using a semiempirical

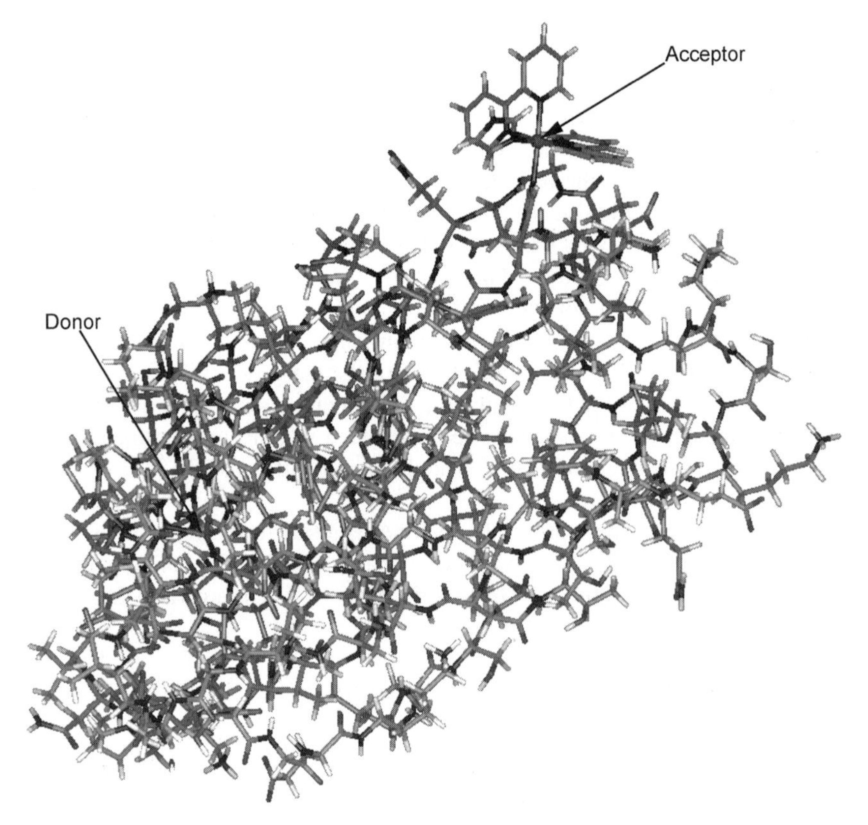

Figure 1. Ru-modified azurin system.

extended Hückel model. The results of pruning of His-126 Ru-modified azurin are shown in Fig. 2.

The pruning procedure naturally leaves donor and acceptor complexes intact, and a number of amino acids that make up the tunneling bridge which connects Ru and Cu ions are identified. In this particular case, two stretches of the protein backbone provide the connection. The two stretches form "molecular wires" along which an electron tunnels between donor and acceptor. The two stretches were identified because for each one the connection to the redox site is strong on

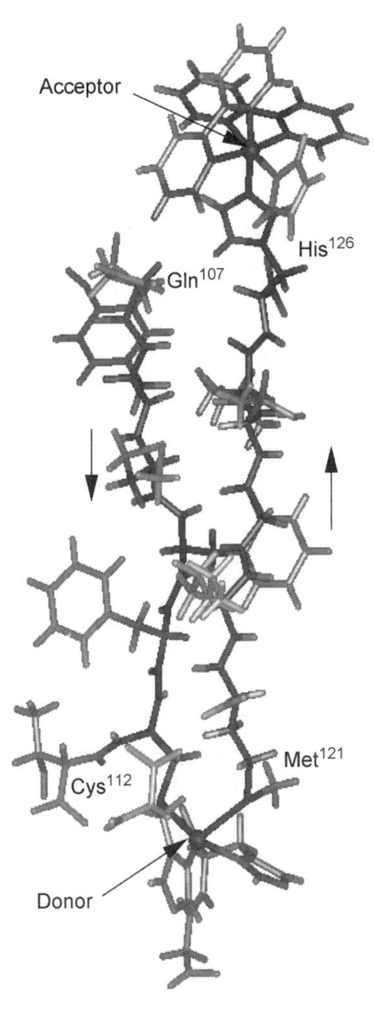

Figure 2. Pruned Ru-modified azurin system.

one end and weak on the other: The Met residue of one wire is much more weakly coupled to the Cu ion than the Cys residue of the other, and the His residue of the Met stretch is much more strongly coupled to Ru than is the Gln residue of the Cys stretch. The actual relative importance of these two paths can only be described by a more accurate calculation, which could correctly describe the relative strengths of (a) weak "through space" coupling of Gln to the Ru complex and (b) Met coupling to Cu center.

The *atoms* of the pruned molecule in Fig. 2, through which electron tunneling occurs, are shown in varying degrees of shaded color. Each atom is characterized by the total tunneling current J_a^+ (see Section II.C). Typically, a good-quality color visualization provides a detailed picture of the distribution of the tunneling in the system (see, for example, Refs. 26, 49, and 50). The analysis of the interatomic fluxes J_{ab} is particularly useful for understanding the structure of the tunneling flow in the system.

The direction of the tunneling current, indicated by the arrows, is related to quantum mechanical phase. In our calculations all wave functions can be made real, and all the information about the quantum mechanical phase is now contained in the sign of the wave functions. The latter determines the direction of the tunneling currents. The destructive interference, for example, shows up as currents in opposite directions, which can lead, under appropriate conditions, to cancellation of the currents. The total (net) flux, as we saw in the preceding theoretical discussion, is related to the coupling matrix element. Thus, the mutually compensating currents imply a decrease of electronic coupling due to destructive quantum interference. Such is the case shown in Fig. 2.

The calculations shown in Fig. 2 were performed at a low-level electronic structure description. Typically, as in this case, we find that the relevant part of the molecule is rather small and that therefore more accurate, ab initio calculations are possible on the pruned molecule. The results of such calculations on a model system, which is derived from our pruned His-126 molecule, is the subject of the next subsection. The goal now is to find out exactly how electrons tunnel in molecular structures shown in Fig. 2.

B. Tunneling Transition in $(His)_2(Met)Cu^{1+}$-(Cys)-(Gly_5)-(His)$Ru^{3+}bpy_2Im$

Here, the method described in the previous theoretical section is applied to study electron tunneling in a system which is a slight modification of that shown in Fig. 2.

The model is shown in Fig. 3. It contains two transition metal complexes, as in the pruned system, which are connected by a stretch of polypeptide chain. Notice that the tunneling current in Fig. 2 is localized in the backbone atoms of the protein. Therefore the nature of the side chains in this case is not important.

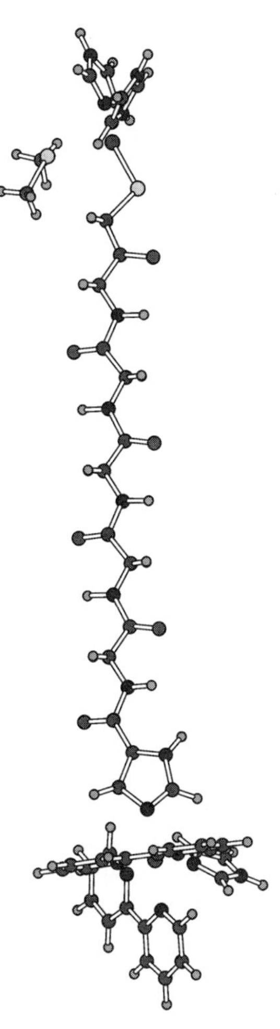

Figure 3. Model charge transfer system for which ab initio calculations are described in the text: $(His)_2(Met)Cu(I/II)$-(Cys)-$(Gly)_5$-(His)Ru(III/II)bpy_2Im.

Electron donor and acceptor in the calculations are, respectively, Cu(I) of the blue copper center in the reduced form, and the Ru^{3+} ion in the -$HisRu^{3+}bpy_2Im$ complex. The donor and acceptor complexes are coordinated to the opposite ends of the polypeptide (Gly_5) chain. The geometry of the polypeptide was chosen, for convenience of visualization, to be planar with all atoms lying in the same plane except for the hydrogen atoms of each of the α-carbons; the length of this system as shown in Fig. 3 is about 35 Å.

Since the two ends of the polypeptide chain are not equivalent, in the absence of an external field the donor and acceptor diabatic states, D–Bridge–A^+ and D^+–Bridge–A, have different energies, E_d and E_a, and therefore no charge transfer can occur. (E_d and E_a are total electronic energies of the system.) In order to induce charge transfer, the two states need to be brought into resonance. In electron transfer reactions, such a resonance occurs in the course of thermal fluctuations of the polar medium surrounding donor and acceptor complexes [8].

To mimic the effect of the polar environment, in this calculation an additional charge Q was placed along the axis of the molecule, and was varied to find a value at which donor and acceptor states have approximately the same energy. The same effect could be achieved by varying configuration of one or several water molecules positioned around donor and acceptor ions. At the transition state value $Q = Q^{\dagger}$, $E_d(Q^{\dagger}) = E_a(Q^{\dagger})$, and the charge is completely delocalized between the donor and acceptor sites. The exact value of $Q^{\dagger}$ is very difficult to determine because the two energies $E_d(Q)$ and $E_a(Q)$, which in our calculations are of the order of 10^4–10^5 eV, should be brought into resonance within a small value of their coupling matrix element, $|E_d(Q^{\dagger}) - E_a(Q^{\dagger})| \leq |T_{DA}|$, where T_{DA} in our and similar systems [6, 7] is of the order of 10^{-2}–10^{-4} cm^{-1}. It is easy, however, to compute an approximate value of the transition state $Q^{\dagger}$ by examining the Mulliken populations of both donor and acceptor ions as a function of Q. The transition between two states in which an electron is localized on donor or acceptor is very sharp [31].

The two approximate many-electron diabatic states $|D\rangle$ and $|A\rangle$ corresponding to localization of the tunneling charge on the donor and acceptor, respectively, can now be calculated for a fixed external charge Q that is sufficiently close to $Q^{\dagger}$. This can be done by iterating the Hartree–Fock wave function from two initial starting points: one that corresponds to the donor configuration, which is taken from calculations for $Q_d < Q^{\dagger}$, and the other that corresponds to the acceptor configuration, which is taken from calculations for $Q_a > Q^{\dagger}$. For a fixed Q, one finds two stable Hartree–Fock solutions, which correspond to $|D\rangle$ and $|A\rangle$ states. Typically, the energies of the two states can be made close within a few hundred wave numbers. (Notice that ideally these energies should be the same.)

The calculations of diabatic states were performed using the Gaussian program with the LanL2DZ basis set [66]. The structure of the tunneling flow calculated using Eq. (3.25) is shown in Fig. 4. The most prominent feature in the flow is the presence of vortices (Fig. 5) [31]. It turns out that these vortices are of the same general nature as those initially discovered by Onsager and Feynman in superfluid helium [67, 68] and later found in many other quantum systems: superconductors, plasma, spin systems, wave fronts, and others. The analog most directly related to this finding is the discovery of vortices by McCullough and Wyatt in quantum fluxes in reactive scattering [69–71]. The

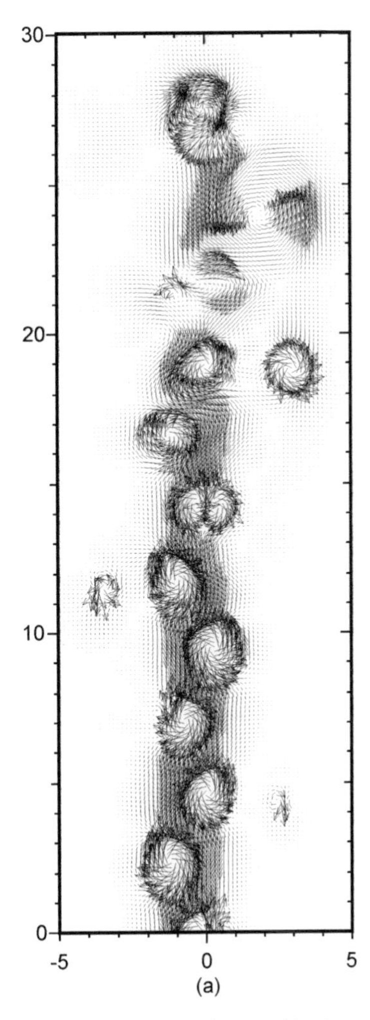

Figure 4. Quantum-mechanical flux in a tunneling transition in a model system shown in Fig. 3.

mathematical nature of the vortices is identical in all the mentioned cases. In all cases, the topology of the quantum flux is controlled by the nodes of the wave function.

C. Quantized Vortices in the Tunneling Flow

It turns out that biorthogonalization mixes the canonical orbitals in our system in such a way that only one pair of orbitals is a major contributor to the tunneling

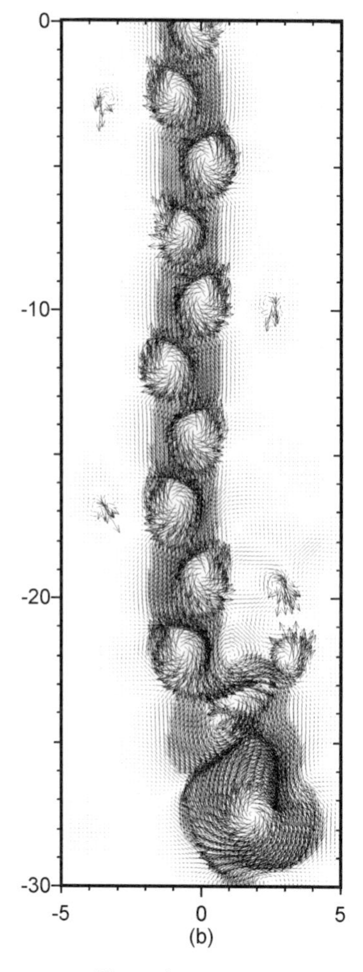

Figure 4. *(Continued).*

current. We will call the orbitals of this pair φ_0^D and φ_0^A. For this pair, the overlap is of the order 10^{-4}, while for others it is less than 0.1% different from 1.0. This pair of orbitals can be taken as a basis of an effective one-electron (or, one-hole) description of the tunneling process. (This is a so-called single tunneling orbital approximation, STOA.) Although such a one-particle picture is what might have been expected, it is worth while to point out that the transferring orbitals that we find are some linear combinations of *all* canonical orbitals. Therefore, the picture that we obtain is far from a trivial statement that the highest occupied molecular orbital (HOMO; or some other single canonical orbital) of the Hartree–Fock (HF) state describes the tunneling electron.

It follows from the above discussion that the current density $\vec{J}(\vec{r})$ shown in Figs. 4 and 5 can be described by a one-electron wave function $\psi(\vec{r}) = (\varphi_0^D(\vec{r}) + i\varphi_0^A(\vec{r}))/2^{1/2}$. Both $\varphi_0^D(\vec{r})$ and $\varphi_0^A(\vec{r})$ have nodes, because they are some excited

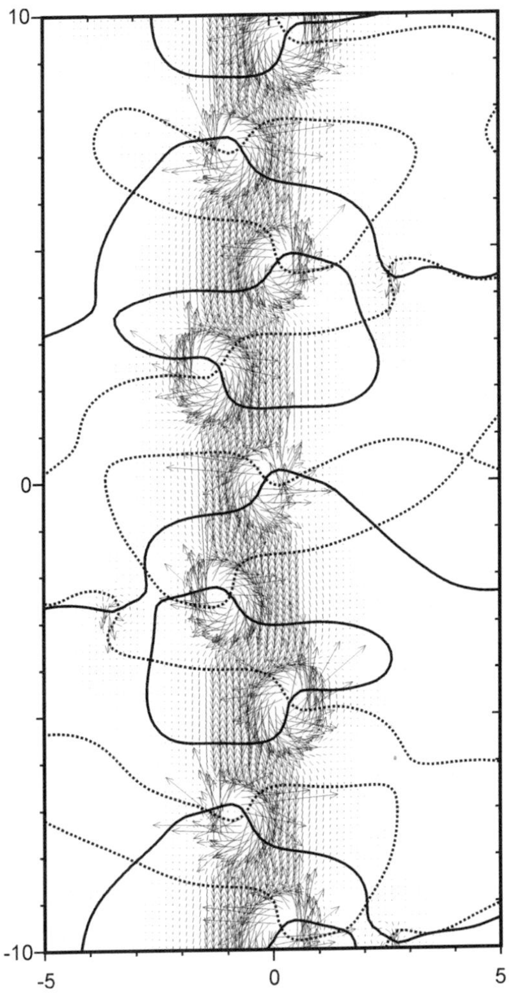

Figure 5. Middle part of the peptide system of Fig. 3 with flux and nodal lines of donor (broken lines) and acceptor orbitals (solid lines)—that is, nodal lines of real and imaginary parts of the wave function of the tunneling electron, respectively. The crossing points of the broken and solid lines correspond to zeros of the complex wave function of the tunneling electron and are the origins of the vortices.

states of the HF Hamiltonian. The nodes of donor and acceptor orbitals correspond some surfaces in three-dimensional (3D) space. Intersection of these surfaces with the xy plane form lines, which are shown in Fig. 5. In 3D space these surfaces intersect to form (most likely, but not necessarily) closed loops. The intersection of these loops with the xy plane of the figure are the crossing points of the nodal lines of donor and acceptor orbitals. These are the centers of vortices. In Figure 5, they are shown as crossing points of solid and dotted lines, which indicate nodes of real and imaginary parts of the wave function of the tunneling electron, respectively.

The nodal lines of complex $\psi(\vec{r})$ are defined by the condition $\varphi_0^D(\vec{r}) = \varphi_0^A(\vec{r}) = 0$. Such nodal lines are topological defects in the sense that on these lines the phase of the wave function ψ is not defined. Around these lines, the tunneling flux exhibits rotational motion (this phenomenon was first described by Dirac in his famous monopole paper [72]), forming structures similar to those found (for example) in superfluid liquids. The velocity field around the vortex lines is quantized. If we write

$$\psi = \rho^{1/2} e^{i\phi} \tag{4.1}$$

where $\rho(\vec{r})$ is the local density of the flux, then, in general,

$$\vec{J}(\vec{r}) = \rho(\vec{r})\left(\frac{\hbar}{m}\right)\nabla\phi(\vec{r}) \tag{4.2}$$

and

$$\oint \frac{\vec{J}(\vec{r})}{\rho(\vec{r})} d\vec{r} = \frac{\hbar}{m}\oint \nabla\phi(\vec{r}) d\vec{r} = \frac{2\pi\hbar}{m} n \tag{4.3}$$

where m is the electron mass and n is an integer. If the integration contour runs around a vortex line, then $n \neq 0$. Although in principle n can be any integer, the most likely value of the circulation number n is one. Values greater than one correspond to multiple crossings of the nodal lines, which are possible, but unlikely. Thus, all our vortices are quantized by the above condition with $n = 1$. Each vortex in Fig. 5 corresponds to a nodal point of the wave function.

The ratio $\vec{J}(\vec{r})/\rho(\vec{r})$ represents the local velocity in the quantum flux, $\vec{v}(\vec{r})$. The above relations show that around each of the vortex lines there is a quantized circulation of the flux, and in the vicinity of *any* vortex line the velocity changes with the distance from the center of the vortex $r_\perp$ as

$$v = \frac{\hbar}{m r_\perp} \tag{4.4}$$

In this sense, all vortices presented in Fig. 5 are similar. At the center, $v \to \infty$, however, the total current is finite because the density of the quantum flux $\rho(\vec{r})$ approaches zero as $r_\perp^2$.

For the velocity of the quantum flux, we can write a "Maxwell" equation,

$$\text{curl } \vec{v}(\vec{r}) = \vec{\Omega}(\vec{r}) \tag{4.5}$$

where

$$\vec{\Omega}(\vec{r}) = \frac{2\pi\hbar}{m} \sum_i \int d\tau \, \vec{u}_i(\tau)\delta(\vec{r} - \vec{R}_i(\tau)) \tag{4.6}$$

where $\vec{u}_i(\tau) = d\vec{R}_i(\tau)/d\tau$ is a tangential vector to a nodal (vortex) line, $\vec{R}_i(\tau)$, and τ is a coordinate along the line.

The stream lines of the tunneling flow in Fig. 5 are the so-called Bohmian trajectories. The whole ensemble of such trajectories forming the flow presents a pictorial hydrodymanic interpretation of quantum mechanics, advocated by David Bohm [52]. It has been shown that under certain conditions the Bohmian trajectories can be chaotic [73]. One of the conditions for the chaotic flux is that more than two eigenstates are mixed in the wave function. In the tunneling case, only two quantum states are involved, and therefore the structure of the quantum flow is not chaotic.

From a practical point of view, the structure of the tunneling flow is an important factor which controls the magnitude of electronic coupling in the system (see discussion in the next section) and with it, the rate of the reaction. This relationship is quite natural because the structure of the tunneling flow reflects the interference effects in electronic transition.

Perhaps the most intriguing question, however, pertains to the possibility of observing the effects related to tunneling vortices experimentally. Since the tunneling current is a source of magnetic field, in principle, it can be measured in nuclear magnetic resonance (NMR) experiments.

As seen from Figs. 4 and 5, the vortices are localized on individual atoms, and therefore we do not expect that deformations of the molecular structure resulting from its thermal motions [45] will destroy the vortices. The thermal motion, of course, will destroy the coherence of the wave function (2.1), but the coherence of individual tunneling jumps (occurring at the transition state of the electron transfer reaction) will not be affected [31]. One may ask, If the atoms move *during* the tunneling transition, will it affect the structure of the tunneling flow? We checked the sensitivity of the matrix element, and that of the tunneling flow to thermal displacements of the atoms, and found that in this case the matrix element changes no more than 15–20%, and the structure of the flow qualitatively remains virtually the same as that shown in Figs. 4 and 5. Thus, thermal motions of the molecular structure shown in Fig. 3 can be viewed as occurring adiabatically during the electron tunneling jump.

Summarizing, the tunneling flow (current) along a polypeptide chain has a nontrivial topology which involves vortices. The structure of the flow reflects the complexity of the dynamics of tunneling electrons in such systems. The analogy with quantum liquids suggests a qualitative picture of biological electron tunneling in which chains of atoms of the protein matrix form a network of molecular tubes connecting donor and acceptor, over which the "quantum electron liquid" can flow. Electrons can tunnel in such tubes only when there is a fluctuational quantum mechanical resonance between the initial and final states [8]. Thus, although always connected, these tubes can be thought of as to open up only infrequently, allowing gradual (and incoherent) leakage of electrons from donor to acceptor. The overall process then resembles quantum percolation.

D. Transfer Matrix Element

The transfer matrix element between donor and acceptor diabatic states could, in principle, be calculated as [19]

$$T_{DA} = H_{DA} - E_0 S_{DA} \tag{4.7}$$

where $\hat{H}$ is the Hamiltonian of the system at the transition state configuration, $\hat{H}(Q^{\dagger})$, $S_{DA} = \langle D|A\rangle$ is the overlap of the two states, and $E_0 = \langle D|\hat{H}|D\rangle = \langle A|\hat{H}|A\rangle$ is their common resonance energy. The application of the above formula for large systems is impeded, however, by numerical difficulties. The problem is that the two terms in Eq. (4.7) by themselves are large numbers, of the order of $10^4\,\mathrm{cm}^{-1}$, and they almost completely cancel each other because T_{DA} is typically of the order of 10^{-2}–$10^{-4}\,\mathrm{cm}^{-1}$. (In particular, the resonant energy of two states is difficult to calculate accurately.) Therefore, the straightforward method of Eq. (4.7) is expected to fail when the coupling matrix element is small.

In Fig. 6, the results of calculating the transfer matrix element using Eq. (3.16), are shown. The total flux was calculated through a plane perpendicular to the axis connecting donor and acceptor ions. The position of the dividing surface was varied in the range of 10 Å from the middle part of the bridge to check the accuracy of the calculations. When the wave functions of diabatic states satisfy the Schrödinger equation exactly in the configuration space, the total flux is guaranteed to be independent of the position of the dividing surface. As seen in Fig. 6, the total flux through the dividing surface fluctuates a small amount, but still provides an accurate estimate of the coupling matrix element. The most remarkable feature of the data in Fig. 6 is the magnitude of the flux itself. A value as small as $10^{-4}\,\mathrm{cm}^{-1}$ for the transfer matrix element obtained by this method is several

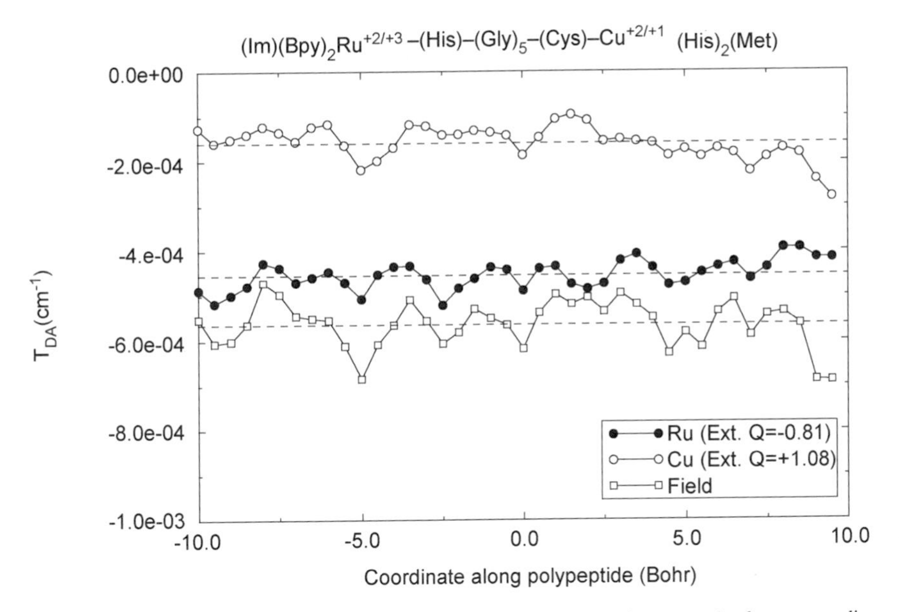

Figure 6. Total flux through a dividing surface between donor and acceptor (a plane perpendicular to the longest axis of the molecule) as a function of the position of the surface. The horizontal axis indicates the coordinate along the longest axis of the molecule, in Bohrs, as in Fig. 4, where the surface crosses the polypeptide. The vertical axis is the total flux, which is equal to the transfer matrix element, measured here in wave numbers.

orders of magnitude smaller than those that could be obtained so far by other methods in the literature (see, for example, Refs. 74 and 75).

The distance dependence of the tunneling matrix element was examined by performing calculations on systems with a varying number of Gly segments in the bridge. The dependence in this case, as expected, is exponential with a β value of about $1\,\text{Å}^{-1}$ (Fig. 7). This is in agreement with what is observed in experiments [4–7].

E. Probing the Nature of the Transition State

The transition state is the resonance between donor and acceptor electronic states. Such a resonance is achieved in the course of thermal fluctuations of the real system. The fluctuations of the polar medium cause the shifts of the potentials of the donor and acceptor complexes. Since fluctuations are random, the two off-resonant levels can move into resonance by moving one level up, or by moving another level down, or by moving one level up and another, simultaneously, down, until their energies match. In fact, all such cases can be realized, so the transition state is not uniquely defined. The question is, How different can these possible transition states in a real system be? It is clear that

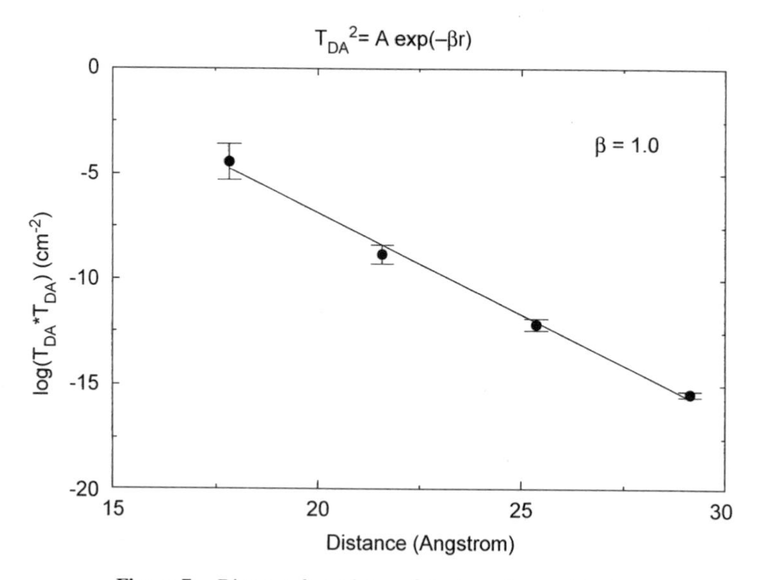

Figure 7. Distance dependence of the tunneling matrix element.

in all cases the position of the pair of resonating states with respect to other states in the system will be different. Also different will be the barrier that the electron tunnels through, and therefore different will be the coupling matrix element for each individual transition state.

We modeled three different ways of bringing two levels in resonance, as discussed above, by varying the external charges and external electric fields. Figure 6 shows the extent of variation of the coupling matrix element in these cases. Surprisingly, for this system we find [34] that the coupling changes by a maximum factor of three. These variations should be appropriately taken into account in the rate expression, by averaging the square of the coupling matrix element, to which the rate is proportional. Additional effects due to the dynamic nature of the protein medium [24, 45–47] are discussed later in this chapter.

F. Electron Transfer or Hole Transfer? Exchange Effects

Electron transfer is associated with superexchange coupling via virtual states in which both donor and acceptor complexes are oxidized; the hole transfer is due to virtual excited states in which both donor and acceptor are reduced. In the latter case a (virtual) hole is present in the protein medium, while in a former case an additional electron is present. Of course, a combination of electron and hole transfer is also possible—a process in which virtual states of both types are involved.

Experimentally, these two cases are impossible to distinguish, because in both cases the initial and final states of the system are the same. Yet, it is interesting to know which type of coupling is actually operating in real systems.

An interesting example that provides some insights into the above issue and shows the effects of the many-electron nature of tunneling transitions is a simple model system, H–(He–He–. . .–He–He)–H^+, in which electron tunneling occurs between two protons, across a chain of three He atoms with interatomic distance 2 Å.

Figure 8 displays the spatial distribution of the tunneling current along the molecular axis. Notice that electron tunneling occurs through the centers of He

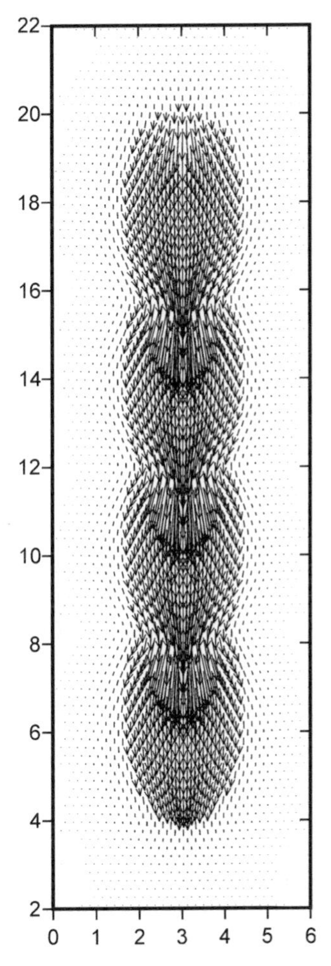

Figure 8. Spatial distribution of tunneling current in H–He–He–He–H^+ system calculated with numerical grid molecular orbitals.

atoms. At first sight, this is counterintuitive, because the $1s$ orbital of a He atom is doubly occupied, and the next available orbital for an additional (tunneling) electron lies above the vacuum level (electron affinity of He is negative). This would imply that the direct tunneling through He atoms should be more difficult than through the vacuum, and therefore a tunneling electron should try to avoid He atoms. The stream lines of the current around each He atom should then look like the lines of the magnetic field expelled from a superconducting sphere. Yet, the calculation shows that electrons move right through the centers of He atoms. This can actually be attributed to the exchange between the tunneling electron and electrons with the same spin on $1s$ orbitals of He atoms. Such a process can also be interpreted as a hole transfer.

This calculation also indicates that any kind of substance is likely to serve as a more suitable medium than vacuum for electrons to tunnel through! Water in proteins, for example, can definitely facilitate the electronic coupling if cavities or gaps filled with water are present along the tunneling path [50, 78].

V. FUTURE WORK. DISCUSSION

The method of tunneling currents, along with other recent approaches to electron tunneling in proteins [20, 21, 23, 24] provides a rigorous theoretical and computational framework for the analysis of biological electron transfer. Further development of this theory is related to (a) the many-electron nature of the problem and (b) the computational techniques of first principles calculations. Below, some of the unresolved theoretical issues are discussed.

A. Polarization Cloud Dynamics. Beyond Hartree–Fock Methods

When an electron tunnels through the protein medium, it moves in the sea of other electrons. The tunneling barrier is, to a large extent, due to interaction of the tunneling electron with other electrons in the medium. There is also an opposite effect, of course: the polarization of the background electrons by the tunneling electron. The energy of the tunneling electron and its tunneling velocity [50] are not very different from those of the electrons of the medium occupying the external orbitals. This means that these electrons will be quick enough to dynamically respond to a moving tunneling charge. Hence, together with the tunneling charge per se, there will be an electron polarization cloud moving in the medium. The compound nature of this tunneling quasi-particle object is rather complex. An additional complication is due to a virtual inseparability of the tunneling electron from those in the polarization cloud because of the electron exchange effect. (A tunneling electron can exchange with the electrons of a nearby atom of the medium as it passes that atom.)

How can one quantitatively describe such a complex tunneling object? The answer to this question determines how accurately one can calculate the tunneling couplings in proteins from first principles theory.

At the present state of theory, the $|D\rangle$ and $|A\rangle$ states are calculated at the HF or similar SCF (DFT) level. This means that the polarization cloud is "tuned" either to donor or acceptor position of the tunneling electron only. The effect of the polarization cloud smoothly dragged on by the tunneling charge is not captured at this level of description. This means that the effective tunneling potential is not quite correctly reproduced in the HF calculations.

To explore this effect, one could modify the description of $|D\rangle$ and $|A\rangle$ states and to include SCF *excited* states that correspond to a tunneling electron/hole localized on the intermediate atoms of the medium, with other electrons adjusted to it. (Notice that this is not the usual superexchange between the zeroth-order states. Most of these interactions are already included in the HF treatment of $|D\rangle$ and $|A\rangle$ states!) The importance of the correlation effects could be examined within this model.

B. New Analytical Methods for the Electron Tunneling Problem

A formal solution of the above polarization/exchange/correlation problem would involve including all excited states (singlets, doublets, etc.) in $|D\rangle$ and $|A\rangle$ states (full CI). As is well known, however, this is practically impossible, because the number of such states is too large. One intermediate solution is suggested above. To make further progress, a deeper theoretical understanding and perhaps a different approach to the problem will be required.

The problem is more general and is related to understanding correlated dynamics in many-electron systems. The long-distance tunneling is the simplest (in some sense) example when only two electronic states are mixed. One possible approach is to apply the advanced analytical methods, such as the many-body Green's functions method, to the problem. The present formulation could provide some advantages, because in our approach the tunneling coupling and other important characteristics are expressed in terms of the tunneling current, which is given by the matrix element of the current density operator; the latter can be expressed in terms of the standard Green's function $-i\langle T\psi(x)\psi^{+}(x')\rangle$ [57] and can be studied using methods of many-body perturbation theory of Fermi systems.

C. Gaussian Basis Versus Real-Space (Grid) Calculations

The advantage of using Gaussian-type basis functions (GTF) in quantum chemistry calculations is well known and is derived from their analytic property which allows for an efficient numerical evaluation of four center integrals in the potential energy expression. The disadvantage of GTF is in their incorrect asymptotic behavior at large distances. Although global properties, such as total energy, may not be very sensitive to long-tail behavior of GTF, the local properties, which depend on the value of the wave function at a given point, can be in significant error when calculated with GTF or with any other finite basis set.

In the calculation of electron tunneling in proteins, the above property of the Gaussian functions can result in a problem. In many cases, electron tunneling in proteins involves "through-space" jumps along the tunneling path, when the distance between atoms is substantially greater than in a usual chemical bond. Indeed, such cases are quite typical (e.g., Ref. 50). The amplitude of the transition is determined by the exponentially small tail of the wave function on the neighboring atom and needs to be evaluated with particular care. The application of standard basis sets in this case is expected to produce poor results.

One solution to such a problem would be to use a large number of diffusive GTF; however, in this case one needs to know in advance between which atoms a more extended basis set should be used. Even when a large number of GTF is used, it is still not clear how accurate the results will be in practice; moreover, such basis sets are not transferable from one molecule to another. Another solution would be to use Slater-type atomic orbitals, which would provide a correct asymptotic behavior. However, *ab initio* calculations are difficult to carry out in this case because of inefficiency of four-center integral calculations.

A third possible solution to this problem is to carry out calculation of the molecular orbitals directly on a real-space grid and to avoid completely any expansion in the atomic basis sets. With a rapid increase of computational power, the idea of directly solving quantum chemistry equations on the numerical grid is gaining much attention and holds significant promise. The molecular orbitals represented on the spatial grid points have the greatest flexibility and can take the proper value at each grid point, both in the core and in the interatomic (or tail) region. The accuracy of the grid-based molecular orbitals is limited only by the grid spacing.

The first such calculation has been carried out recently on a simple model system [33]. The calculation clearly demonstrated the advantages of the real-space approach over standard atomic basis functions. However, much further work needs to be done to make this technique practical. The main problem is the large amount of numerical data required to describe all molecular orbitals on the spatial grid. A combined method, in which the tunneling orbitals are represented on the grid (it is these orbitals that need to be represented most accurately in the tunneling calculations), while other core orbitals are calculated using usual Gaussian functions, could be developed and implemented in such a calculation.

D. DFT Calculations of Tunneling Currents

We have recently demonstrated that Density Functional Theory (DFT) can be employed for the description of electron tunneling in molecular model systems

[33]. The calculation employed the full approximation scheme of the multigrid technique [76, 77] to solve both the Kohn–Sham equations and the Poisson equation, and it yielded excellent results both for the spatial distribution of tunneling currents and for the tunneling matrix element. Since redox sites of the proteins typically contain transition metal ions, the DFT formulation of the tunneling theory is of considerable interest.

The problem here is that the Kohn–Sham orbitals are not exactly equivalent to the usual molecular orbitals of the Slater determinants. As we have demonstrated, the tunneling orbitals are identified in a biorthogonalization procedure, by unitary transformation of the canonical molecular orbitals of diabatic states $|D\rangle$ and $|A\rangle$. One can do such transformations on the Slater-determinant wave functions; similar transformations of the Kohn–Sham orbitals can be carried out as well.

Given that DFT is the most accurate method for the description of transition metal complexes, the application of DFT in the problem of long-distance tunneling in redox proteins would be of great value. Numerous applications of the DFT-based theory of tunneling currents will be possible: cytochrome c oxidase, bc1 complex, Ru-modified proteins, PRC, and others.

E. Protein Dynamic Effects: Sensitivity of Tunneling Paths and Ab Initio Calculation of Electron–Phonon Coupling

As increasingly accurate first-principles methods for calculating the electronic couplings of redox sites become available, one can more reliably evaluate the role of protein dynamics in the long-distance electron tunneling. Previous calculations have been performed mostly at the semiempirical level—for example, in Refs. 24 and 45. Only a few *ab initio* studies (e.g., Refs. 79 and 80) have been reported in the past.

There are two effects to consider: (1) The static or inhomogeneous effect. Electrons in different molecules tunnel through different barriers, because of different instantaneous configurations of the proteins. To evaluate this effect, one needs to average $|T_{DA}|^2$ over thermal configurations of the protein and to examine sensitivity of the pathways. (2) Due to the dynamic nature of the medium, the tunneling electrons can exchange energy with vibrations of the protein, which results in inelastic tunneling [46].

Quantitatively, inelastic tunneling is described by the probability $P(\varepsilon)$ that the tunneling electron will exchange energy ε with the protein when it makes the tunneling jump from donor to acceptor. The longer the distance between donor and acceptor, the larger the probability that such an energy exchange will occur. The ET reaction rate is now given by [46]

$$k = \sum_{\varepsilon} P(\varepsilon) k_0(\Delta G^0 + \varepsilon) \tag{5.1}$$

where k_0 is the standard nonadibatic Marcus–Levich–Dogonadze expression [8, 9] calculated for an averaged coupling, $\langle T_{DA}^2 \rangle$, and for a shifted driving force. The problem is to accurately evaluate $P(\varepsilon)$.

In Ref. 46 we showed that $P(\varepsilon)$ is the Fourier transform of a normalized autocorrelation function of the coupling matrix element $\langle T_{DA}(t)T_{DA}(0)\rangle/|T_{DA}|^2$. Variations of T_{DA} along the dynamic trajectory of the protein can be obtained in a similar way as described in Ref. 45 and by using our new methods for evaluation of T_{DA}.

The dynamic effects are due to high-frequency vibrations of the bridging medium. There can be one- or multiphonon effects. These effects become observable in the rate when the probability $P(\varepsilon)$ becomes sufficiently large for the values of ε which are comparable with the usual activation energy of the reaction. (For example, one or several quanta of hydrogen bonds, which are typical elements of the paths, will make a drastic change in the reaction rate.) The sensitivity of T_{DA} to configuration of the medium is expressed in terms of the electron–phonon coupling constant (derivative of the coupling with respect to configuration changes of coordinates of the medium). The calculation of $P(\varepsilon)$ for specific proteins is a challenging problem that needs to be addressed in the future.

Another interesting dynamical issue is the sensitivity of the pathways to slow and large-amplitude configurational changes of the protein. The question is, Does the (dynamic) inhomogeneity of the protein structure randomize the pathways to an extent that individual paths determined for a fixed configuration of the protein become meaningless?

The positive answer to the last question would provide support of Dutton's concept of the unstructured, effective dielectric protein medium [2]. The negative answer would lend more support to Gray's pathways model [5]. Although it is clear that the answer will depend on the system, it would be interesting to study individual cases and clearly show the existence of these two limits. The issue of the sensitivity of the tunneling paths is related to one of the most fundamental biological questions in this area: whether, indeed, there are specific, evolutionary designed and optimized tunneling routes between redox centers in proteins [2].

VI. CONCLUSION

The method of tunneling currents provides a rigorous framework for the analysis of long-distance electron tunneling. It allows one to gain new insights into the complex subject of tunneling in many-electron systems and provides a practical method of *ab initio* calculation of small tunneling matrix elements. Electron tunneling pathways in proteins can be reliably mapped with this method. A lot of interesting work lays ahead. The picture developed so far is only a self-consistent Hartree–Fock picture. Correlation effects—in particular, dynamic correlations—which describe the polarization cloud around the tunneling electron/hole can in principle be described in this theory. It is tempting to think that the description of such a complex phenomenon as tunneling in a many-electron system can be quantitatively reduced to an effective quasi-one-electron/hole picture. If so, the reduced description would be possible which should greatly simplify the theory without losing accuracy. Further research, however, is needed to address this and other interesting issues.

Acknowledgments

I would like to acknowledge the contributions of my students and post-docs: Iraj Daizadeh, Daniel Katz, Eunjoo Lee, Dmity Medvedev, Margaret Chung, Jessica Swanson, John Gehlen, Eugene Heifets, Jianxin Guo, Jian Wang, Jens Antony, Jongseob Kim, Yury Georgievskii, and Jim Snyder. I would like also to acknowledge many stimulating discussions of the subject of the present chapter with Marshall Newton. Encouragement by Rudy Marcus and Harry Gray at various stages of our work has been a great help. This work was supported by the National Institutes of Health (GM54052-02) and by the Sloan and Beckman Foundations.

References

1. (a) D. Devault, *Quantum Mechanical Tunneling in Biological Systems*, Cambridge University Press, Cambridge, 1984. (b) *Electron and Proton Transfer in Chemistry and Biology*, A. Muller et al., eds., Elsevier, Amsterdam, 1992. (c) *Long Range Electron Transfer in Biology*, P. Bertrand, ed., Springer Series in Structure and Bonding, Vol. 75, Springer, New York, 1991.
2. C. C. Page, C. C. Moser, Xiaoxi Chen, and P. L. Dutton, Natural engineering principles of electron tunnelling in biological oxidation–reduction. *Nature* **402**, 47–52 (1999).
3. C. C. Moser, J. M. Keske, K. Warncke, R. S. Farid, and L. P. Dutton, Nature of biological electron-transfer. *Nature* **355**, 796 (1992).
4. A. Di Bilio, C. Dennison, H. B. Gray, B. E. Ramires, G. Sykes, and J. Winkler, Electron transfer in Ru-modified plastocyain. *J. Am. Chem. Soc.* **120**, 7551 (1998).
5. H. B. Gray and J. R. Winkler, Electron-transfer in ruthenium-modified proteins. *Annu. Rev. Biochem.* **65**, 537–561 (1996).
6. R. Langen, I. Chang, J. P. Germanas, J. H. Richards, J. R. Winkler, and H. B. Gray, Electron tunneling in proteins: Coupling through a beta-strand. *Science* **268**, 1733 (1995).
7. D. R. Casimiro, J. H. Richards, J. R. Winkler, and H. B. Gray, Electron-transfer in ruthenium-modified cytochromes-C-sigma-tunneling pathways through aromatic residues. *J. Phys. Chem.* **97**, 13073 (1993).
8. R. A. Marcus and N. Sutin, Electron transfers in chemistry and biology. *Biochim. Biophys. Acta* **811**, 265 (1985).
9. A. M. Kuznetsov, *Charge Transfer in Physics, Chemistry, and Biology*, Gordon and Breach, Amsterdam, 1995.
10. M. Bixon and J. Jortner, eds., Electron transfer—from isolated molecules to biomolecules. *Adv. Chem. Phys.* **106**, 107 (1999).

11. M. Wikstrom, *Curr. Opin. Struct. Biol.* **8**, 480–488 (1998).
12. M. Pollak and B. Shklovskii, eds., *Hopping Transport in Solids*, North-Holland, New York, 1991.
13. A. K. Felts, W. T. Pollard, and R. A. Friesner, Long-range electron transfer: A mechanism for anomalous distance dependence. *J. Chem. Phys.* **99**, 2929–2940 (1995).
14. A. Okada, V. Chernyak, and S. Mukamel, Solvent Reorganization in long-range electron transfer: Density matrix approach. *J. Phys. Chem.* **102**, 1251 (1998).
15. D. N. Beratan and S. S. Skourtis, Electron transfer mechanisms. *Curr. Opin. Chem. Biol.* **2**, 235 (1998).
16. V. Mujica, A. Nitzan, Y. Mao, W. Davis, M. Kemp, A. E. Roitberg, and M. Ratner, Electron transfer in molecules and molecular wires: Geometry dependence, coherent transfer, and control. *Adv. Chem. Phys.* **107**, 403 (1999).
17. E. Eggers, M. E. Michel-Beyerle, and B. Giese, *J. Am. Chem. Soc.* **120**, 12950 (1998); B. Giese, *Acc. Chem. Res.* (2000), in press.
18. D. DeVault and B. Chance, Studies of photosynthesis using a pulsed laser. *Biophys. J.* **6**, 825 (1966).
19. (a) M. D. Newton, Quantum chemical probes of electron-transfer kinetics—the nature of donor–acceptor interactions. *Chem. Rev.* **91**, 767 (1991); (b) M. D. Newton, Control of electron transfer kinetics. *Adv. Chem. Phys.* **106**, 303 (1999); (c) M. D. Newton, Modeling donor/acceptor interactions: Combined roles of theory and computation. *Int. J. Quant. Chem.* **77**, 255 (2000).
20. H. M. McConnel, *J. Chem. Phys.* **35**, 508 (1961).
21. S. Larsson, Electron transfer in chemical and biological systems. Orbital rules for nonadiabatic transfer. *J. Am. Chem. Soc.* **103**, 4034 (1981); Electron transfer in proteins, *J. Chem. Soc. Faraday Trans.* **2**, 1375 (1983); Electron transfer in proteins, *Biochim. Biophys. Acta* **1365**, 294 (1998).
22. J. N. Onuchic, D. N. Beratan, J. R. Winkler, and H. B. Gray, Electron-tunneling pathways in proteins. *Science* **258**, 1740 (1992).
23. S. S. Skourtis and D. Beratan, Theories of structure-function relationships for bridge-mediated electron transfer reactions. *Adv. Chem. Phys.* **106**, 377 (1999).
24. (a) J. J. Regan and J. N. Onuchic, Electron-transfer tubes. *Adv. Chem. Phys.* **107**, 497 (1999). (b) I. A. Balabin and J. N. Onuchic, *Science* **290**, 114 (2000).
25. A. A. Stuchebrukhov, Tunneling currents in electron transfer reactions in proteins. *J. Chem. Phys.* **104**, 8424–8432 (1996).
26. A. A. Stuchebrukhov, Tunneling currents in electron transfer reactions in proteins. II. Calculation of tunneling matrix element and tunneling currents using non-orthogonal basis sets. *J. Chem. Phys.* **105**, 10819–10829 (1996).
27. A. A. Stuchebrukhov, Tunneling currents in proteins: Non-orthogonal atomic basis sets and Mulliken population analysis. *J. Chem. Phys.* **107**, 6495 (1997).
28. A. A. Stuchebrukhov, Tunneling currents in long-distance electron transfer reactions. III. Many-electron formulation. *J. Chem. Phys.* **108**, 8499 (1998).
29. A. A. Stuchebrukhov, Tunneling currents in long-distance electron transfer reactions. IV. Many-electron formulation. Interatomic Currents and Mulliken population analysis. *J. Chem. Phys.* **108**, 8510 (1998).
30. E. N. Heifets, I. Daizadeh, J. X. Guo, and A. A. Stuchebrukhov, Electron tunneling in quasi-one-dimensional resonant molecular systems. Ab initio study. *J. Phys. Chem.* **102**, 2847 (1998).

31. Iraj Daizadeh, Jian-xin Guo, and Alexei Stuchebrukhov, Vortex structure of the tunneling flow in long-range electron transfer reactions. *J. Chem. Phys.* **110**, 8865 (1999).
32. A. A. Stuchebrukhov, Ab initio calculations of long-distance electron tunneling in metalloorganic systems of biological origin. *Int. J. Quant. Chem.* **77**, 16 (2000).
33. Jian Wang and Alexei Stuchebrukhov, DFT calculations of tunneling current in molecules. *Int. J. Quant. Chem.* **80**(4), 591 (2000) (Proceedings of Sanibel 2000).
34. Jongseob Kim and Alexei Stuchebrukhov, Ab initio calculations of long-distance electron tunneling in a model peptide system. *J. Phys. Chem.* **104**, 8606 (2000).
35. S. Goedecker, Linear scaling electronic structure methods. *Rev. Mod. Phys.* **71**, 1085 (1999).
36. A. A. Stuchebrukhov, Dispersion relations for electron and hole transfer in donor–bridge–acceptor systems. *Chem. Phys. Lett.* **225**, 55–61 (1994).
37. A. A. Stuchebrukhov, On the non-orthogonal basis set calculation of the bridge-mediated electronic matrix elements. *Chem. Phys. Lett.* **265**, 643–648 (1997).
38. I. Daizadeh, J. N. Gehlen, and A. A. Stuchebrukhov, Calculation of electronic tunneling matrix elements in proteins. Comparison of exact and approximate one-electron methods for Ru-modified proteins. *J. Chem. Phys.* **106**, 5658 (1997).
39. J. W. Evenson and M. Karplus, *J. Chem. Phys.* **96**, 5272 (1992); *Science* **262**, 1247 (1993).
40. M. Gruschus and A. Kuki, *J. Phys. Chem.* **97**, 5581 (1993).
41. P. Siddarth and R. A. Marcus, *J. Phys. Chem.* **94**, 2985, 8430 (1993); *ibid.* **97**, 2400, 13078 (1993).
42. M. Ratner, *J. Phys. Chem.* **94**, 4877 (1990); M. Kemp, A. Roitberg, V. Mujica, T. Wanta, and M. A. Ratner, *J. Phys. Chem.* **100**, 8349 (1996).
43. D. J. Katz and A. A. Stuchebrukhov, A new expression for superexchange matrix element in long-distance electron transfer reactions. *J. Chem. Phys.* **109**, 4960 (1998).
44. I. Kurnikov and D. N. Beratan, Ab initio based effective Hamiltonians for long-range electron transfer: Hartree–Fock analysis. *J. Chem. Phys.* **105**, 9561 (1996).
45. I. Daizadeh, E. S. Medvedev, and A. A. Stuchebrukhov, Effect of protein dynamics on biological electron transfer. *Proc. Natl. Acad. Sci. USA* **94**, 3703–9708 (1997).
46. E. S. Medvedev and A. A. Stuchebrukhov, Inelastic tunneling in long-distance electron transfer reactions. *J. Chem. Phys.* **107**, 3821 (1997).
47. E. S. Medvedev and A. A. Stuchebrukhov, Dynamic effects in long-distance biological electron transfer reactions. *IUPAC J.* **70**(11), 2201–2210 (1998).
48. J. N. Gehlen, I. Daizadeh, A. A. Stuchebrukhov, and R. A. Marcus, Tunneling matrix element in Ru-modified blue copper proteins: Pruning the protein in search of electron transfer pathways. *Inorg. Chim. Acta* **243**, 271–282 (1996).
49. Margaret Chung, Iraj Daizedeh, Alexei Stuchebrukhov, and Paul Heelis, Electron transfer pathways in E. coli DNA Photolyase: Trp306 to FADH. *Biophys. J.* **76**, 1241 (1999).
50. Dmitry Medvedev and Alexei Stuchebrukhov, Electron transfer tunneling pathways in bovine heart cytochrome c oxidase. *J. Am. Chem. Soc.* **122**, 6571 (2000).
51. Jens Antony, Dmitry Medvedev, and Alexei Stuchebrukhov, Theoretical study of electron transfer between the photolyase catalytic cofactor FADH and DNA thymine dimer. *J. Am. Chem. Soc.* **122**, 1057 (2000).
52. *Bohmian mechanics and Quantum Theory: An Apprisal*, J. Cushing, A. Fine, and S. Goldstein, eds., Kluver Academic, Dordtecht, Holland, 1996.
53. R. Cave and M. D. Newton, Generalization of the Mulliken–Hush treatment for the calculation of electron transfer matrix elements. *Chem. Phys. Lett.* **249**, 15 (1996).

54. R. Cave and M. D. Newton, Calculation of electronic coupling matrix elements for ground and excited state electron transfer reactions: comparison of the generalized Mulliken–Hush and block diagonalization methods. *J. Chem. Phys.* **106**, 9213 (1997).
55. M. Shephard, M. N. Paddon-Row, and K. D. Jordan, Electronic coupling through saturated hydrocarbon bridges. *Chem. Phys.* **176**, 289 (1993).
56. M. N. Paddon-Row and M. Shephard, Through bond orbital coupling, the parity rule and the design of superbridges which exibit greatly enhanced electronic coupling. *J. Am. Chem. Soc.* **119**, 5355 (1997).
57. A. A. Abrikosov, L. P. Gorkov, and I. E. Dzialoshinskii, *Methods of Quantum Field Theory in Statistical Physics*, Prentice-Hall, New York, 1963.
58. A. Szabo and N. S. Ostlund, *Modern Quantum Chemistry*, Macmillan, New York, 1982.
59. A. T. Amos and G. G. Hall, *Proc. R. Soc. Lond.* **A263**, 483 (1961).
60. H. King, R. E. Stanton, H. Kim, R. E. Wyatt, and R. G. Parr, *J. Chem. Phys.* **47**, 1936 (1967).
61. A. F. Voter and W. A. Goddard, III, *Chem. Phys.* **57**, 253 (1981).
62. M. D. Newton, *J. Phys. Chem.* **92**, 3049 (1988).
63. M. D. Newton, K. Ohta, and E. Zhong, *J. Phys. Chem.* **95**, 2317 (1991).
64. Yu. Georgievski, J. Kim, and A. A. Stuchebrukhov, *J. Chem. Phys.* (to be published).
65. R. Bader, *Atoms in Molecules : A Quantum Theory*, Oxford, 1990.
66. M. J. Frisch, G. W. Trucks, H. B. Schlegel, P. M. W. Gill, B. G. Johnson, M. A. Robb, J. R. Cheeseman, T. A. Keith, G. A. Petersson, J. A. Montgomery, K. Raghavachari, M. A. Al-Laham, V. G. Zakrzewski, J. V. Ortiz, J. B. Forseman, J. Cioslowski, B. B. Stefanov, A. Nanayakkara, M. Challacombe, C. Y. Peng, P. Y. Ayala, W. Chen, M. W. Wong, J. L. Andres, E. S. Replogle, R. Gomperts, R. L. Martin, D. J. Fox, J. S. Binkely, D. J. Defrees, J. Baker, J. P. Stewart, M. Head-Gordon, C. Gonzalez, and J. A. Pople, *Gaussian 94*, Gaussian, Inc., Pittsburgh, PA, 1995.
67. L. Onsager, *Nuovo Cimento* **6** (Suppl. 2), 249 (1949).
68. R. P. Feynman, *Phys. Rev.* **94**, 262 (1954).
69. E. A. McCullough, Jr. and R. E. Wyatt, *J. Chem. Phys.* **54**, 3578 (1971).
70. J. O. Hirschfelder, A. C. Cristoph, W. E. Palke, *J. Chem. Phys.* **61**, 5435 (1974).
71. J. O. Hirschfelder, C. J. Goeble, and L. W. Bruch, *J. Chem. Phys.* **61**, 5456 (1974).
72. P. A. M. Dirac, *Proc. R. Soc. Lond.* **A133**, 60 (1931).
73. R. H. Parmenter and R. W. Valentine, *Phys. Lett. A.* **201**, 1 (1995).
74. C. X. Liang and M. D. Newton, *J. Phys. Chem.* **96**, 2855 (1992); *ibid.* **97**, 3199 (1993).
75. R. Cave and M. D. Newton, *Chem. Phys. Lett.* **249**, 15 (1996).
76. A. Brandt, S. McCormick, and J. Ruge, Multigrid methods for differential eigenproblems. *SIAM J. Sci. Stat. Comput.* **4**, 244 (1983).
77. A. Brandt, Multi-level adaptive solutions to boundary-value problem. *Math. Comput.* **31**, 333 (1977).
78. T. M. Henderson and R. J. Cave, An ab initio study of specific solvent effects on the electronic coupling element in electron transfer reactions. *J. Chem. Phys.* **109**, 7414–7423 (1998).
79. L. W. Ungar, M. D. Newton, and G. A. Voth, Classian and quantum simulation of electron transfer through a polypeptide. *J. Phys. Chem.* **103**, 7367 (1999).
80. I. Kurnikov, L. D. Zusman, M. G. Kurnikova, R.S. Farid, and D. N. Beratan, Structural fluctuations, spin, reorganization energy, and tunneling energy control of intramolecular electron transfer. *J. Am. Chem. Soc.* **119**, 5690 (1997).

MAGNETIC FIELD INFLUENCE ON DYNAMICS OF SINGLET–TRIPLET CONVERSION

VLADIMIR I. MAKAROV

Institute of Chemical Kinetics and Combustion, Novosibirsk, Russia; and Department of Chemistry, University of Puerto Rico, San Juan, Puerto Rico

IGOR V. KHMELINSKII

Universidade do Algarve, Faro, Portugal

CONTENTS

Advances in Chemical Physics, Volume 118, Edited by I. Prigogine and Stuart A. Rice.
ISBN 0-471-43816-2

I. INTRODUCTION

Magnetic field is a weak perturbation, which, however, can essentially change excited state dynamics of a molecule. Studies of magnetic field effects on molecular fluorescence intensity and decay rate, after excitation into a well-defined vibronic or rovibronic state, are extremely informative. They are used to elucidate magnetic effect mechanisms in the intramolecular dynamics of the excited states, as well as to study correlations between intramolecular dynamics, intramolecular coupling, and level structure. Magnetic field effects were reported for various gaseous systems, which are listed in Table I. The observed effects were explained using the level anticrossing (LAM), direct (DM), and indirect (electron-spin and nuclear-spin decoupling) (IM) mechanism theories [19–23]. Later we shall discuss these mechanisms in detail; here, however, we have to explain the meaning of terms DM and IM, because the frequently used definition of these mechanisms [21, 22] differs from the one used here. In the literature, DM corresponds to a direct magnetic-field-induced coupling of the discrete levels of the fluorescent state with dissociative continuum or rovibronic quasi-continuum of the neighboring "dark" nonemitting state, while IM

TABLE I
Molecular Systems whose Fluorescence is Sensitive to External Magnetic Field [1–18]

LAM	H_2CO	D_2CO	$(COH)_2$	C_2H_2	NO
DM	I_2	NO_2	SO_2(fluor.)	CS_2	Cl_2CS
IM	H_2CO	D_2CO	$(COH)_2$	$(COF)_2$	C_2H_2
LAM	HCOOH				
DM	HCOOH	H_3CCOH	NO		
IM	Pyrimidine	*s*-Triazine	SO_2(phos.)	Pyrazine	

describes such coupling between the discrete levels of both the fluorescent and the "dark" states [21,22]. In both cases mentioned, the magnetic-field-induced interaction couples the fluorescent-state level with the dark-state levels in the first order of the perturbation theory. Thus, both mechanisms have the same physical nature. In the present review, we shall understand DM as the magnetic-field-induced coupling of the fluorescent state and neighboring dark-state levels, created in the first order of the perturbation theory, while IM will be considered as such interaction created in higher orders of the perturbation theory. Note that a well-known electron spin and nuclear-spin decoupling mechanism (ENSDM) [23] can be considered as IM, because in the frameworks of ENSDM, the magnetic-field-induced interaction arises in higher orders of the perturbation theory applied to the level of the fluorescent state.

Study of the singlet–triplet (S–T) conversion mechanism is an important photochemical problem, because frequently, a photochemical process involves transition to a triplet excited state and the subsequent chemical transformations of this state. Magnetic field affecting S–T conversion creates a unique possibility to study the S–T conversion mechanism in detail. New knowledge about such mechanisms can be obtained from the excited system response to the field action. As follows from Table I, magnetic-field-induced fluorescence quenching in H_2CO [5], D_2CO [5], $(COH)_2$ [3] $(COF)_2$ [19], C_2H_2 [8] pyrazine [14], pyrimidine [15], and *s*-triazine [16] systems may be described using the IM (ENSDM) theory; that is, in these cases the S–T conversion dynamics is field-dependent. The IM is also applicable to the SO_2 (*phosphorescence*) [3], although here magnetic field induces coupling between different triplet electronic states. For S–T conversion of singlet excited states of the $(COH)_2$ [3,4,20,21,24–36], $(COF)_2$ [19,37–41], pyrazine [14,42–48], pyrimidine [15,49], and *s*-triazine [16,50] molecules, a comprehensive magnetic quenching study has been undertaken. The respective authors, using magnetic, radio-frequency, and microwave frequency fields, had achieved a virtually complete understanding of the S–T conversion process in these systems. In the present review we shall focus our attention on the S–T conversion processes induced by the magnetic field, considering also in brief the LAM and DM theories and the respective experimental data, which can be explained using these theories.

Magnetic field affects the S–T conversion through the indirect mechanism (ENSDM), understood as the field influence upon the S–T conversion, at $\zeta_t > 1$. The ratio ζ_t is defined as

$$\zeta_t = \frac{\langle V_{ST} \rangle}{\langle \gamma_t \rangle} \tag{1.1}$$

where $\langle \gamma_t \rangle$ is the average triplet level width and $\langle V_{ST} \rangle$ is the averaged S–T coupling interaction in zero magnetic field. Magnetic field B mixes fine and hyperfine sublevels of the triplet state; thus the effective density of triplet levels, coupled by the S–T interaction to a singlet level, will increase providing higher f_{IM} factor values. We define the latter as

$$f_{IM} = \frac{\rho_t(B)}{\rho_t(0)} \tag{1.2}$$

where $\rho_t(B)$ is the field-dependent effective density of triplet levels. Thus, the probability to find the excited system in a triplet state will also increase at higher f_{IM} values. Such a probability growth will create an additional molecular fluorescence quenching for the nonzero $\{\gamma_t\}$ values. Thus, in the simplest case the f_{IM} factor determines the magnetic field effect. The system excited-state dynamics depends on the field, if the Zeeman interaction, coupling the hyperfine and fine triplet-state sublevels, is smaller than the energy of the triplet-level fine splitting [19]. We shall consider this problem in detail later, as the analysis of the nature of the triplet-level fine splitting is a very important question for the detailed understanding of the IM, as well as for its correct usage in experimental data analysis.

Note that triplet states of polyatomic nonlinear molecules may be described using the Hund case (b) coupling scheme [51, 52]. Further on, we shall always analyze this coupling scheme only. In the general case of the Hund case (b) system, the fine splitting energy of the triplet levels is defined by anisotropic spin–spin, isotropic spin–rotational, and spin–orbit interactions [52]. In a typical case of a rigid molecule, the spin–spin splitting energy ($\lambda \sim 0.1$–$0.2\,\text{cm}^{-1}$) is much larger than the other interactions mentioned; that is, the fine splitting energy is primarily determined by the anisotropic spin–spin interaction. However, for the high J' or N' values, the energy of the fine splitting is mainly defined by the spin–rotational interaction, being much lower than the spin–spin interaction energy [19, 37–41]. Here J' is the total angular momentum of the singlet excited level, $N' = R + L + l$, where R is the rotational angular momentum, L is the electronic angular momentum, and l is the oscillatory angular momentum. Moreover, the fine splitting energy is linearly dependent on the $(2J' + 1)^{-1}$ value for the singlet excited level considered [41, 47, 49]. The total angular momentum J' of the singlet excited level is equal, with good accuracy, to the rotational angular momentum of the singlet-state levels. Such dependence has been observed experimentally in studies of the magnetic field effect in function of the field strength, for individual rotational levels of oxalylfluoride [41], pyrazine [47], pyrimidine [49], and *s*-triazine [16, 50]. The phenomenon observed has been explained qualitatively by an interaction between rotational and low-frequency vibrational motions of the molecule. For higher J', such an effect averages the anisotropic spin–spin interaction to zero. Understanding and correct usage of the averaging mechanism is very important in the experimental data analysis. For example, the averaging effect discussed can significantly change the optically detected EPR (OD EPR) spectra of gaseous systems, determined by the fine and hyperfine structure of the triplet levels [37–41]. The OD EPR spectra have been studied experimentally for $(COF)_2$ excited to the 0_0^0 transition band, as well as to the $J' = 4$ rotational level of the same band of the $\widetilde{A}^1A_u \leftarrow \widetilde{X}^1A_g$ transition, at room temperature and in a cooled molecular jet, respectively [37–41].

Thus, in the present review we shall first consider the general questions related to the magnetic field influence on the excited state dynamics, and then use the theoretical models developed to analyze experimental data for the systems presented in Table I. Our attention will be focused on the field-induced S–T conversion processes for the lowest singlet excited states of the acetylene, oxalylfluoride, pyrazine, pyrimidine, and *s*-triazine molecules.

II. THEORETICAL BACKGROUND

The S–T conversion is a fundamental problem of photochemistry and chemical physics. The extensively developed radiationless theory is used to analyze the S–T conversion process [53–89]. We shall later consider the S–T interaction schemes, taking into account the selection rules for perturbations coupling the levels of singlet and triplet states. However, for now we shall limit ourselves to the analysis of the simplest interaction scheme, shown in Fig. 1, for which an exhaustive theoretical analysis exists. In the dipole approximation, there are allowed optical transitions between the ground state $|x\rangle$ and the first singlet excited state $|s\rangle$ levels, while the transitions between the ground and the dark-state levels $|q\rangle$ are forbidden. In case of the S–T conversion, the dark-state levels $|q\rangle$ correspond to the triplet-state levels $|t\rangle$. The $|s\rangle$ and $|t\rangle$ levels are coupled by an intramolecular interaction V_{ST}. The intramolecular interaction V_{ST} can be determined by the first-order spin–orbit perturbation (SO), as well as by the second-order vibronic–spin–orbit, (VSO) and rotational–orbit–spin–orbit (ROSO) perturbations. Intramolecular interactions causing higher-order perturbations will be neglected. Following the Van Vleck theory, the second-order interactions may be defined as [90–92]

$$V_{ST}^{(2)} = \sum_f \frac{\langle s|V_1|f\rangle \times \langle f|V_2|t\rangle}{E_s - E_f} + \sum_g \frac{\langle s|V_2|g\rangle \times \langle g|V_1|t\rangle}{E_s - E_g} \qquad (2.1)$$

where $|f\rangle$ and $|g\rangle$ are levels of intermediate singlet and triplet states, respectively. Since in a typical case, the intermediate singlet and triplet states are high-lying electronic states, with the energy higher than that of the excited states considered, the excited-state dynamics may be evaluated using an interaction scheme

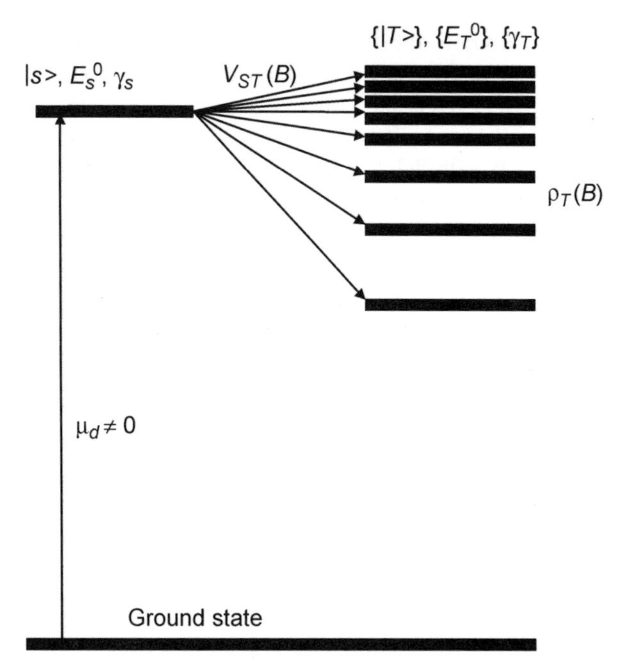

Figure 1. Interaction scheme describing the S–T conversion process in the first order of the perturbation theory.

presented in Fig. 1. Note that the intermediate states only determine the scale of the $V_{ST}^{(2)}$ interaction. The total effective Hamiltonian of the system considered may be represented by

$$\widehat{H} - \frac{i}{2}(\Gamma) = \widehat{H}_0 - \frac{i}{2}(\gamma) + V_{ST} \tag{2.2}$$

where

$$\widehat{H}_0 \cdot |j\rangle = E_j^0 \cdot |j\rangle; \ \widehat{H} \cdot |n\rangle = \varepsilon_n^0 \cdot |n\rangle \tag{2.3}$$

Here, $|j\rangle = |s\rangle$, $|t\rangle$, or $|q\rangle$ is an eigenfunction of the zero-order Hamiltonian, $|n\rangle = \sum C_{nj} \cdot |j\rangle$ is the eigenwave function of the total Hamiltonian, (γ) is the diagonal matrix of the $|j\rangle$ level widths, (Γ) is the diagonal matrix of the $|n\rangle$ level widths, E_j^0 are the eigenvalues of the zero-order Hamiltonian, ε_n^0 are the eigenvalues of the total Hamiltonian, and ρ_t is the effective level density of the triplet state.

Since the excited-state dynamics has been extensively studied in the framework of such an interaction scheme [53–89], we shall consider this problem only very briefly. Note that to observe the real dynamics of an excited state, the condition $|\Delta\omega_{exc}| \gg |V_{ST}|$ should be satisfied, limiting the capability to prepare a nonstationary excited state. To analyze our problem qualitatively, it is very useful to introduce the following parameters:

$$\eta_t = \rho_t \cdot \langle V_{ST} \rangle, \qquad \chi_t = \rho_t \cdot \langle \gamma_t \rangle, \qquad \zeta_t = \frac{\langle V_{ST} \rangle}{\langle \gamma_t \rangle} \tag{2.4}$$

where $\langle \rangle$ designate an averaged value of the corresponding parameter. The meaning of the η_t and χ_t parameters is quite clear: The first determines the number of the triplet-state levels, which are in the "interaction" resonance with a singlet-state level considered; the second defines the character of the neighboring triplet spectrum. For each of these parameters, three different limit cases may be considered: (1a) $\eta_t \ll 1$ is a resonance case; (1b) $\eta_t \sim 1$ is an intermediate case; (1c) $\eta_t \gg 1$ is a statistical limit; (2a) $\chi_t \ll 1$ corresponds to the discrete triplet spectrum; (2b) $\chi_t \sim 1$ corresponds to an intermediate case spectrum; (2c) $\chi_t \gg 1$ corresponds to a quasi-continuum triplet spectrum. The meaning of the ζ_t parameter may be understood, if we note that the time evolution of the initially prepared state can be represented in the following form [53–89]:

$$P(t) = |\langle \Psi(0) | \Psi(t) \rangle|^2 = \sum_n |a_n|^4 \exp(-\Gamma_n t) + \sum_{m,n} |a_m|^2 \cdot |a_n|^2 \exp(-\Gamma_{mn} t) \cdot \cos\left[\frac{2\pi}{h} \Delta\varepsilon^0_{mn} t\right] \tag{2.5}$$

where

$$|\Psi(t)\rangle = \sum_n a_n \cdot |n\rangle \cdot \exp\left[-\frac{i}{h}\left(\varepsilon^0_n - \frac{\Gamma_n}{2}\right)t\right] \tag{2.6}$$

$$|a_n|^2 = \frac{|C_{ns}|^2}{\sum_n |C_{ns}|^2}; \quad \Gamma_n = |C_{ns}|^2 \gamma_s + \sum_n |C_{nt}|^2 \gamma_t \tag{2.7}$$

$$\Gamma_{mn} = \frac{\Gamma_m + \Gamma_n}{2}; \quad \Delta\varepsilon^0_{mn} = \varepsilon^0_m - \varepsilon^0_n \tag{2.8}$$

Equation (2.5) comprises two terms: The first one determines the decay of the quasi-stationary states of the total system Hamiltonian, and the second defines quantum interference effect of these states. The latter is important, if the condition $\langle\Gamma_{mn}\rangle \ll (2\pi/h)\langle\Delta\varepsilon_{mn}^0\rangle$ is satisfied. $\langle\Gamma_{mn}\rangle\sim\langle\gamma_t\rangle$ with good accuracy at $\eta_t>1$. Here $\min(\Delta\varepsilon_{mn}^0) = 2V_{ST}$. Thus, the ζ_t parameter determines the character of the excited-state dynamics: (3a) For $\zeta_t\ll 1$ the quantum-state interference processes are insignificant; in this case, the simplest kinetic approach can be used, with the rate constant of the radiationless process obtained using the "golden" Fermi rule $k_{nr} = (2\pi/h)\rho_t V_{ST}^2$; (3b) at $\zeta_t\sim 1$ the quantum state interference processes become observable; this is most complicated case to analyze; (3c) for $\zeta_t\gg 1$, the quantum-state interference processes are significant; in this case the decay defined by Eq. (2.5) may be represented with good accuracy by a biexponential function:

$$P(t) = A_f \cdot \exp\left(-\frac{t}{\tau_f}\right) + A_s \cdot \exp\left(-\frac{t}{\tau_s}\right) \tag{2.9}$$

with the well-known relationships

$$\frac{1}{\tau_f} = \frac{2\pi}{h}\rho_t V_{ST}^2 = \frac{2\pi}{h}\frac{\eta_t^2}{\rho_t}; \quad \frac{A_f}{A_s} = (\rho_t V_{ST})^2 \tag{2.10}$$

very useful in the analysis of experimental data [75, 77]. We shall use these equations later.

As follows from Fig. 1, the interaction scheme studied includes many different parameters, with only the level widths being the external ones, dependent on the gas pressure. In this case, the magnetic field may be treated as another variable external perturbation, capable of altering the excited-state dynamics. As was mentioned above, various mechanisms are used to analyze the magnetic field effects: levels anticrossing (LAM), direct (DM), and indirect mechanisms (IM). Perturbations induced by magnetic field are described by the Zeeman operator, to be analyzed in detail later.

The LAM applies when the conditions $\eta_t\ll 1$ and $V_{ST}\neq 0$ are satisfied. Since the g factors of singlet- and triplet-state levels are different, the energy gap between the coupled levels is field-dependent, causing field dependence of the fluorescence quantum yield. For the two coupled levels, the energy gap may be represented by

$$\Delta E_{ST}^0 = E_S^0 - E_T^0 \pm \Delta g\mu_B B \tag{2.11}$$

As follows from this equation, the absolute value of ΔE_{ST}^0 has a minimum of 0, which corresponds to the maximal mixing of the $|s\rangle$ and $|t\rangle$ levels. Note that the equation for the fluorescence quantum yield contains the $|\Delta E_{ST}^0|^2$ value only. Hence, LAM operates on the diagonal matrix elements of the Zeeman operator only, thus the field dependence of the fluorescence quantum yield should demonstrate the "resonance" effects.

The interaction scheme of Fig. 1 should be used for the DM, with the $|s\rangle$ and $|q\rangle$ levels of the same multiplicity, and one should substitute the V_{ST} perturbation by the magnetic-field-induced perturbation of $V_Z(B)$. In this case, an interaction induced by the magnetic field is created in the first order of the perturbation theory. All qualitative considerations presented above may equally be used to analyze the DM theory, though in this case we have to remember that all the parameters and constants mentioned above become field-dependent, excluding, of course, the zero-order parameters (γ_s, γ_q, E_s^0, E_q^0, ρ_q, χ_q). Note also that at $\zeta_q(B) \gg 1$, the fast component lifetime of Eq. (2.10) is a field-dependent parameter, $\tau_f(B) = (h/2\pi)(\rho_q V_Z^2(B))^{-1}$. The magnetic quenching (MQ) of NO_2 [93], SO_2 [94], and CS_2 [95] fluorescence has been explained using the DM theory, based on the fast-component lifetime in all these cases being field-dependent.

In the general case, the IM takes into account the magnetic-field-induced coupling between the $|s\rangle$ and $|q\rangle$ levels in the higher orders of the perturbation theory. The appropriate interaction scheme contains a sequence of levels belonging to intermediate dark levels. At least one of these levels should be coupled by intramolecular interactions with the $|s\rangle$ level, while every one of them should be coupled by the magnetic-field-induced interaction both to each other and to the $|q\rangle$ level. Magnetic field quenching of the SO_2 phosphorescence [96] has been explained using the general approach of the IM theory, where the interaction scheme includes three different electronic states ($\tilde{a}\,^3B_1$, $\tilde{b}^3A_2$, and $^3B_2$). The levels of the $\tilde{a}^3B_1$ phosphorescent state are coupled by an intramolecular interaction with the quasi-resonance levels of the $\tilde{b}^3A_2$ state, while the magnetic field induces coupling between the levels of the $\tilde{b}^3A_2$ state and the levels of the low-lying $^3B_2$ state. The field-induced interaction in this case is created in the second order of the perturbation theory as regards the levels of the phosphorescent state. Apparently, at present, this is the only example of the field-induced luminescence quenching treated using the general IM approach. Another very typical case of the IM is the field effect on the S–T dynamics. Structure of the triplet-state levels, the S–T intramolecular interaction nature, matrix elements of the Zeeman perturbation, and detailed interaction schemes need to be considered to analyze the latter problem in detail. Matrix elements of the Zeeman perturbation are also important in the analysis of the LAM and of the DM.

A. Structure of the Triplet-State Levels

This problem is a very important step towards complete analysis of the S–T coupling mechanism and matrix elements of the Zeeman operator. At this step, we get the eigenstates of the effective Hamiltonian [90–92], diagonalized by intramolecular perturbations in the framework of the same electronic state, which result from the Born–Oppenheimer approximation. Thus, we shall now determine the diagonal in terms of the electronic quantum number matrix elements of the effective Hamiltonian. Since all the molecular systems that will be considered here may be reasonably represented by a symmetric top, we shall further analyze the symmetric top case only. Also, given that all the systems studied are polyatomic nonlinear molecules (note that C_2H_2 becomes nonlinear in electronic excited states), the electronic angular momentum of these systems is close to zero; thus the Hund case (b) approximation is applicable to all these molecules with a good accuracy.

This problem has been studied extensively in the literature [97–109], and it was shown that the fine structure of the rotational triplet-state levels is defined mainly by the anisotropic dipole–dipole spin–spin, isotropic spin–orbit, and isotropic spin–rotational interactions. In this case, the spin–orbit interaction is usually much smaller than the other interactions mentioned. In the typical case, the main term determining the hyperfine structure of the fine triplet-state sublevels is the isotropic Fermi interaction, which is usually much weaker than the interactions determining the energy of the fine splitting of the triplet-state levels. Thus, to analyze the S–T coupling and magnetic-field-induced coupling mechanisms, we can limit ourselves to the fine sublevels of the triplet-state rotational levels, and then extend the analysis, taking into account the hyperfine structure of the triplet-state fine sublevels.

Wave functions of the triplet-state fine sublevels may be written as [107, 109, 110]

$$|\psi_{fine}\rangle = |\varepsilon, \upsilon, JNKMS\rangle \tag{2.12}$$

where the triplet-state fine sublevel is described by $J = N - \widetilde{S}$ (total angular momentum quantum number), with $N = R + L + l$, R being the rotational, L the electronic, and l the vibrational angular momentum quantum numbers; K is the projection of N on the symmetric top axis, M is the projection of J on the space-fixed axis, and S is the spin angular momentum quantum number, respectively. Here, J is defined using the $\widetilde{S} = -S$ reverse spin angular momentum, because for the angular momentum J the commutative relationships have different signs in the space-fixed coordinate system and in the molecule-fixed reference frame [90–92, 108]. In Eq. (2,12), ε stands for the quantum numbers defining the electronic state, and υ stands for the quantum numbers defining the vibrational state. The energy of the fine sublevels relative to the triplet-state rotational sublevel considered may be determined as [107, 109, 110]

$$
\begin{aligned}
E_1(N+1,K) = E_0(N,K) &+ \left[\chi\frac{K^2}{(N+1)^2}+\mu\right](N+1) \\
&- 2\lambda\frac{(N+1)^2-K^2}{(N+1)(2N+3)}
\end{aligned} \tag{2.13}
$$

$$
E_2(N,K) = E_0(N,K) - \chi\frac{K^2}{N(N+1)} - 2\lambda\frac{K^2}{N(N+1)} \tag{2.14}
$$

$$
E_3(N-1,K) = E_0(N,K) - \left[\chi\frac{K^2}{N^2}+\mu\right]N - 2\lambda\frac{N^2-K^2}{N(2N-1)} \tag{2.15}
$$

where λ is the anisotropic constant of the spin–spin interaction, and μ and χ are isotropic constants of the spin–rotational interaction. Note that the condition $\lambda \gg \mu$, χ is true in a typical case.

To take into account the hyperfine splitting of the triplet-state fine sublevels, we shall limit ourselves to the analysis of the J-coupling scheme [110], in the frameworks of which the total angular momentum of the studied system is defined as follows: $F = J - \widetilde{I}$, where $\widetilde{I} = -I$ is the reverse nuclear angular momentum of the studied system. The wave functions describing the hyperfine components may be represented as

$$
|\psi_{hf}\rangle = |\varepsilon, \upsilon, FJNKM_FSI\rangle \tag{2.16}
$$

where M_F is the projection of F on the space-fixed axis. In our analysis of the hyperfine splitting of the fine sublevels of the triplet state, we shall take into account the isotropic Fermi interaction only, which is determined by the $a_F(\mathbf{S}\cdot\mathbf{I})$ operator. Matrix elements of this operator may be determined using the projection method [111–114]; thus the energy of the F-hyperfine components may be represented as

$$
\begin{aligned}
E_F = a_F(\mathbf{S}\cdot\mathbf{I}) = a_F &\times [J(J+1) - N(N+1) + S(S+1)] \\
&\times \frac{[F(F+1) - J(J+1) - I(I+1)]}{4J(J+1)}
\end{aligned} \tag{2.17}
$$

Thus, using the approximations mentioned, we can determine the hyperfine and fine components of the triplet state in the symmetric top system.

Although the hyperfine splitting of the rotational levels is insignificant for a singlet state, we have to remember that each rotational level of the $|\,\varepsilon, \upsilon, NKM\rangle$

state consists of the hyperfine sublevels of the $| \varepsilon, \upsilon, FNKM_FI\rangle$ state and that sublevels of different states with the same F may be coupled by some intramolecular interaction.

B. The S–T Coupling Mechanisms

Typically, the most significant intramolecular interactions determining coupling between levels of the singlet and triplet states are the first-order spin–orbit interaction and the second-order vibronic–spin–orbit and rotational–orbit–spin–orbit interactions [97–109]. The selection rules for these interactions have been analyzed by Stevens and Brand [97] and by Howard and Schlag [108]. Here we shall briefly consider this problem, using an approach developed by these authors. For our treatment, we have to determine operators of the vibronic (nonadiabatic), rotational–orbit (Coriolis), and spin–orbit interactions. We also have to define the basic electron–spin wave functions of the singlet and triplet states, as well as redetermine the Hund case (b) wave functions, using linear combinations of the Hund case (a) wave functions. This is useful because, in the latter case, spin angular momentum is coupled to the molecular rotation, thus both the M and P projections of J on the space-fixed and molecule-fixed axes are defined; that is, the transformation properties of the Hund case (a) wave functions are completely defined under transition from the space-fixed coordinate system to the molecule-fixed reference frame.

1. Vibronic Interaction Operator

The vibronic interaction will be represented in the Herzberg–Teller approximation [97, 109]:

$$H_{ev}(q,Q) = \sum_i \left[\frac{\partial H(q,Q)}{\partial Q_i}\right]_{Q_{i0}} (Q_i - Q_{i0}) \tag{2.18}$$

where q and Q are the electronic and nuclear coordinates, respectively, with Q_i being the ith nuclear coordinate and Q_{i0} being the same in equilibrium configuration.

2. Operator of the Coriolis Interaction

To begin with, we shall define this operator for the singlet state, and then use the operator form obtained for the triplet state as well. Using the theory developed by Van Vleck, the operator of the Coriolis interaction may be defined as [90–92]

$$\begin{aligned} H_{Cor} = A_r \cdot \left[(J_a - L_a - l_a)^2 - J_a^2\right] \\ + B_r \cdot \left[(J_b - L_b - l_b)^2 - J_b^2\right] + C_r \cdot \left[(J_c - L_c - l_c)^2 - J_c^2\right] \end{aligned} \tag{2.19}$$

For a singlet state, the condition $J_i = N_i$ is satisfied, where N_i is the ith projection of N on the ith axis of the molecule-fixed coordinate system, and A_r, B_r, and C_r are the rotational constants with respect to corresponding axes of the molecule-fixed reference frame. Thus, the Coriolis operator may be written as

$$H_{Cor} = -2A_rJ_a \cdot (L_a + l_a) - 2B_rJ_b \cdot (L_b + l_b) - 2C_rJ_c \cdot (L_c + l_c) \qquad (2.20)$$

Since we are only interested by the coupling induced by this perturbation between different electronic states, the operator may be reduced to

$$H_{Cor} = -2A_rJ_a \cdot L_a - 2B_rJ_b \cdot L_b - 2C_rJ_c \cdot L_c \qquad (2.21)$$

Further on, Eq. (2.21) will be used. We shall specially note the instances where the vibrational angular momentum is accounted for.

3. Spin–Orbit Perturbation Operator

Using the theory developed by Howard and Schlag [108], the operator of the spin–orbit interaction for the case studied may be written as

$$\begin{aligned} H_{SO} &= H_{SO(1)} + H_{SO(2)} \\ &= \lambda_1 \sum_q (-1)^q l^{(1)}_{1(q)} S^{(1)}_{1(-q)} + \lambda_2 \sum_q (-1)^q l^{(1)}_{2(q)} S^{(1)}_{2(-q)} \end{aligned} \qquad (2.22)$$

Here, subscripts 1 and 2 refer to the first and the second electron, λ_i are the spin–orbit constants, $l^{(1)}_{i(q)}$ are the qth components of the orbital angular momentum operator represented in the spherical tensor form, and $S^{(1)}_{i(-q)}$ are the qth components of the spin angular momentum operator represented in the spherical tensor form. We shall also use the $l^{(1)}_{\pm\,(q)}$ and $S^{(1)}_{\pm\,(q)}$ components, defined as

$$l^{(1)}_{\pm\,(q)} = \frac{\lambda_1 l^{(1)}_{1(q)} \pm \lambda_2 l^{(1)}_{2(q)}}{2} \qquad (2.23)$$

$$S^{(1)}_{\pm\,(q)} = \frac{1}{2}(-1)^q \left(S^{(1)}_{1(q)} \pm S^{(1)}_{2(q)} \right) \qquad (2.24)$$

Thus (1.24) may be rewritten in another form:

$$H_{SO} = \sum_q (-1)^q l^{(1)}_{-(q)} S^{(1)}_{-(-q)} + \sum_q (-1)^q l^{(1)}_{+(q)} S^{(1)}_{+(-q)} \qquad (2.25)$$

The latter form of the spin–orbit interaction operator is the most suitable for analysis of the S–T coupling problem, and it will be used further on.

4. *Basic Electron–Spin Wave Functions. Representation of the Hund Case (b) Wave Functions by Linear Combination of the Hund Case (a) Wave Functions*

Previously, while discussing the theory developed by Van Vleck [90–92], we introduced $\widetilde{S} = -S$, the reverse spin angular momentum operator whose eigen-wave functions may be represented as [108]

$$|\widetilde{SS_z}\rangle = (-1)^{S+S_z}|S, -S_z\rangle \tag{2.26}$$

A diagonal basis for the $(\widetilde{S}_1 + \widetilde{S}_2)^2$ and $(\widetilde{S}_{1z} + \widetilde{S}_{2z})$ operators may be represented as

$$\begin{aligned}
|00\rangle &= (2)^{-1/2}(\alpha_1\beta_2 - \beta_1\alpha_2)\\
|11\rangle &= (\alpha_1\alpha_2)\\
|10\rangle &= (2)^{-1/2}(\alpha_1\beta_2 + \beta_1\alpha_2)\\
|00\rangle &= (\beta_1\beta_2)
\end{aligned} \tag{2.27}$$

where $\alpha = |1/2,1/2\rangle$ and $\beta = |1/2,-1/2\rangle$.

The wave functions in the case (b) after Hund may be represented by linear combinations of the Hund case (a) wave functions, using the general angular momentum theory [108]. The required representation may be written as

$$|\varepsilon, \upsilon, JNKMS\rangle = \sum_{P,\Sigma}(S - \Sigma JP|SJNK) \cdot |\varepsilon, \upsilon, JPMS - \Sigma\rangle \tag{2.28}$$

The latter equation may be rewritten as

$$\begin{aligned}
|\varepsilon, \upsilon, JNKMS\rangle &= (2N+1)^{1/2}\\
&\times \sum_{P,\Sigma}(-1)^{J+P} \cdot \begin{pmatrix} J & N & S \\ P & -K & -\Sigma \end{pmatrix}|\varepsilon, \upsilon, JPMS\Sigma\rangle
\end{aligned} \tag{2.29}$$

where Σ is the projection of S on the symmetric top axis, and the tall brackets represent the $3j$ symbol. This representation will find much use later on.

5. *Matrix Elements of the First-Order Perturbations*

Since only the spin–orbit interaction can couple levels of the singlet and triplet states, this section will be devoted to the selection rules for the matrix elements of (a) vibronic interactions, acting between singlet and singlet-state levels and triplet and triplet-state levels, (b) Coriolis interactions, acting between singlet and singlet-state levels and triplet and triplet-state levels, and (c) spin–orbit interactions, acting between singlet and triplet state levels.

The vibronic perturbation operator acts on the vibronic part of the system wave functions only, thus the matrix elements of interest for the singlet-state levels may be given by

$$\begin{aligned}&\langle \varepsilon, \upsilon, NKM|H_{ev}(q,Q)|\varepsilon', \upsilon', N'K'M'\rangle \\ &\quad = \sum_i \langle \varepsilon| \left[\frac{\partial H(q,Q)}{\partial Q_i}\right]_{Q_{i0}} |\varepsilon'\rangle \cdot \langle \upsilon|(Q_i - Q_{i0})|\upsilon'\rangle \times \delta_{NN'}\delta_{KK'}\delta_{MM'}\end{aligned} \tag{2.30}$$

and for the triplet-state levels by

$$\begin{aligned}&\langle \varepsilon, \upsilon, JNKMS|H_{ev}(q,Q)|\varepsilon', \upsilon', J'N'K'M'S'\rangle \\ &\quad = (2N+1)\sum_i \sum_{P,\Sigma,P',\Sigma'} (-1)^{J+P+J'+P'} \times \langle \varepsilon| \left[\frac{\partial H(q,Q)}{\partial Q_i}\right]_{Q_{i0}} |\varepsilon'\rangle \\ &\quad \times \langle \upsilon|(Q_i - Q_{i0})|\upsilon'\rangle \times \begin{pmatrix} J & N & S \\ P & -K & -\Sigma \end{pmatrix} \\ &\quad \times \begin{pmatrix} J' & N' & S' \\ P' & -K' & -\Sigma' \end{pmatrix} \cdot \delta_{JJ'}\delta_{PP'}\delta_{MM'}\delta_{SS'}\delta_{\Sigma\Sigma'}\end{aligned} \tag{2.31}$$

As follows from these equations, we have to take into account the vibronic species of the states studied, namely, their symmetry, while considering the coupling induced by the vibronic interaction between different electronic states. Hence, the selection rules for the quantum numbers of the different angular momenta, along with projections of these momenta, may be given by

$$\Delta J = 0;\ \ \Delta N = 0;\ \ \Delta K = 0;\ \ \Delta M = 0;\ \ \Delta S = 0 \tag{2.32}$$

For the Coriolis perturbation, the matrix elements can be written for the singlet-state levels as

$$\begin{aligned}&\langle \varepsilon, \upsilon, NKM|H_{Cor}|\varepsilon', \upsilon', N'K'M'\rangle \\ &\quad = -2\sum_q A_q \cdot \langle \varepsilon|L_q|\varepsilon'\rangle \cdot \langle \upsilon|\upsilon'\rangle \cdot \begin{pmatrix} N & 1 & N' \\ K & -q & -K' \end{pmatrix} \cdot \delta_{NN'}\delta_{MM'}\end{aligned} \tag{2.33}$$

and for the triplet-state levels:

$$
\begin{aligned}
&\langle \varepsilon, \upsilon, JNKMS|H_{Cor}|\varepsilon', \upsilon', J'N'K'M'S'\rangle \\
&\quad = -2(2N+1)\sum(-1)^{J+P+J''+P'} \cdot A_q \cdot \langle \varepsilon|L_q|\varepsilon'\rangle \cdot \langle \upsilon|\upsilon'\rangle \\
&\quad \times \begin{pmatrix} J & N & S \\ P & -K & -\Sigma \end{pmatrix} \cdot \begin{pmatrix} J & 1 & J' \\ P & -q & -P' \end{pmatrix}
\end{aligned}
\tag{2.34}
$$

Here $A_q = A_r, B_r, C_r$. Using the rule of sums and properties of the $3j$ symbols, we can obtain the selection rules for the singlet-state levels:

$$
\Delta N = 0; \qquad \Delta K = 0, \pm 1; \qquad \Delta M = 0; \qquad \Delta S = 0 \tag{2.35}
$$

and for the triplet-state levels:

$$
\Delta J = 0; \qquad \Delta N = 0, \pm 1; \qquad \Delta K = 0, \pm 1; \qquad \Delta M = 0; \qquad \Delta S = 0 \tag{2.36}
$$

The $\langle \varepsilon|L_q|\ \varepsilon'\rangle$ matrix elements determine additional selection rules based on the orbital symmetry of the interacting states.

Matrix elements of the spin–orbit interaction can be determined using the spin–orbit operator (2.25), and the wave functions can be defined by (2.29). Note that the symmetric part of the spin–orbit operator (2.25) should be omitted, because any matrix element of the spin part of this symmetric operator between a singlet and a triplet wave function (2.27) is always zero. Thus, taking into account the condition

$$
\langle 1q|S^{(1)}_{-(-q)}|0q'\rangle = \delta_{-qq'} \tag{2.37}
$$

the required spin–orbit interaction operator and its matrix elements may be determined as

$$
H_{SO} = \sum_q (-1)^q l^{(1)}_{-(q)} S^{(1)}_{-(-q)} \tag{2.38}
$$

and

$$
\begin{aligned}
&\langle \varepsilon, \upsilon, NKM|H_{SO}|\varepsilon', \upsilon', JN'K'M'S\rangle \\
&\quad = \sum_q (-1)^{N+K}(2N'+1)^{1/2} \begin{pmatrix} N & N' & 1 \\ K & -K' & q \end{pmatrix} \times \langle \varepsilon|l^{(1)}_{-(q)}|\varepsilon'\rangle\langle \upsilon|\upsilon'\rangle
\end{aligned}
\tag{2.39}
$$

respectively. We note from (2.39) that the spin–orbit interaction matrix elements in question are nonvanishing, if the conditions

$$\Delta N = 0, \pm 1 \quad \text{and} \quad \Delta K = 0, \pm 1 \tag{2.40}$$

are true. In this case, we disregarded the selection rules which follow from the $\langle \varepsilon | l^{(1)}_{-(q)} | \varepsilon' \rangle$ matrix element analysis. If these selection rules were taken into account, the selection rules for the rotational states would be simplified. For example, if $\langle \varepsilon | l^{(1)}_{-(0)} | \varepsilon' \rangle \neq 0$, the selection rules (2.40) will be reduced to

$$\Delta N = 0, \pm 1 \quad \text{and} \quad \Delta K = 0 \tag{2.41}$$

while for $\langle \varepsilon | l^{(1)}_{-(\pm 1)} | \varepsilon' \rangle \neq 0$, they will be given by

$$\Delta N = 0, \pm 1 \quad \text{and} \quad \Delta K = \pm 1 \tag{2.42}$$

Later we shall consider such cases for the particular systems studied.

6. *Matrix Elements of the Second-Order Perturbations*

As was noted above, the most interesting perturbations of the second order are the vibronic–spin–orbit (VSO) and the rotational–orbit–spin–orbit (ROSO) interactions. We may analyze the matrix elements of these interactions using Eq. (2.1), and hence define matrix elements for the vibronic, Coriolis, and spin–orbit perturbations. Thus, for the VSO interaction, the selection rules for the rotational quantum numbers are dependent on the vibronic symmetry of the intermediate vibronic states, which participate in the matrix elements of the electron–vibrational interaction. In this case, one of the following sets of selection rules for the rotational quantum numbers applies:

$$\Delta N = 0, \pm 1; \qquad \Delta K = 0 \tag{2.43}$$

$$\Delta N = 0, \pm 1; \qquad \Delta K = \pm 1 \tag{2.44}$$

$$\Delta N = 0, \pm 1; \qquad \Delta K = 0, \pm 1 \tag{2.45}$$

Later we shall consider these selection rules in detail for the real systems studied.

For the ROSO perturbation, the selection rules are dependent on the matrix elements of the orbital angular momentum operator, yielding one of the following sets of conditions:

$$\Delta N = 0, \pm 1; \qquad \Delta K = 0 \tag{2.46}$$

$$\Delta N = 0, \pm 1; \qquad \Delta K = \pm 2 \tag{2.47}$$

$$\Delta N = 0, \pm 1; \qquad \Delta K = 0, \pm 2 \tag{2.48}$$

Further on, we shall consider these selection rules in detail for the specific systems studied.

C. Matrix Elements of the Zeeman Perturbation

Matrix elements of the Zeeman operator have been considered in publications [22, 23, 39, 116] where the angular momenta coupling schemes were analyzed for the case (a) after Hund, and also in the so-called "low"- and "high"-field limits of the case (b) after Hund. To analyze this problem, we have to note that the nonlinear polyatomic molecules belong to the Hund case (b) angular momenta coupling scheme; that is, their fine splitting energy of the triplet-state levels is much smaller than the energy interval between the rotational levels of the triplet vibronic state. Therefore, we shall limit ourselves to the analysis of the Hund case (b) systems. To begin with, we shall disregard the hyperfine splitting of the triplet-state levels as the first approximation; we shall, however, include this phenomenon later. We shall also focus our attention on the low- and high-field limits [39], which we have to define. In the low-field limit, the coupling energy of different angular momenta is much higher than that of the Zeeman interaction: min $(\Delta E_{fine}) \gg V_Z(B)$. In the high-field limit, the coupling energy of different angular momenta is much lower than that of the Zeeman interaction: max $(\Delta E_{fine}) \ll V_Z(B)$; thus all the angular momenta studied are quantized independently with respect to external field. To study the field-induced matrix elements, the Zeeman operator should be defined.

1. Zeeman Interaction Operator

In the space-fixed coordinate system, the Zeeman operator may be defined as [22, 23, 116]:

$$H_Z^{(s)}(B) = -\vec{B} \cdot \left(\vec{\mu}_L^{(s)} + \vec{\mu}_S^{(s)}\right) \tag{2.49}$$

where

$$\vec{\mu}_L^{(s)} = g_{orb}\beta \vec{L}^{(s)} \quad \text{and} \quad \vec{\mu}_S^{(s)} = g_{spin}\beta \vec{S}^{(s)} \tag{2.50}$$

Here, β is the Born magneton and $g_{spin} \simeq 2g_{orb}$. Using spherical tensor notations, Eq. (1.51) may be rewritten as

$$H_Z^{(s)}(B) = -\beta \sum_Q (-1)^Q \left[g_{orb} B_Q^{(s1)} \cdot L_{-Q}^{(s1)} + g_{spin} B_Q^{(s1)} \cdot S_{-Q}^{(s1)}\right] \tag{2.51}$$

where the superscript $s1$ denotes a spherical tensor of the first rank, defined in the space-fixed coordinate system, $Q=0, \pm 1$. If an external magnetic field is directed along the Z-axis of the space-fixed reference frame, Eq. (1.53) may be rewritten as

$$H_Z^{(s)}(B) = -\beta\left[g_{orb}B_0^{(s1)} \cdot L_0^{(s1)} + g_{spin}B_0^{(s1)} \cdot S_0^{(s1)}\right] \tag{2.52}$$

$$H_Z^{(s)}(B) = -\beta B \cdot \left(g_{orb}L_0^{(s1)} + g_{spin}S_0^{(s1)}\right) \tag{2.53}$$

Further on, we shall use the latter form of the Zeeman operator.

Since in the low-field limit different angular momenta are coupled to the molecular motion, it is very useful to determine the Zeeman operator in the molecule-fixed coordinate system. Transformation of Eq. (2.52) from the space-fixed coordinate system to the molecule-fixed reference frame can be represented using the techniques of spherical tensor transformation:

$$B_c^{(s1)} \cdot T_c^{(s1)} = B \sum_{a,b,c,d} (1a, 0c|10) \cdot (1b, 1d|00) \cdot D_{ab}^{(1)}(\gamma, \theta, \varphi) \cdot T_d^{(1)} \tag{2.54}$$

$$B_c^{(s1)} \cdot T_c^{(s1)} = B \sum_q (10, 00|10) \cdot (1q, 1-q|00) \cdot D_{0q}^{(1)}(\gamma, \theta, \varphi) \cdot T_{-q}^{(1)} \tag{2.55}$$

$$B_c^{(s1)} \cdot T_c^{(s1)} = \frac{B}{\sqrt{3}} \sum_q (-1)^q \cdot D_{0q}^{(1)}(\gamma, \theta, \varphi) \cdot T_{-q}^{(1)} \tag{2.56}$$

because the conditions $a=c=0$ and $b=-d=q$ are satisfied. Here $D_{0q}^{(1)}(\gamma, \theta, \varphi)$ is the rotational matrix, γ, θ, φ are the Euler angles, defining position of the molecule-fixed coordinate system with respect to the space-fixed coordinate system. Using Eq. (2.56), the Zeeman operator in the molecule-fixed reference frame can be written as

$$H_Z^{(m)} = -\frac{\beta B}{\sqrt{3}} \sum_q (-1)^q \cdot D_{0q}^{(1)}(\gamma, \theta, \varphi) \cdot \left(g_{orb}L_{-q}^{(1)} + g_{spin}S_{-q}^{(1)}\right) \tag{2.57}$$

2. *Low-Field Limit*

Since min $(\Delta\ E_{fine}) \gg V_Z(B)$, the different angular momenta are coupled to the molecular motion. Thus, to define the matrix elements of the Zeeman perturbation, we may conveniently use the Zeeman operator in the form of Eq. (2.57). Note that this operator includes the $D^{(1)}_{0q}(\gamma, \theta, \varphi)$ rotational matrix, whose matrix elements are undetermined for the Hund case (b) states, because the transformation properties of the Hund case (b) wave functions are undefined for the transition from the space-fixed to the molecule-fixed reference frame. Therefore, we have to represent the Hund case (b) wave functions by the Hund case (a) wave functions [see Eq. (2.29)]. For the Hund case (a), the matrix elements of the $D^{(1)}_{0q}(\gamma, \theta, \varphi)$ operator are defined [22, 23, 116]. Thus, the matrix elements required may be written as

$$
\begin{aligned}
\langle \varepsilon, \upsilon, JNKMS | H^{(s)}_Z(B) | \varepsilon', \upsilon', J'N'K'M'S' \rangle &= -\frac{\beta B}{\sqrt{3}} \\
\times \sum_{P,\Sigma,P',\Sigma',q} (-1)^{J+S+P+P'+q} \cdot &\begin{pmatrix} J & N & S \\ P & -K & -\Sigma \end{pmatrix} \begin{pmatrix} J & N & S \\ P & -K & -\Sigma \end{pmatrix} \\
\times \left\langle \varepsilon, \upsilon, JPMS\Sigma \middle| D^{(1)}_{0q}(\gamma, \theta, \varphi) \cdot \left(g_{orb} L^{(1)}_{-q} + g_{spin} S^{(1)}_{-q} \right) \middle| \varepsilon', \upsilon', J'P'M'S'\Sigma' \right\rangle
\end{aligned}
\tag{2.58}
$$

Here,

$$
\begin{aligned}
&\left\langle \varepsilon, \upsilon, JNKMS \middle| D^{(1)}_{0q}(\gamma, \theta, \varphi) \cdot \left(g_{orb} L^{(1)}_{-q} + g_{spin} S^{(1)}_{-q} \right) \middle| \varepsilon', \upsilon', J'N'K'M'S' \right\rangle \\
&\quad = (-1)^{M-P} \cdot [(2J+1)(2J'+1)]^{1/2} \cdot \begin{pmatrix} J & 1 & J' \\ M & 0 & -M' \end{pmatrix} \begin{pmatrix} J & 1 & J' \\ P & q & -P' \end{pmatrix} \\
&\quad \times \left\langle \varepsilon, S\Sigma \middle| \left(g_{orb} L^{(1)}_{-q} + g_{spin} S^{(1)}_{-q} \right) \middle| \varepsilon', S'\Sigma' \right\rangle \cdot \langle \upsilon | \upsilon' \rangle
\end{aligned}
\tag{2.59}
$$

where

$$
\left\langle \varepsilon, S\Sigma \middle| \left(g_{orb} L^{(1)}_{-q} \right) \middle| \varepsilon', S'\Sigma' \right\rangle = g_{orb} \langle \varepsilon | L^{(1)}_{-q} | \varepsilon' \rangle \delta_{SS'} \delta_{\Sigma\Sigma'}
\tag{2.60}
$$

and

$$
\begin{aligned}
\left\langle \varepsilon, S\Sigma \middle| \left(g_{spin} S^{(1)}_{-q} \right) \middle| \varepsilon', S'\Sigma' \right\rangle \cdot \langle \upsilon | \upsilon' \rangle &= g_{spin} \langle S\Sigma | S^{(1)}_{-q} | S'\Sigma' \rangle \delta_{\varepsilon\varepsilon'} \delta_{\upsilon\upsilon'} \\
&= g_{spin} (-1)^{S-\Sigma} \cdot \begin{pmatrix} S & 1 & S' \\ -\Sigma & -q & \Sigma' \end{pmatrix} \cdot \delta_{\varepsilon\varepsilon'} \delta_{\upsilon\upsilon'} \delta_{SS'}
\end{aligned}
\tag{2.61}
$$

Thus, selection rules for the matrix elements of the Hund case (a) wave functions may be given as

$$\Delta J = 0, \pm 1; \qquad \Delta P = 0, \pm 1; \qquad \Delta M = 0; \qquad \Delta \Sigma = 0, \pm 1; \qquad \Delta S = 0 \tag{2.62}$$

while the respective selection rules for the matrix elements of the Hund case (b) wavefunctions may be given as

$$\Delta N = 0, \pm 1; \qquad \Delta K = 0, \pm 1; \qquad \Delta M = 0; \qquad \Delta S = 0 \tag{2.63}$$

for $S = 0$, and

$$\Delta J = 0, \pm 1; \qquad \Delta K = 0, \pm 1, \pm 2; \qquad \Delta M = 0; \qquad \Delta S = 0 \tag{2.64}$$

for $S \neq 0$. Note that the orbital part of the Zeeman operator can mix different electronic states, while its spin part can mix levels of the same vibronic state only at $S \neq 0$. Note that only the spin part of the Zeeman operator is of interest in the analysis of the field-induced S–T processes. Further on, we shall use matrix elements of the spin part of the Zeeman operator to study the problem discussed.

3. *High-Field Limit*

At max $(\Delta E_{fine}) \ll V_Z(B)$, the N and S angular momenta are quantized independently with respect to external magnetic field; that is, in this case we may use the Zeeman operator defined in the space-fixed coordinate system (2.53), the Wigner–Eckart theorem, and properties of the reduced matrix elements [116]. Thus, matrix elements of the Zeeman operator may be written as

$$\begin{aligned}
&\langle \varepsilon, \upsilon, JNKMS | H_Z^{(s)}(B) | \varepsilon', \upsilon', J'N'K'M'S' \rangle \\
&\quad = -\beta B \langle \varepsilon, \upsilon, JNKMS | g_{orb} L_0^{(s1)} + g_{spin} S_0^{(s1)} | \varepsilon', \upsilon', J'N'K'M'S' \rangle \\
&\quad = (-1)^{J-M} \begin{pmatrix} J & 1 & J' \\ M & 0 & -M' \end{pmatrix} \\
&\qquad \times \beta B \left\langle \varepsilon, \upsilon, JNKS \left\| g_{orb} L_0^{(s1)} + g_{spin} S_0^{(s1)} \right\| \varepsilon', \upsilon', J'N'K'S' \right\rangle \\
&\quad = (-1)^{J-M} \beta B \begin{pmatrix} J & 1 & J' \\ M & 0 & -M' \end{pmatrix} [(2J+1)(2J'+1)]^{1/2} \\
&\qquad \times \begin{bmatrix} (-1)^{N+S+J'+1} \begin{pmatrix} N & J & S \\ J' & N' & 1 \end{pmatrix} \delta_{SS'} \langle L \rangle F_{\upsilon\upsilon'} + \\ +(-1)^{N'+S'+J+1} \begin{pmatrix} S & J & N \\ J' & S' & 1 \end{pmatrix} \delta_{NN'} \delta_{\varepsilon\varepsilon'} \delta_{\upsilon\upsilon'} \langle S \rangle \end{bmatrix}
\end{aligned} \tag{2.65}$$

where

$$\langle L\rangle = \langle \varepsilon, NK||L^{(s1)}||\varepsilon', N'K'\rangle$$
$$= (-1)^{N-M_N} \frac{\langle \varepsilon, NKM_N|L_0^{(s1)}|\varepsilon', N'K'M_N'\rangle}{\begin{pmatrix} N & 1 & N' \\ -M_N & 0 & M_N' \end{pmatrix}}$$
$$= (-1)^{N-M_N} \frac{\left\langle \varepsilon, NKM_N \middle| \sum_q D_{0q}^{(1)}(\gamma, \theta, \varphi) L_{-q}^{(1)} \middle| \varepsilon', N'K'M_N' \right\rangle}{\begin{pmatrix} N & 1 & N' \\ -M_N & 0 & M_N' \end{pmatrix}}$$
$$\times \sum_q (-1)^{N-M_N} [(2N+1)(2N'+1)]^{1/2} \begin{pmatrix} N & 1 & N' \\ -K & q & K' \end{pmatrix} \langle \varepsilon | L_{-q}^{(1)} | \varepsilon' \rangle \tag{2.66}$$

$$\langle S\rangle = \langle S||S^{(s1)}||S'\rangle = (-1)^{S-M_S} \frac{\langle SM_S|S_0^{(s1)}|S'M_S'\rangle}{\begin{pmatrix} S & 1 & S' \\ -M_S & 0 & M_S' \end{pmatrix}} \tag{2.67}$$

$$\langle SM_S|S_0^{(s1)}|S'M_S'\rangle = [S(S+1)(2S+1)]^{1/2}; \qquad S = S' \tag{2.68}$$

$$F_{\upsilon\upsilon'} = \langle \upsilon|\upsilon'\rangle \text{ is the Franck–Condon factor} \tag{2.69}$$

As follows from these equations, the selection rules for the orbital part of the Zeeman operator may be given by

$$\Delta N = 0, \pm 1; \qquad \Delta K = 0, \pm 1; \qquad \Delta M_N = 0; \qquad \Delta S = 0; \qquad \Delta M_S = 0 \tag{2.70}$$

while for the spin part of this operator the selection rules are

$$\begin{matrix} \Delta N = 0; & \Delta K = 0; & \Delta M_N = 0; & \Delta S = 0; \\ \Delta M_S = 0; & \Delta\varepsilon = 0; & \Delta\upsilon = 0 & \end{matrix} \tag{2.71}$$

that is, in the high-field limit the spin part of the Zeeman operator does not mix any levels of the molecular states studied, whereas the orbital part of this operator can mix different levels of the same electronic state, as well as levels of different electronic states. However, for the problem of the field-induced S–T conversion, only the spin part of the Zeeman operator is of interest and shall be used in the following analysis.

4. *Systems with a Nuclear Angular Momentum*

When the system studied contains any nuclei with nonzero nuclear angular momentum, we have to include the hyperfine structure of the triplet-state levels. Also, we have to keep in mind that the levels of the singlet state are described by the total angular momentum $\vec{F}$, which results from interaction of the rotational angular momentum $\vec{N}$, and the nuclear angular momentum $\vec{I}$. The wave functions used in this case may be represented as

$$|\varepsilon, \upsilon, FNKM_FM_N\rangle \tag{2.72}$$

$$|\varepsilon, \upsilon, FJNKM_FMS\rangle \tag{2.73}$$

for the singlet and triplet levels, respectively. Further on, we shall assume that the magnetic field mixes the F components of the triplet, but not of the singlet state. In the low-field limit ($E_{hyperfine} \gg E_{Zeeman}$), the selection rules shown above apply, with the additional conditions

$$\Delta F = 0, \pm 1; \qquad \Delta M_F = 0 \tag{2.74}$$

If the coupling energy of I and J is smaller than the interaction energy of these moments with the magnetic field, then the F components are completely mixed by the field B. It means that J and I are quantized independently with respect to external magnetic field. In this case, $\Delta I = 0$ and $\Delta M_I = 0$, where M_I is the projection of I on the space-fixed axis coinciding with the external magnetic field direction. The condition $M_F = M_I + M_J = M_I + M_N + M_S$ is satisfied for a transition from the case of completely coupled moments to the case of completely decoupled moments at $B \neq 0$.

Effective density of triplet levels, coupled to a given singlet level, can increase by a factor of $f_{IM} = (2\langle I\rangle + 1) f(N, K)$ in the presence of a magnetic field. Here, N and K belong to the triplet levels, $f(N, K)$ depends on the energies of fine and N-, K-splitting of the triplet levels, with $\langle I\rangle$ being the average value of the quantum number of the total nuclear angular momentum. This momentum is averaged over different realizations of the I value, with the respective statistical weights. Note that $f(N, K)$ can change from 8/3 to $(K + 1)$ in function of the triplet level K-structure, with N-splitting being always much larger than the hyperfine and fine splitting. For the triplet state of our systems, corresponding to a prolate symmetric top, $f(N, K) = 8/3$ [19]. Further on, this conclusion will be useful in the analysis of the field effects in glyoxal, formaldehyde, oxalylfluoride, and acetylene molecules.

D. Interaction Schemes Analyzed With and Without a Magnetic Field

In the present review, we shall consider the magnetic field-induced S–T conversion in molecules shown in Table II. Here, OST and PST denote oblate symmetric top and prolate symmetric top, respectively. Thus, to analyze the S–T interaction schemes in these system both at $B = 0$ and $B \neq 0$, we need to consider the selection rules for the orbital and vibronic parts of the S–T interactions mentioned above. For now, we can limit ourselves to the C_{2v}, C_{2h}, D_{2h}, and D_{3h} point groups. In the present section, we shall disregard the nuclear spin statistics, discussing this issue briefly later. As was noted, we have to analyze the first-order SO (the orbital part only), and the second-order VSO (combination of the matrix elements of nonadiabatic perturbation and the orbital part of the SO perturbation); and ROSO interactions (combination of orbital parts of both the Coriolis and SO perturbations); that is, the matrix elements

$$\langle \varepsilon | L_q^{(1)} | \varepsilon' \rangle; \tag{2.75}$$

$$\langle \varepsilon | V_{ev} | \varepsilon'' \rangle \langle \varepsilon'' | L_q^{(1)} | \varepsilon' \rangle; \qquad \langle \varepsilon | L_q^{(1)} | \varepsilon'' \rangle \langle \varepsilon'' | V_{ev} | \varepsilon' \rangle \tag{2.76}$$

$$\langle \varepsilon | L_p^{(1)} | \varepsilon'' \rangle \langle \varepsilon'' | L_q^{(1)} | \varepsilon' \rangle; \qquad \langle \varepsilon | L_q^{(1)} | \varepsilon'' \rangle \langle \varepsilon'' | L_p^{(1)} | \varepsilon' \rangle \tag{2.77}$$

should be considered at $B = 0$. We shall only consider the general statements, applicable to the possible interaction schemes at $B = 0$ and $B \neq 0$ without presenting detailed spectroscopic data for the systems studied. Later, analyzing

TABLE II
The Molecular Systems Analyzed

Compound Name	Symmetry Group	Geometry
Formaldehyde	C_{2v}	PST
Formaldehyde, D-substituted	C_{2v}	PST
Glyoxal	C_{2h}	PST
Oxalylfluoride	C_{2h}	PST
Acetylene	C_{2h}	PST
Pyrazine	D_{2h}	OST
Pyrimidine	C_{2v}	OST
s-Triazine	D_{3h}	OST

magnetic fluorescence quenching in these systems, we shall consider the respective spectroscopy in detail and will analyze the implications of the nuclear angular momentum.

The S–T coupling should be considered between the $\widetilde{A}^1A_2$ and $\widetilde{A}^3A_2$ states for the formaldehyde molecule [52]. Components of the orbital tensor are transformed by irreducible representations $A_2(q=0)$, B_1, and $B_2(q=\pm 1)$ of the C_{2v} point group. Thus, the spin–orbit interaction cannot couple directly the levels of the $\widetilde{A}^1A_2$ and $\widetilde{A}^3A_2$ states. However, these states can be coupled in the second order of the perturbation theory. (i) Noting that the vibronic modes of formaldehyde are transformed by the irreducible representations a_1, b_1, and b_2 of the C_{2v} point group, matrix elements of the VSO interaction may be written as

$$\begin{aligned}&\left\langle A_2\left|\begin{pmatrix}a_1\\b_1\\b_2\end{pmatrix}\right|\begin{pmatrix}B_2\\B_1\end{pmatrix}\right\rangle\left\langle\begin{pmatrix}B_2\\B_1\end{pmatrix}\left|\begin{pmatrix}A_2\\B_1\\B_2\end{pmatrix}\right|A_2\right\rangle\\&\left\langle A_2\left|\begin{pmatrix}A_2\\B_1\\B_2\end{pmatrix}\right|\begin{pmatrix}B_2\\B_1\end{pmatrix}\right\rangle\left\langle\begin{pmatrix}B_2\\B_1\end{pmatrix}\left|\begin{pmatrix}a_1\\b_1\\b_2\end{pmatrix}\right|A_2\right\rangle\end{aligned}\tag{2.78}$$

Thus, the transitions with $\Delta K=\pm 1$ are allowed for the VSO perturbation. (ii) In the case of the ROSO interaction, the matrix elements

$$\left\langle A_2\left|\begin{pmatrix}A_2\\B_1\\B_2\end{pmatrix}\right|\begin{pmatrix}A_2\\B_2\\B_1\end{pmatrix}\right\rangle\left\langle\begin{pmatrix}A_1\\B_2\\B_1\end{pmatrix}\left|\begin{pmatrix}A_2\\B_1\\B_2\end{pmatrix}\right|A_2\right\rangle\tag{2.79}$$

should be used, with the selection rules $\Delta K=0,\pm 2$ (see above). Thus, in the case of the VSO interaction, the effective density of triplet-state levels coupled to a singlet state is $\rho_{eff}=6\rho_{vib}$, while for the ROSO interaction it is $\rho_{eff}=9\rho_{vib}$, where ρ_{vib} is the density of the triplet-state vibronic levels. For a given vibronic state of the triplet state, the interaction schemes between chosen singlet-state levels and the corresponding triplet rovibronic levels are shown in Fig. 2 for both S–T coupling mechanisms. Note that for both cases the complete interaction scheme is a superposition of the $n=\rho_{vib}$ schemes shown, each corresponding to a different vibronic triplet state.

Since for the formaldehyde the energy of the N- and K-splitting is much larger than that of the fine splitting of the triplet-state levels, it is sufficient to consider the magnetic-field-induced coupling of the fine triplet sublevels of the same (N,K)-level only. Thus, the level interaction schemes of Fig.2 would transform into ones of Fig. 3. Recalling that the energy of Zeeman interaction in the

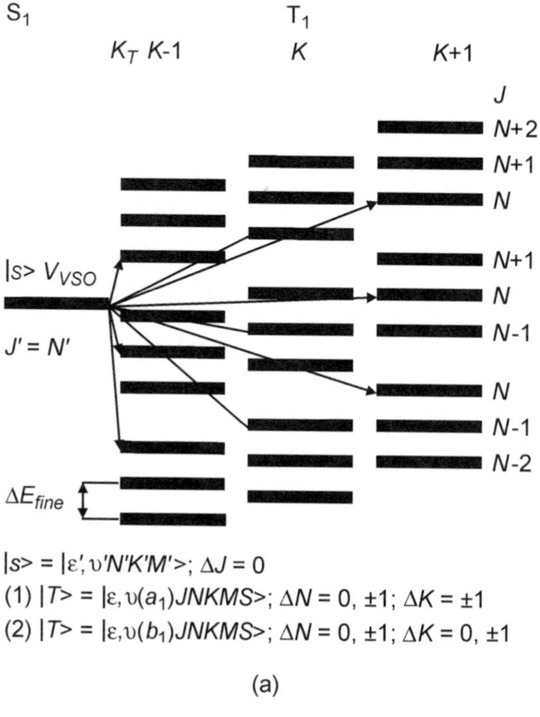

Figure 2. Interaction scheme for the VSO [(a) $B=0$, (c) $B\neq 0$]: (1) $|\ \varepsilon,\ \upsilon\ (a_1),\ JNKMS\rangle$, $\Delta N=0,\pm 1$, $\Delta K=\pm 1$; (2) $|\ \varepsilon,\ \upsilon\ (b_1),\ JNKMS\rangle$, $\Delta N=0,\pm 1$, $\Delta K=0,\pm 1$), and ROSO [(b) $B=0$, (d) $B\neq 0$]: (1) $|\ \varepsilon,\ \upsilon\ (a_1),\ JNKMS\rangle$ $\Delta N=0,\pm 1$, $\Delta K=0,\pm 1$; (2) $|\ \varepsilon,\ \upsilon\ (b_1),\ JNKMS\rangle$, $\Delta N=0,\pm 1$, $\Delta K=0,\pm 1$) S–T coupling mechanisms of a PST. The schemes are presented for a single vibronic triplet state only. Interaction schemes between a given singlet level and rovibronic levels corresponding to different vibronic levels of the triplet state are obtained by superposition of several interaction schemes similar to the one presented. The schemes of Fig. 3 become the schemes of Fig. 4 at $V_{ST}(0) \ll V_Z\ (B)$.

$$|\varepsilon, \upsilon, NKM_N M_S S\rangle = \sum_{J=|N-S|}^{N+S} |\varepsilon, \upsilon, JNKMS\rangle \times \langle JM|NKM_N M_S\rangle.$$

experimental fields and the energy of fine splitting are much larger than that of the S–T coupling, we can start by diagonalizing the system Hamiltonian over the Zeeman interaction, using this procedure on the mixed levels of the triplet state. Note that the Zeeman interaction in the low-field limit is weaker than the fine splitting, while in the high-field limit the field-induced interaction is stronger than the fine splitting; on the other hand, the fine splitting energy is always higher than that of the S–T coupling, because the Franck–Condon factor reduces significantly the S–T coupling matrix elements. Next, we can analyze the S–T

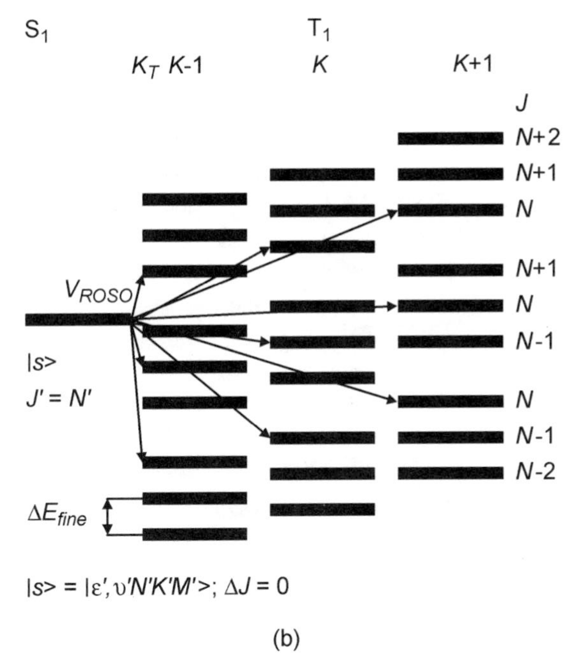

(b)

Figure 2. *(Continued).*

interaction between the new mixed levels of the triplet state, and the singlet level can be considered. Further on, this will be the procedure used. In this framework the interaction schemes of Fig. 3 will transform into ones of Fig. 4. Thus the interaction schemes, where the excited-state dynamics should be analyzed using the interactions coupling levels in the first and higher orders of the perturbation theory, may be reduced to a scheme, where the interaction couples levels in the first order of the perturbation theory. In the latter scheme, the effective S–T interaction $V_{ST}(B)$ becomes field-dependent.

For glyoxal and oxalylfluoride (C_{2h}), the singlet excited state $\widetilde{A}\,^1A_u$ can be coupled directly to the levels of the $\widetilde{a}\,^3A_u$ triplet state by the SO perturbation, also, the same states can be coupled in the second order of the VSO and ROSO perturbations. For the SO interaction, the matrix element

$$\langle A_u|A_g|A_u\rangle \tag{2.80}$$

does not vanish, yielding the selection rules $\Delta K = \pm 1$. Both these molecules have vibrational modes, which are transformed by the $5a_g + 2a_u + b_g + 4b_u$ irreducible representation of the C_{2h}-point group. Thus, for the VSO interaction,

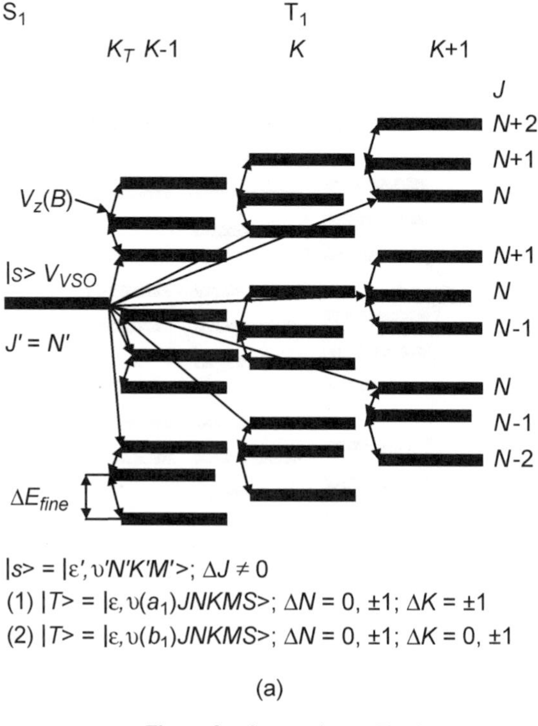

Figure 3. See caption to Fig. 2.

the only non vanishing matrix elements are

$$\left\langle A_u \middle| \begin{pmatrix} a_g \\ b_g \end{pmatrix} \middle| \begin{pmatrix} A_u \\ B_u \end{pmatrix} \right\rangle \left\langle \begin{pmatrix} A_u \\ B_u \end{pmatrix} \middle| \begin{pmatrix} A_g \\ B_g \\ B_g \end{pmatrix} \middle| A_u \right\rangle$$
$$\left\langle A_u \middle| \begin{pmatrix} A_g \\ B_g \end{pmatrix} \middle| \begin{pmatrix} A_u \\ B_u \end{pmatrix} \right\rangle \left\langle \begin{pmatrix} A_u \\ B_u \end{pmatrix} \middle| \begin{pmatrix} a_g \\ b_g \end{pmatrix} \middle| A_u \right\rangle \tag{2.81}$$

yielding $\Delta K = \pm 1$ and $\Delta K = 0, \pm 1$ as the respective selection rules for the vibronic modes of a_g and b_g. The ROSO matrix elements can be represented as

$$\left\langle A_u \middle| \begin{pmatrix} A_g \\ B_g \\ B_g \end{pmatrix} \middle| \begin{pmatrix} A_u \\ B_u \end{pmatrix} \right\rangle \left\langle \begin{pmatrix} A_u \\ B_u \end{pmatrix} \middle| \begin{pmatrix} A_g \\ B_g \\ B_g \end{pmatrix} \middle| A_u \right\rangle \tag{2.82}$$

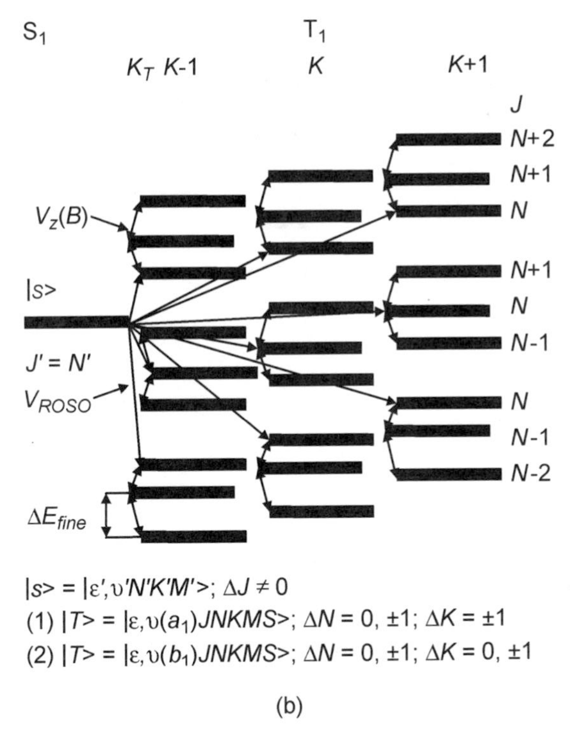

Figure 3. *(Continued).*

giving the selection rules $\Delta K = 0, \pm 2$. Thus, effective level density for both molecules in function of the interaction may be written as (SO) $\rho_{eff} = 6\rho_{vib}$, (VSO, a_g) $\rho_{eff} = 6\rho_{vib}$ (a_g) + $9\rho_{vib}(b_g)$, (VSO, b_g) $\rho_{eff} = 9\rho_{vib}$ $(a_g) + 6\rho_{vib}$ (b_g), (ROSO) $\rho_{eff} = 9\rho_{vib}$. Taking into account the selection rules obtained here and in Sections II.B and II.C, we can obtain interaction schemes at $B = 0$ and $B \neq 0$, very similar to those shown in Figs. 2–4. However, for glyoxal and oxalylfluoride, we have to consider additional interaction schemes describing direct SO interaction and VSO interaction for the a_g and b_g vibrational modes.

For the $\widetilde{A}^1A_u(C_{2h})$ singlet excited state of acetylene, there are at least six triplet states in the energy region of interest: $\widetilde{a}^3B_u$ (*trans*), $\widetilde{b}^3A_u$(*trans*), $\widetilde{c}^3B_u$(*trans*), $\widetilde{a}^3B_2$(*cis*), $\widetilde{b}^3A_2$(*cis*), and $\widetilde{c}^3A_2$(*cis*). Further on, we shall limit ourselves to the *trans* configuration of the acetylene molecule in the singlet excited and triplet states. The vibrational modes of acetylene in the excited *trans* states are $3a_g + a_u + 2b_u$. Thus, the first and second matrix elements relevant to

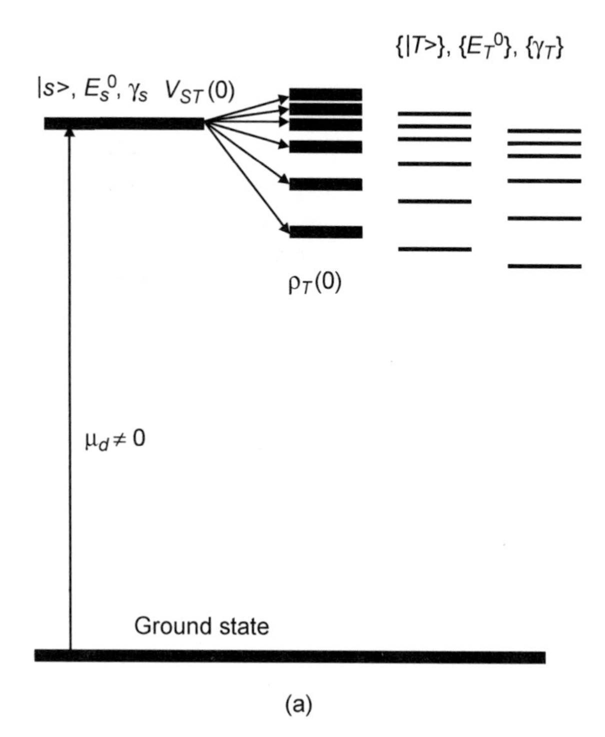

Figure 4. See legend to Fig. 2

the S–T interactions may be represented as

$$\begin{aligned}
&\left\langle A_u \left| \begin{pmatrix} A_g \\ B_g \\ B_g \end{pmatrix} \right| \begin{pmatrix} A_u \\ B_u \end{pmatrix} \right\rangle - \text{SO} \\
&\langle A_u | a_g | A_u \rangle \left\langle A_u \left| \begin{pmatrix} A_g \\ B_g \\ B_g \end{pmatrix} \right| \begin{pmatrix} A_u \\ B_u \end{pmatrix} \right\rangle - \text{VSO} \\
&\left\langle A_u \left| \begin{pmatrix} A_g \\ B_g \\ B_g \end{pmatrix} \right| \begin{pmatrix} A_u \\ B_u \end{pmatrix} \right\rangle \left\langle \begin{pmatrix} A_u \\ B_u \end{pmatrix} | a_g | \begin{pmatrix} A_u \\ B_u \end{pmatrix} \right\rangle - \text{VSO} \\
&\left\langle A_u \left| \begin{pmatrix} A_g \\ B_g \\ B_g \end{pmatrix} \right| \begin{pmatrix} A_u \\ B_u \end{pmatrix} \right\rangle \left\langle \begin{pmatrix} A_u \\ B_u \end{pmatrix} \left| \begin{pmatrix} A_g \\ B_g \\ B_g \end{pmatrix} \right| \begin{pmatrix} A_u \\ B_u \end{pmatrix} \right\rangle - \text{ROSO}
\end{aligned} \tag{2.83}$$

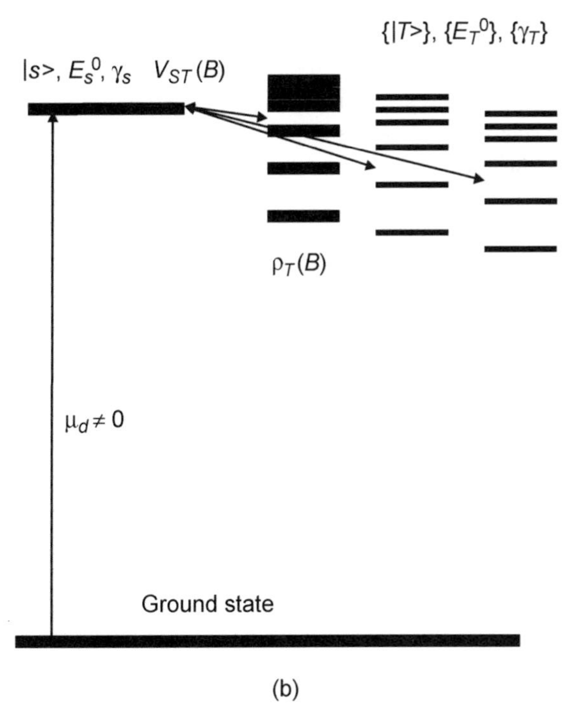

Figure 4. *(Continued).*

Thus, the selection rules in this case can be written as

$$
\begin{array}{lllll}
\Delta N = 0, \pm 1; & \Delta K = \pm 1 & \text{and} & \Delta N = 0, \pm 1; & \Delta K = 0 - \mathrm{SO} \\
\Delta N = 0, \pm 1; & \Delta K = \pm 1 & \text{and} & \Delta N = 0, \pm 1; & \Delta K = 0 - \mathrm{VSO} \\
\Delta N = 0, \pm 1; & \Delta K = \pm 1 & \text{and} & \Delta N = 0, \pm 1; & \Delta K = 0 - \mathrm{VSO} \\
\Delta N = 0, \pm 1; & \Delta K = 0, \pm 2 - \mathrm{ROSO} & & &
\end{array}
\tag{2.84}
$$

Using the same procedure, we can obtain the effective level density and consider interaction schemes at $B \neq 0$ and $B = 0$ for acetylene. We shall present the results later, because this problem has been analyzed earlier. The above systems are nearly-PST molecules, where energy of the K-splitting is much larger than that of the fine splitting. For the diazine and triazine systems, which correspond to nearly oblate (diazines) and oblate (s-triazine) symmetric tops, having close values of the A_r, B_r, C_r (or A_r, B_r) rotational constants, the energy of the K-splitting can be close to that of the fine splitting. In this case, magnetic field can also mix sublevels corresponding to different K-levels, causing extra complications. In the present section, we shall consider the general approach in the simplest case of the pyrazine molecule: D_{2h}-point group; close to the OST geometry. The analysis for pyrimidine and s-triazine is similar, and will be discussed in detail later.

The first singlet excited and triplet states of the pyrazine molecule are $\widetilde{A}\,{}^1B_{3u}$ and $\widetilde{a}\,{}^3B_{3u}$ respectively, with the y-axis directed along the N–N bond. Components of the axial vector are transformed by the B_{g1}, B_{g2}, and B_{g3} irreducible representations of the D_{2h}-point group. The molecule studied has 24 vibrational modes, which can be presented as follows: $5a_g + 2a_u + 4b_{1g} + 2b_{1u} + b_{2g} + 4b_{2u} + 2b_{3g} + 4b_{3u}$. The SO interaction does not couple levels of the $\widetilde{A}\,{}^1B_{3u}$ and $\widetilde{a}\,{}^3B_{3u}$ states, which can only be coupled by the second-order VSO and ROSO interactions. Thus, the nonvanishing matrix elements of these interactions may be represented as

$$\left\langle B_{3u}\middle|\begin{pmatrix} b_{1g} \\ b_{2g} \\ b_{3g} \end{pmatrix}\middle|\begin{pmatrix} B_{2u} \\ B_{1u} \\ A_u \end{pmatrix}\right\rangle\left\langle\begin{pmatrix} B_{2u} \\ B_{1u} \\ A_u \end{pmatrix}\middle|\begin{pmatrix} B_{1g} \\ B_{2g} \\ B_{3g} \end{pmatrix}\middle|B_{3u}\right\rangle - \text{VSO}$$
$$\left\langle B_{3u}\middle|\begin{pmatrix} B_{1g} \\ B_{2g} \\ B_{3g} \end{pmatrix}\middle|\begin{pmatrix} B_{2u} \\ B_{1u} \\ A_u \end{pmatrix}\right\rangle\left\langle\begin{pmatrix} B_{2u} \\ B_{1u} \\ A_u \end{pmatrix}\middle|\begin{pmatrix} b_{1g} \\ b_{2g} \\ b_{3g} \end{pmatrix}\middle|B_{3u}\right\rangle - \text{VSO} \qquad (2.85)$$
$$\left\langle B_{3u}\middle|\begin{pmatrix} B_{1g} \\ B_{2g} \\ B_{3g} \end{pmatrix}\middle|\begin{pmatrix} B_{2u} \\ B_{1u} \\ A_u \end{pmatrix}\right\rangle\left\langle\begin{pmatrix} B_{2u} \\ B_{1u} \\ A_u \end{pmatrix}\middle|\begin{pmatrix} B_{1g} \\ B_{2g} \\ B_{3g} \end{pmatrix}\middle|B_{3u}\right\rangle - \text{ROSO}$$

As follows from these expressions, selection rules for the interactions studied and the effective level density of the triplet states are

$$\begin{array}{lll} \Delta N = 0, \pm 1; & \Delta K = 0; & \rho_{eff} = 3\rho_{vib}(b_{1g}) - \text{VSO} \\ \Delta N = 0, \pm 1; & \Delta K = \pm 1; & \rho_{eff} = 6\rho_{vib}(b_{2g}) - \text{VSO} \\ \Delta N = 0, \pm 1; & \Delta K = \pm 1; & \rho_{eff} = 6\rho_{vib}(b_{3g}) - \text{VSO} \\ \Delta N = 0, \pm 1; & \Delta K = 0, \pm 2; & \rho_{eff} = 9\rho_{vib}(b_{3g}) - \text{ROSO} \end{array} \qquad (2.86)$$

Note that for this molecule, the values of the rotational constants A_r, B_r, C_r are close; thus the energy of the K-splitting given by

$$E(K) = \left(\frac{B_r + C_r}{2} - A_r\right)K^2 \qquad (2.87)$$

can be smaller than that of fine splitting of the triplet-state levels; that is, here the magnetic field can also couple levels with different K values. The interaction schemes at $B=0$ and $B \neq 0$ can be presented as shown in Fig. 5 for the VSO and ROSO interactions respectively. This case differs from that of a PST by the $f(N, K)$ factor value, which starts at 8/3, and may achieve the value of $K+1$ for formaldehyde, glyoxal, oxalylfluoride, and acetylene [19, 39–41]. The latter happens if the fine splitting energy of the triplet-state levels is larger than $((B_r+C_r)/2-A_r)N^2$, where N is the quantum number of the rotational angular momentum of the triplet-state level. We shall analyze this issue later using real molecular parameters.

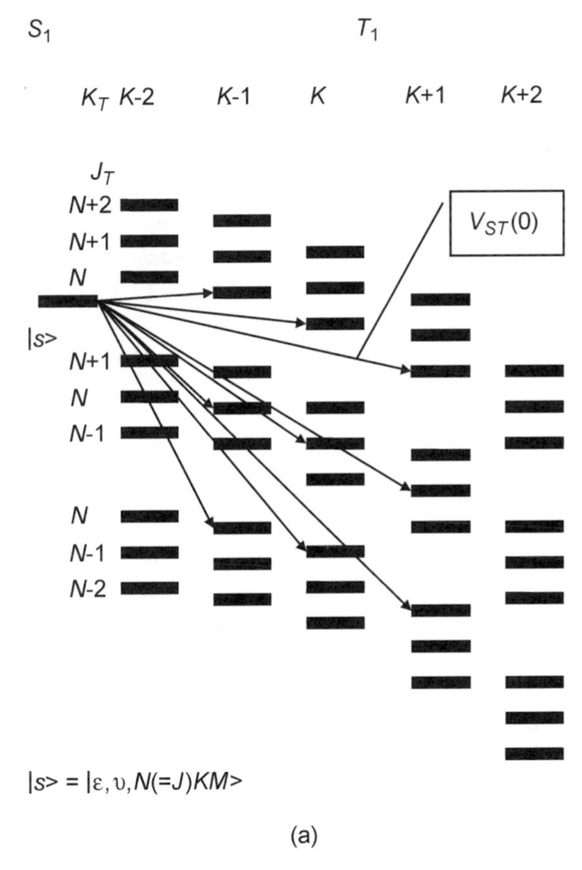

Figure 5. Interaction scheme for the VSO and ROSO S–T coupling mechanisms of an oblate symmetric top: (a) $B=0$, $\Delta J=0$. (b) $B=0$, $\Delta J \neq 0$. Both schemes are presented for a single vibronic triplet state only.

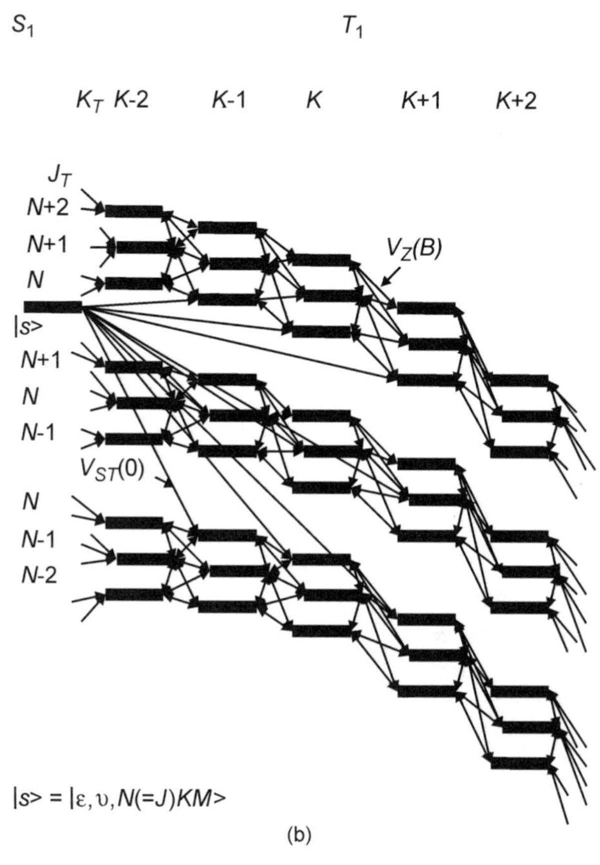

(b)

Figure 5. *(Continued).*

The last question we would like to address in general has been mentioned above—namely, representation of new triplet-state levels and of matrix elements coupling these, using the levels of the singlet state at $B \neq 0$.

E. Triplet-State Wave Functions and S–T Coupling Matrix Elements in the Presence of a Magnetic Field

Since the matrix element of the S–T interaction is much smaller than the fine splitting energy of the triplet-state levels, and the Zeeman interaction energy is close to that of the fine splitting in the field range typically used, we shall start with determining the new triplet-state levels by diagonalizing the molecular Hamiltonian over the Zeeman interaction while excluding the S–T interaction. As the second step, we can determine the matrix elements coupling the new triplet states with a singlet-state level. Such procedure has been implemented for a PST with $A_r \gg B_r$, C_r [19]. There, the energy of the fine splitting of the triplet-state levels was determined by the spin–rotational interaction, because the anisotropic spin–spin interaction was averaged to zero for high values of N within the approximations used. We shall consider this averaging mechanism in detail later, demonstrating briefly the procedure used [19], because we are going to use the same approximation. As follows from Eqs. (2.13)–(2.15) for $N \gg 1$ and omitting the terms for the anisotropic spin–spin interaction we obtain

$$\Delta E_1 = \Delta = |E_1(N+1,K) - E_2(N,K)| \approx |E_2(N,K) - E_3(N=1,K)| \tag{2.88}$$

For a molecule that can be described by a PST geometry, it is sufficient to analyze the magnetic field-induced coupling of fine components belonging to the same rotational triplet level only. The new triplet state wavefunctions may be written as [19]

$$|T_f^{(k)}\rangle = \sum_i^3 C_{fi}^{(k)}(B)|T_i^{(0)}\rangle \tag{2.89}$$

where $|T_f\rangle$ is the triplet state level wave function at $B \neq 0$, $|T_i^{(0)}\rangle$ is the triplet state level wave function at $B=0$, $k=1$, 2, 3 corresponds to the interaction schemes shown in Fig. 6, with $C_{fi}(B)$ determined as

$$\begin{aligned}
|C_{11}^{(k)}(B)| &= \frac{\Delta}{\sqrt{\Delta^2 + 2V_Z^2(B)}} \\
|C_{12}^{(k)}(B)| = |C_{13}^{(k)}(B)| &= \frac{V_Z(B)}{\sqrt{\Delta^2 + 2V_Z^2(B)}} \\
|C_{21}^{(k)}(B)| = |C_{31}^{(k)}(B)| &= \frac{V_Z(B)}{\sqrt{\Delta^2 + 2V_Z^2(B)}} \\
|C_{22}^{(k)}(B)| = |C_{33}^{(k)}(B)| &= \frac{V_Z^2(B)}{(\Delta + S(B))\sqrt{\Delta^2 + 2V_Z^2(B)}} \\
|C_{23}^{(k)}(B)| = |C_{32}^{(k)}(B)| &= \frac{V_Z^2(B)}{(\Delta - S(B))\sqrt{\Delta^2 + 2V_Z^2(B)}}
\end{aligned} \tag{2.90}$$

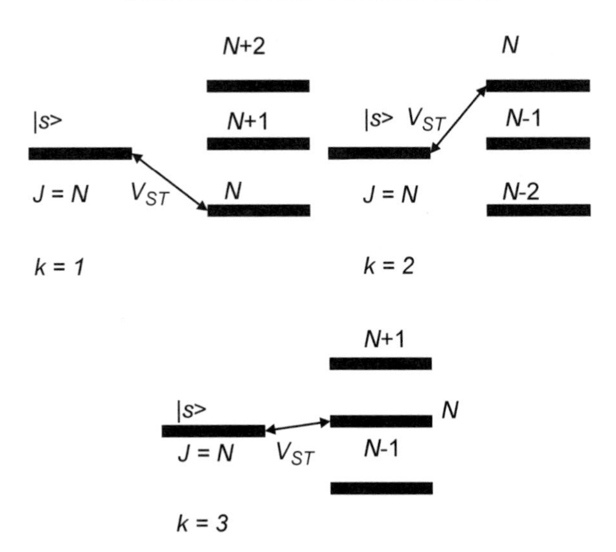

Figure 6. Interaction schemes corresponding to $k = 1,2$, and 3.

Thus, the required matrix elements may be represented as

$$\begin{aligned}
|\langle s|V_{ST}|T_{11}\rangle| &= V_{ST}\left|C_{13}^{(1)}(B)\right| = V^{(2)} \\
|\langle s|V_{ST}|T_{21}\rangle| &= V_{ST}\left|C_{23}^{(1)}(B)\right| = V^{(4)} \\
|\langle s|V_{ST}|T_{31}\rangle| &= V_{ST}\left|C_{33}^{(1)}(B)\right| = V^{(3)} \\
|\langle s|V_{ST}|T_{12}\rangle| &= V_{ST}\left|C_{12}^{(2)}(B)\right| = V^{(2)} \\
|\langle s|V_{ST}|T_{22}\rangle| &= V_{ST}\left|C_{22}^{(2)}(B)\right| = V^{(3)}
\end{aligned} \tag{2.91}$$

$$\begin{aligned}
|\langle s|V_{ST}|T_{32}\rangle| &= V_{ST}\left|C_{32}^{(2)}(B)\right| = V^{(4)} \\
|\langle s|V_{ST}|T_{13}\rangle| &= V_{ST}\left|C_{11}^{(3)}(B)\right| = V^{(1)} \\
|\langle s|V_{ST}|T_{23}\rangle| &= V_{ST}\left|C_{21}^{(3)}(B)\right| = V^{(2)} \\
|\langle s|V_{ST}|T_{33}\rangle| &= V_{ST}\left|C_{31}^{(3)}(B)\right| = V^{(2)}
\end{aligned} \tag{2.92}$$

As we can see from Eqs. (2.91) and (2.92), four different B-dependent S–T coupling matrix elements can be identified; also, $V^{(1)} \to 0$ at $B \to \infty$.

In the case of an OST, where $A_r = B_r \approx C_r$, the solution is much more complex and can only be analyzed numerically. In this case, different interaction schemes may be realized at $B \neq 0$, dependent on the relative values of the fine splitting Δ and on $(C_r - B_r)\, N^2$. The interaction schemes at $B = 0$ and $B \neq 0$ are

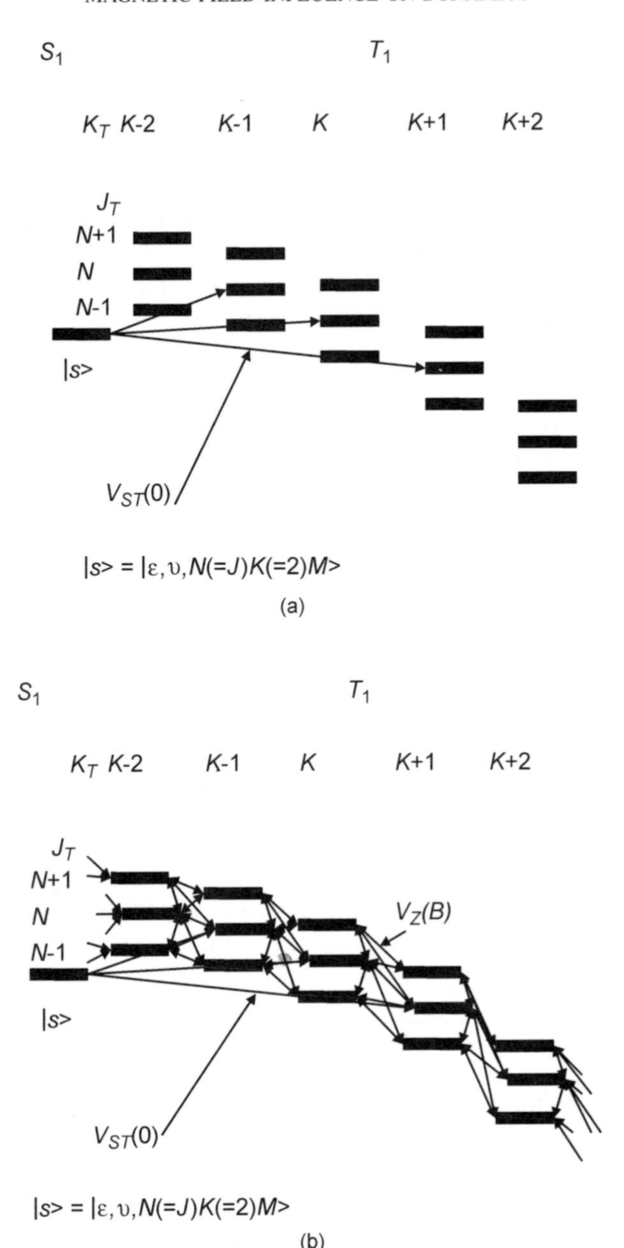

Figure 7. Interaction scheme determining the S–T coupling of an oblate symmetric top $\Delta \gg (C-B)N^2$: (a) $B=0$. (b) $B \neq 0$.

shown in Figs. 7a and 7b, respectively, for $\Delta \gg (C_r - B_r)\, N^2$. As we see from Fig. 7b, the secular equation defining eigenvalues of the new triplet-state levels at $B \neq 0$ has the order $3(N+1)$, thus we can't determine its roots analytically in a general case. We shall analyze this issue in detail while considering diazine and triazine.

Note that the detailed analysis of excited-state dynamics in the frameworks of the DM theory had been published by Lin and Fujimura [117]. Therefore, the present review will only consider the systems where the magnetic-field-induced processes are due to the indirect electron-spin- and nuclear-spin-decoupling mechanism.

III. MAGNETIC FIELD INFLUENCE ON EXCITED-STATE DYNAMICS IN OXALYLFLUORIDE, ACETYLENE, DIAZINE, AND TRIAZINE

We shall not discuss the MFE in spectroscopy and excited-state dynamics of the formaldehyde and glyoxal molecules, because the respective analysis was done for glyoxal [118], with the formaldehyde system being quite similar. However, we shall make use of the experimental results and the respective conclusions.

A. Oxalylfluoride

This molecule has two configurations, *trans* and *cis*, which can be distinguished spectroscopically [119]. Each has three electronic states in the spectral region of interest: $\widetilde{X}\,^1A_g, \widetilde{A}\,^1A_u, \widetilde{a}\,^3A_u$ and $\widetilde{X}\,^1A_1, \widetilde{A}\,^1A_2, \widetilde{a}\,^3A_2$ for the *trans* and *cis* configurations, respectively. Further on, we shall focus our attention on the *trans* configuration only, belonging to the C_{2h}-point group. The reason is that the *cis* configuration is less abundant:

$$\frac{n_{cis}}{n_{trans} + n_{cis}} \approx \exp(-1) \div \exp(-1.5)$$

due to the lower by about 200–300 cm^{-1} ground-state energy of the *trans* configuration [119], where n_{cis} and n_{trans} are ground-state populations of the respective conformations. Moreover, the absorption and laser-induced fluorescence (LIF) spectra of oxalylfluoride measured at room and cooled-jet conditions were found to mainly contain the bands and rotational lines corresponding to the *trans* configuration [19, 41, 120–127].

Absorption and LIF spectra of oxalyl fluoride are well known at lower spectral resolution, 0.1–0.2 cm^{-1} [19, 121–127], and at higher spectral resolution of 0.020–0.025 cm^{-1} for the 0_0^0 band [41, 120].

1. *Field Dependence*

MFE on the integrated LIF intensity and decay for different vibrational bands of the low-resolution LIF spectra has been studied by Makarov, Abe, and Hayashi [19] with 30-ns time resolution. Magnetic field reduced the integrated fluorescence intensity, as well as the decay lifetime and amplitude. The band intensities decrease steeply at lower fields $B < 0.3\,\mathrm{T}$, saturating at $B \gtrsim 0.3\,\mathrm{T}$. The magnetic field fluorescence quenching was band-dependent, while the band contours remained unchanged. Field dependences of the MFE, $R(B) = I_{fl}(B)/I_{fl}(0)$, measured for different bands of the LIF spectrum were studied in detail [19]. The typical half-width value $B_{1/2}$ of such dependences was within the 116- to 167-G range, while of the saturation value $R(\infty)$ of the field effect decreased with growing excess energy above the vibrationless level of the $\widetilde{A}\,^1A_u$ singlet state studied. The excess energy ΔE_{excess} dependence of $R(\infty) \approx R(0.3\,\mathrm{T})$ saturates at higher ΔE_{excess}. The same dependence of the $B_{1/2}$ is irregular, with the $B_{1/2}$ values distributed in an apparently random fashion in the 116- to 167-G range.

2. *Pressure Dependence*

Pressure dependences of $R(0.3\,\mathrm{T})$ are known for the $5_0^1 7_1^1$-[19, 128], and for the 0_0^0 bands [39]. Magnetic field quenching was found to be a collision-induced process for the 0_0^0-band [39], while for the $5_0^1 7_1^1$-band, it was observed even at the $P = 0.3\,\mathrm{mTorr}$ pressure of $(COF)_2$. This effect distinguishes the mechanisms of field-induced quenching of the two levels, to be discussed later.

3. *Oxalylfluoride Fluorescence Decay*

Fluorescence decay was measured for the 0_0^0 bands [19, 39–41] and for the $5_0^1 7_1^1$ bands [19, 128]. Decays were measured for the individual rotational lines of the 0_0^0-band in a supersonic collimated molecular beam [41]. The decay of the 0_0^0-band fluorescence at $B = 0$ could be fitted by an exponential function, with a biexponential approximation required at $B \neq 0$. The fast decay component lifetime $\tau_f^{-1} = (2.36 \pm 0.19) \times 10^7\,\mathrm{s}^{-1}$ is field-independent, whereas the slow component lifetime depends on B. The fast component amplitude increases, while that of the slow one decreases with the field. For the $5_0^1 7_1^1$-band fluorescence, the decay could be fitted using the biexponential approximation at $B = 0$ and using a three-exponential approximation at $B \neq 0$:

$$\begin{aligned} I_{fl}(t, B) = {} & A_f(B) \cdot \exp\left(-\frac{t}{\tau_f}\right) + A_{s1}(B) \cdot \exp\left(-\frac{t}{\tau_{s1}(B)}\right) \\ & + A_{s2} \cdot \exp\left(-\frac{t}{\tau_{s2}}\right) \end{aligned} \tag{3.1}$$

with $A_f(B) + A_{s1}(B) + A_{s2} = A_f(0) + A_s(0)$ and $\tau_{s1}(0) = \tau_{s2} = \tau_s(0)$. Here, τ_f is the field-independent fast-component lifetime, $\tau_s(0)$ is the slow-component lifetime,

$A_f(0)$ is the fast-component amplitude, and $A_s(0)$ is the slow component amplitude, all at $B = 0$.

The fluorescence decay studies in supersonic collimated beam showed that the decay of a single rotational line of the 0_0^0 band could be fitted by an exponential function at $B = 0$, with the biexponential approximation required at $B \neq 0$ [41]. The fast component lifetime varies from one rotational line to the other, staying in the 37.8- to 53.5-ns range. At higher B values the slow-component lifetime increases, the fast-component amplitude increases, and the slow-component amplitude decreases, with the integrated signal intensity remaining constant.

Pressure dependences (Stern–Volmer plots) of the reverse lifetime of the slow decay component for the 0_0^0 and $5_0^1 7_1^1$ bands were also taken at $B = 0$ and at $B = 0.3\,\text{T}$ [19, 39, 128]. All the plots for the 0_0^0 band were linear, with the slope and the inverted intercept being $1.57 \times 10^7\,\text{Torr}^{-1}\,\text{s}^{-1}$, $18\,\mu\text{s}$ and $2.24 \times 10^7\,\text{Torr}^{-1}\,\text{s}^{-1}$, $27\,\mu\text{s}$ at $B = 0$ and $B \neq 0$, respectively. Thus, the collisional rate constant of the fluorescence quenching increases with a field, while the collisionless rate constant decreases. In case of the $5_0^1 7_1^1$ band, the plot for $B \neq 0$ is nonlinear in the low-pressure region, with the slope and the inverted intercept at lower pressures of $1.82 \times 10^7\,\text{Torr}^{-1}\,\text{s}^{-1}$, $20\,\mu\text{s}$ and $4.50 \times 10^7\,\text{Torr}^{-1}\,\text{s}^{-1}$, $20\,\mu\text{s}$ at $B = 0$ and $B \neq 0$, respectively; the same parameters estimated in the "high"-pressure range were $1.84 \times 10^7\,\text{Torr}^{-1}\,\text{s}^{-1}$, $20\,\mu\text{s}$ and $2.36 \times 10^7\,\text{Torr}^{-1}\,\text{s}^{-1}$, $1.0\,\mu\text{s}$ at $B \neq 0$ and $B = 0$, respectively. Thus, at $B \neq 0$ the collision-induced rate constant of fluorescence quenching decreases passing from lower to higher pressures, while the collisionless rate constant increases. This phenomenon is very interesting, and it will be explained later using the qualitative model published earlier.

4. *The J-Dependence of the $B_{1/2}$ Parameter Measured for Rotational Levels of the 0_0^0 Band in the Supersonic Collimated Beam*

This dependence is very useful, while its analysis presents a very interesting problem. The $B_{1/2}$ versus $(2J' + 1)^{-1}$ plot is linear (see Fig. 8) [41]; the slope is $0.12\,\text{T}$, with $B_{1/2}$ decreasing at higher J. In the next section, while discussing theoretical models of the MFE in the $(COF)_2$ fluorescence, we shall also explain this phenomenon.

5. *Field Effect on the $(COF)_2$ Phosphorescence*

The integrated intensity of the $(COF)_2$ phosphorescence was found to grow at $B \neq 0$, upon excitation to the 0^0 level [39, 40]. The field and pressure dependences of the MFE defined as $I_{ph}(0)/I_{ph}(B)$ were studied. These dependences and those measured for the 0_0^0-band fluorescence quenching are practically the same. Thus, it has been assumed that an external magnetic field influences the S–T conversion process only [39, 40].

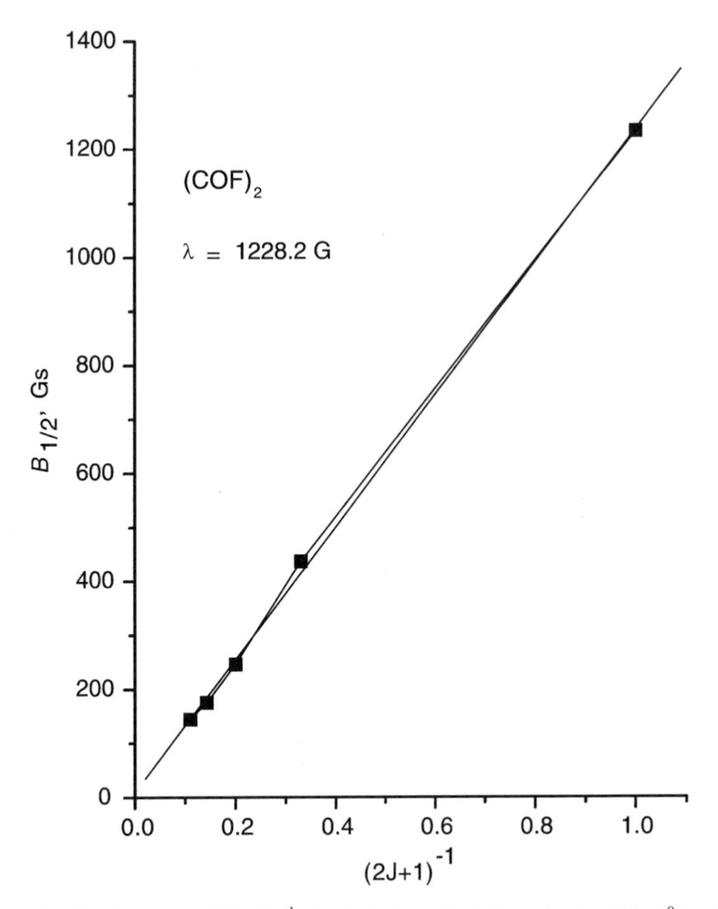

Figure 8. The $B_{1/2}$ versus $(2J+1)^{-1}$ plot for individual rotational levels of the 0^0_0 band of the oxalylfluoride molecule.

6. *OD EPR Effect*

The unresolved OD EPR spectrum measured under excitation of the oxalylfluoride molecule to the 0^0 level was recorded using microwave field in the 3-cm band [37, 38, 40]. Detailed power dependences of the microwave field effects measured in the maximum of the studied signal for the fluorescence intensity decrease and phosphorescence intensity increase are shown in Fig. 9. Fluorescence decay measured in the maximum of this signal has also been studied. It was found that (i) the width of the observed OD EPR signal is about 5.4 mT, (ii) the intensity of the OD EPR signal saturates at higher microwave power, (iii) the intensity of the $(COF)_2$ phosphorescence increases and saturates at higher

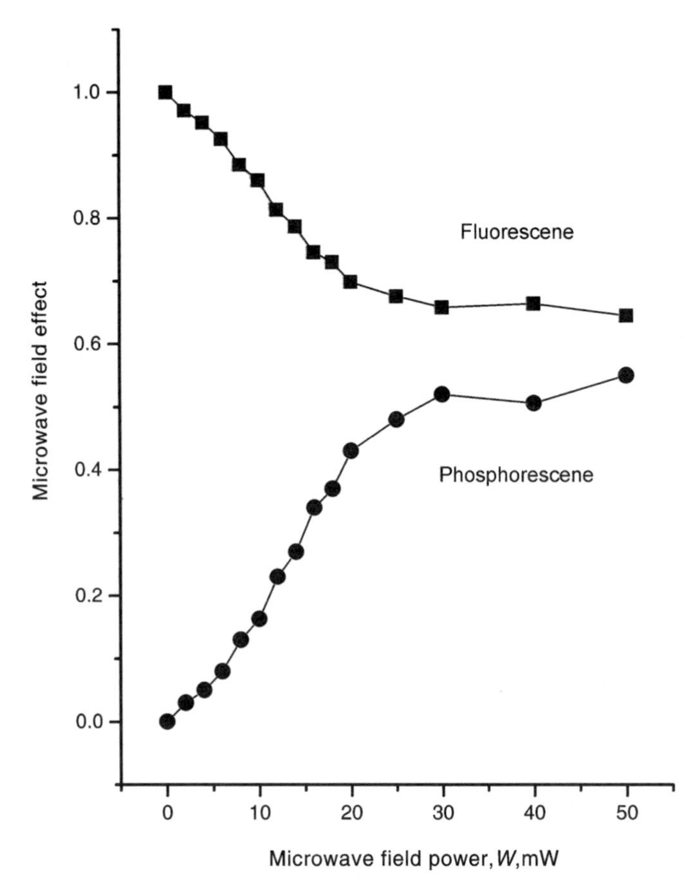

Figure 9. Microwave-power dependence of the OD EPR signal measured for oxalylfluoride fluorescence and phosphorescence in the maximum of the 0_0^0-band LIF spectrum at $B = 0.3295$ T, $P = 30$ mTorr, $T = 300$ K.

microwave power, and (iv) in the presence of a microwave field, both lifetime and amplitude of the slow decay component decrease.

In a supersonic collimated beam, the OD EPR spectrum measured for a single $4_{31} \leftarrow 3_{21}$ rotational line was resolved, with the measured spectrum including a set of lines. We shall analyze this spectrum later using the previously published theoretical approach [41].

7. *Experimental Data Analysis*

The indirect mechanism theory (ENSDM) has been extensively used to analyze the experimental data [19, 37–41]. The S–T coupling between levels of the $0^0(\widetilde{A}^1A_u)$ and $\widetilde{a}^3A_u$ states results from the second-order VSO interaction [39], whose selection rules have been determined above in Eqs. (2.81) and (2.82). It was also shown that for all vibrational bands and rotational lines of the 0_0^0-band, the condition $\eta_t(0) \leq 1$ is satisfied [19, 37–41]. It means that at $B = 0$ there is at least one level of the triplet state in the "interaction" resonance with the singlet level studied. Note also that since the decay of the 0^0-state levels can be fitted by an exponential function [41], we can assume that for all these levels the condition $\eta_t(0) < 1$ should also be satisfied. Given that (i) the fast-component lifetime is field-independent, (ii) the field dependences are saturated, and (iii) the MFE $I_{fl}(B)/I_{fl}(0)$ decreases with an increase of the excess energy above the vibrationless level of the $\widetilde{A}^1A_u$ state, it has been concluded that the observed MFEs on the $(COF)_2$ fluorescence can be explained using the ENSDM [19, 37–41]. Let us consider this mechanism step by step as applied to the problem studied.

8. *Decay*

As we already mentioned, the decay of initially prepared state can be fitted by a biexponential function, when one of the two sets of conditions is met:

$$\Delta\omega_{exc} \gg \langle|V_{ST}(0)|\rangle; \qquad \eta_T(0) = \rho_T(0) \cdot \langle|V_{ST}(0)|\rangle \gg 1; \qquad \zeta_T(0) = \frac{\langle|V_{ST}(0)|\rangle}{\langle\gamma_T\rangle} > 1 \tag{3.2}$$

or

$$\Delta\omega_{exc} \gg \langle|V_{ST}(0)|\rangle; \qquad \eta_T(0) = \rho_T(0) \cdot \langle|V_{ST}(0)|\rangle \leq 1; \qquad \zeta_T(0) = \frac{\langle|V_{ST}(0)|\rangle}{\langle\gamma_T\rangle} > 1; \qquad G \gg 1 \tag{3.2}$$

Here, $\Delta\omega_{exc}$ is the coherent width of the exciting radiation, $\langle|V_{ST}(0)|\rangle$ is the average value of the S–T interaction at $B = 0$, $\rho_T(0)$ is the effective density of the triplet-state levels, $\langle\gamma_T\rangle$ is the average value of the triplet-level width, and G is the number of the singlet-state levels excited. In both these cases, relations (2.10) can be used to analyze the fluorescence decay. In oxalylfluoride, the conditions of (3.2) are applicable, because the condition $A_f(0)/A_s(0) < 1$ is valid for all the levels studied [19, 37–41]. Using this approximation and the theoretical statements shown above, the $V_{ST}(0)$ and $\rho_T(0)$ values were estimated for the 0_0^0 band, both as averaged values and values for individual rotational lines [19, 37–41].

The $B_{1/2}$ versus $(2J' + 1)^{-1}$ dependence is linear [41]. This result can be explained by the averaging effect of the anisotropic spin–spin interaction [37–41], which, together with the isotropic spin–rotational interactions, determines the fine splitting energy of the triplet levels. The anisotropic interaction averaging occurs via interaction between low-frequency vibrational and rotational motions of the studied system, termed Coriolis interaction. The slope of the $B_{1/2}$ versus $(2J'+1)^{-1}$ plot determines the anisotropic spin–spin constant λ [39, 41], if the $J'=N'$ values of the excited singlet state correlate with the corresponding N values of the triplet state. Such correlation is quite accurate for systems in the low-level density limit. Thus, for the 0_0^0 band of oxalylfluoride, $\lambda \approx 0.12$ T. In diazine and triazine, to be considered below, the level density has intermediate values. Hence, the rotational singlet-state levels with a defined $J'=N'$ are coupled to triplet-state levels with $N=N'$, $N' \pm 1$. Here the slope of the $B_{1/2}$ versus $(2J'+1)^{-1}$ plot is proportional to λ, which can be determined from the analysis of the $B_{1/2}$ versus $(2\overline{N}+1)^{-1}$ plot [39, 41], where $\overline{N}$ is the averaged value of the rotational quantum number of the triplet-state levels coupled to the singlet-state level considered.

The condition

$$E(M_N.M_I) = \left(\chi \cdot \frac{K \cdot M_N}{N} + \mu \cdot M_N. + a_{FC} \cdot M_I\right) \cdot M_S \tag{3.3}$$

was used to analyze the OD EPR spectrum [41]. Here, the spin rotational constant $\chi \approx 0$, while $\mu = 0.41$ mT and $a_{FC} = 2.72$ mT.

We have considered the magnetic field quenching of the oxalylfluoride fluorescence in detail, as this analysis demonstrates almost every possible method of the S–T conversion studies using external magnetic and microwave fields.

B. Acetylene

Acetylene is one of the fundamental molecules in chemistry and a key in bridging the gap between diatomic molecules and large polyatomic molecules. The 210- to 240-nm spectral region of acetylene was studied extensively [129–134]. Ingold and King [132] and Innes [133] had shown that the upper state has a C_{2h} nonlinear trans-planar geometry, in contrast to the well-known linear ground state. The acetylene molecule in the singlet excited state is close to a PST, whose rotational levels are described by the total angular momentum J'. For a singlet state, $J'=N'$; thus the relevant quantum numbers are the rotational angular momentum N' and its projection K' on the symmetric top axis. Detailed vibrational and rotation analysis of the system has been presented by Watson, Herman, Van Crean, and Colin [135]. Zeeman quantum beats [136], Zeeman anticrossing [137], Stark anticrossing [138], and the decrease of relative quantum yield below the predissociation threshold [139] show that the levels of the $\widetilde{A}^1A_u$ state are coupled by intramolecular interactions with those of a neighboring "dark" state. Dupre et al. [137, 140] have found a rapid increase in the Zeeman anticrossing density with increasing excitation energy below the threshold. The acetylene predissociation in the spectral region considered has been studied by Haijima, Fuji, and Ito [139, 141]. They found that the fluorescence quantum yield of acetylene is strongly dependent on the rotational quantum number of the excited state in the 46,339 cm^{-1} and 46,673 cm^{-1} spectral regions. Later Suzuki and Hashimoto [142] studied predissociation of acetylene excited to individual rovibronic levels of the $\widetilde{A}^1A_u$ state. They found that the fluorescence quantum yield dependences on the rotational quantum number J' measured for the different vibronic subbands of the $\widetilde{A}^1A_u \leftarrow \widetilde{X}^1A_g$ transition have the same profile as that obtained by Haijima, Fuji, and Ito [139, 141]; that is, the fluorescence quantum yield drops sharply at higher J', while the predissociation yield of the H atom is virtually independent on J'. Abe and Hayashi have studied MFE on the fluorescence quantum yield of acetylene [8]. They found that the fluorescence quantum yield decreases with a field. In our study [143], we showed that the magnetic-field quenching of the $V_1^2K_1^{m=0,1,2}$ subband fluorescence has a collisional nature, because the integrated intensity of the acetylene fluorescence is field-independent in the collisionless conditions.

Collisional quenching of different vibrational levels of the $\widetilde{A}^1A_u$ state has been studied by Stephenson, Blazy, and King [144]. Fluorescence decay of individual rotational levels of the $K' = 1$ subband was measured, and could be fitted by an exponential function with good accuracy [144]. An increase of the vibrational quantum number from 0 to 2 caused the rate constant of the collision-induced quenching of the excited acetylene by the ground-state acetylene to increase from 4.37 to 14.6 μs^{-1} $Torr^{-1}$, with the collisionless constant decreasing from 3.69 to 2.90 μs^{-1}. Note that the decay lifetime dependence on the rotational quantum number is very weak for all the vibrational levels studied. The collisional and collisionless rate constants for different rotational lines of the $V_1^2K_1^{m=0,1,2}$ subbands were also measured [145], comparing well with the data obtained by Stephenson, Blazy, and King for the $V_1^2K_1^1$ subband, with larger differences for the $V_1^2K_1^{m=0,2}$ subbands [144]. Ochi and Tsuchiya obtained decay lifetimes of individual rotational levels of the $\upsilon' = 3$ state lying in the 45,303-cm^{-1} region in the 0.45- to 2.4-μs range [146]. Thus, the decay lifetime is essentially dependent on the rovibronic state studied.

1. *Magnetic Quenching of the Acetylene Fluorescence*

Abe and Hayashi studied magnetic-field quenching of the acetylene fluorescence for individual rotational lines of the $V_0^{n=1-4}$ bands, at $P = 20$ mTorr [8]. Makarov, Cruz, and Quinones studied such quenching for the $V_1^2K_1^{m=0,1,2}$ subbands in the 3- to 598-mTorr pressure range, as well as in supersonic jets [143]. Abe and Hayashi demonstrated that (i) a strong decrease of the fluorescence intensity in function of B is observed at lower fields $B < 0.2$ T for the $n = 1$–4 levels, with low saturation values achieved at $B \gtrsim 0.2$ T, (ii) within a vibrational level, the MQ is more pronounced as the rotational quantum number increases, and (iii) the MQ at high field strengths is more pronounced as the vibrational quantum number increases. Abe and Hayashi analyzed the S–T conversion problem using the indirect ENSDM mechanism. Makarov, Cruz, and Quinones showed for the $V_1^2 K_1^{m=0,1,2}$ subbands that the field effect studied has a collisional nature [143]. The observed data were explained using the IM theory in the limit of low level density.

The IM was discussed above and has been employed for the analysis of the S–T conversion in the $(COH)_2$ and $(COF)_2$ molecules, which belong to the C_{2h}-point group along with the $\widetilde{A}^1A_u$ state of C_2H_2. Note, however, that the problem is much more complex for acetylene as compared to glyoxal and oxalylfluoride, because acetylene has at least six triplet electronic states in the energy range of interest [142], together with the dissociative limit of the $\widetilde{X}^1A_u$ ground state at about (131 ± 1) kcal/mol [147–159].

C. Diazines

The diazine (pyrazine and pyrimidine) spectroscopy and excited-state dynamics have been studied extensively [14–16, 42–50]. The excited-state dynamics in diazines at $B \neq 0$ is much more complex than in acetylene or oxalylfluoride, given that all the azobenzene systems considered are at least very close to an OST symmetry; thus the conditions $A_r \approx B_r$ ($A_r = B_r$ for *s*-triazine) and $(B_r - C_r) \ll B_r$ are satisfied. In this case, an external magnetic field should also mix the rotational levels with different values of K_a, corresponding to the same vibronic level of the triplet state. Besides, these systems include nitrogen ($I_N = 1$) and hydrogen ($I_H = 1/2$) or deuterium ($I_D = 1$) atoms; thus the hyperfine structure of the triplet levels is much more complex than that in acetylene or oxalylfluoride. We also have to take into account the hyperfine structure of the triplet-state levels at $B \neq 0$. We shall briefly consider the experimental data published and propose a qualitative explanation using the above theoretical models, because no detailed theoretical analysis of the excited-state dynamics of these molecules at $B \neq 0$ is available.

1. *Pyrazine*

Rotational dependence of fluorescence quantum yield and decay of pyrazine-h_4 and pyrazine-d_4 vapors at $B < 200$ G has been studied [14, 15, 44, 47, 48]. The profile of the integrated fluorescence dependence on the magnetic field strength changed at higher quantum number J' of the total angular momentum of the excited state. The decay profile varied as well: At higher J', the saturation value of the MFE and the half-width $B_{1/2}$ of the B-dependence decrease; the fluorescence decay at both $B = 0$ and $B \neq 0$ demonstrates quantum beat effects for low J', while at higher J' it can be fitted using the biexponential approximation. Thus, with an increase in J', we can observe a transition from the low-level density limit to the intermediate case. A very significant experimental result is the fact that the $B_{1/2}$ versus $(2J' + 1)^{-1}$ plot can be fitted by a linear function for both the pyrazine-h_4 and pyrazine-d_4 molecules (Fig. 10). The same was

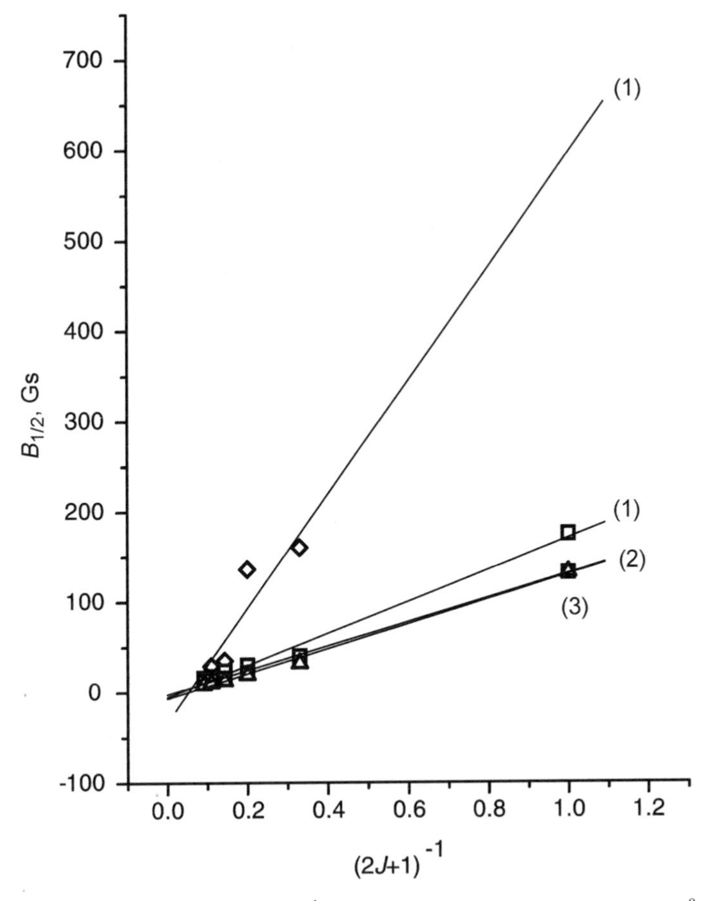

Figure 10. The $B_{1/2}$ versus $(2J+1)^{-1}$ plot for single rotational levels of the 0_0^0 band of pyrazine-h_4, 0_0^0 band of pyrazine-d_4, $6a_0^2$ band of pyrimidine, and 6_0^2 band of *s*-triazine.

observed for the rotational levels of the vibrationless state of oxalylfluoride [41] (see above in Section III.A.8). The dependence observed in the oxalylfluoride has been explained by an averaging effect of the anisotropic spin–spin interaction, which determines the fine splitting energy of the triplet-state levels. The mechanism of this averaging involves interaction between rotational and vibrational molecular motions in the triplet state. This interaction mixes rotational levels of the triplet state with different K_a values. It has been argued that in the triplet state coupled to the singlet state of pyrazine considered, the Coriolis interaction is significant, mixing rotational levels of this triplet state with different K_a values [43–45, 160–175]. Thus, the theory developed for the oxalylfluoride molecule [41] can also be applied to the analysis of the $B_{1/2}$ versus $(2J'+1)^{-1}$ plots obtained for different vibronic states of pyrazine-h_4 and pyrazine-d_4. The slope of this dependence is determined by the anisotropic spin–spin constant in the low-level density systems, while for the intermediate-level density, the slope is proportional to this constant [41].

2. *Pyrimidine*

Magnetic field effects on fluorescence intensity and fluorescence decay of pyrimidinevapor were reported for excitation into individual rotational levels of different vibronic states of the S_1 singlet state [15, 176]. The behavior of pyrimidine fluorescence at $B \neq 0$ is close to that of pyrazine fluorescence for all the vibronic states studied. The $B_{1/2}$ versus $(2J'+1)^{-1}$ plot could be fitted by a linear function (Fig. 10), whose slope is proportional to the anisotropic spin–spin constant, for all the vibronic bands studied: 0_0^0, $6a_0^1$, 12_0^2, and $6a_0^2$.

3. *s-Triazine*

The *s*-triazine molecule is an exact OST. The fluorescence quantum yield and lifetime of the *s*-triazine vapor emitting from an initially prepared level (IPL) of the S_1 state depend on the excited rotational level, namely the quantum yield φ_{fl} and lifetime τ_s of the slowly decaying portion of fluorescence emitted from an IPL respectively decreased and increased at growing J' [16, 50, 177–180]. Transformations of the fluorescence decay profile at growing J' for *s*-triazine are similar to those in pyrazine and pyrimidine [168, 177–180]. The *B*-dependence of the MFE of the integrated fluorescence varies with growing J', starting from $J'=1$, in the same way as for pyrazine and pyrimidine: the $B_{1/2}$ versus $(2J'+1)^{-1}$ plot can be fitted by a linear function, as shown for the 6_0^2 band in Fig. 10 [16, 50]. The slope of this dependence is 0.0656 T. If we assume the same mechanism for the $B_{1/2}$ decrease in triazine as that in oxalylfluoride, pyrazine, and pyrimidine, the slope obtained should be proportional to the anisotropic spin–spin constant.

D. Estimates of the Anisotropic Spin–Spin Constants in Diazine and Triazine

The diazine and triazine systems correspond to the intermediate molecular case; thus, to estimate the anisotropic spin–spin constants, we have to take into account the coupling of the singlet-state rotational level studied to different rotational levels of the triplet state, using the selection rules obtained. Using the averaging mechanism of the anisotropic spin–spin interaction, and defining averaged value of J_T for triplet levels coupled to the singlet one studied, we estimated anisotropic spin–spin constants for pyrazine-h_4 of about 898 G, pyrazine-d_4 586 G, pyrimidine 637 G, and *s*-triazine 661 G. Thus, the study of the J'-dependence of the $B_{1/2}$ can yield estimates of different spectroscopic parameters of the neighboring "dark" triplet state.

IV. CONCLUSIONS

Study of the S–T conversion dynamics in magnetic field can give detailed information about the mechanism of the S–T conversion and the structure of the triplet-state levels lying in the quasi-resonance with the excited singlet levels studied. We demonstrated that for acetylene, oxalylfluoride, and azobenzene systems, where magnetic field affects the S–T conversion dynamics, such information has been obtained, helping to better understand the respective S–T conversion mechanisms.

References

1. W. Steubing, *Verh. Disch. Phys.*, 1181 (1913).
2. R. Solarz, S. Butler, and D. H. Levy, *J. Chem. Phys.* **58**, 5172 (1973).
3. M. Lombardi, R. Jost, C. Michel, and A. Tramer, *Chem. Phys.* **46**, 273 (1980).
4. M. Matsuzaki and S. Nagakura, *Z. Phys. Chem. Neue Folge, Bd.* **101**, 283 (1976).
5. N. I. Sorokin, N. L. Lavrik, G. I. Skubnevskaya, N. M. Bazhin, and Yu. N. Molin, *Nouv. J. Chim.* **4**, 395 (1980).
6. A. Matsuzaki and S. Nagakura, *Bull. Chem. Soc. Japan* **49**, 359 (1976).
7. V. I. Makarov, N. L. Lavrik, G. I. Skubnevskaya, and N. M. Bazhin, *Opt. Spektrosk.* **50**, 290 (1981).
8. H. Abe and H. Hayashi, *Chem. Phys. Lett.* **206**, 337 (1993).
9. Ch. Ottinger and A. F. Vilisov, *J. Chem. Phys.* **100**, 1805 (1994).
10. Y. Fukuda, H. Hayashi, and S. Nagakura, *Chem. Phys. Lett.* **119**, 480 (1985).
11. H. Abe, *MR Chem. Dynamics Team, MR Science Research Group, The Institute of Physical and Chemical Research (RIKEN)*, Japan, February 4–5, p. 1 (1995).
12. M. Sumitani, H. Abe, and S. Nagakura, *J. Chem. Phys.* **94**, 1923 (1991).
13. V. I. Makarov, I. V. Khmelinskii, S. A. Kochubei, and V. N. Ishchenko, *Chem. Phys.*, accepted for publication.
14. Y. Matsumoto, L. H. Spangler, and D. W. Pratt, *J. Chem. Phys.* **80**, 5539 (1984).
15. N. Ohta, T. Takemura, M. Fujita, and H. Baba, *J. Chem. Phys.* **88**, 4197 (1988).
16. N. Ohta and T. Takemura, *J. Chem. Phys.* **93**, 877 (1990).
17. H. Abe and H. Hayashi, *J. Phys. Chem.* **98**, 2797 (1994).
18. J. Nakamura and S. Nagakura, *Chem. Phys. Lett.* **74**, 228 (1980).
19. V. I. Makarov, H. Abe, and H. Hayashi, *Mol. Phys.* **84**, 911 (1985).
20. M. Lombardi, R. Jost, C. Michel, and A. Tramer, *Chem. Phys.* **57**, 341 (1981).
21. M. Lombardi, R. Jost, C. Michel, and A. Tramer, *Chem. Phys.* **57**, 355 (1981).
22. A. Matsuzaki and S. Nagakura, *Helvet. Cem. Acta* **61**, 675 (1978).
23. P. R. Stannard, *J. Chem. Phys.* **88**, 3932 (1978).
24. R. Jost, M. Lombardi, C. Michel, and A. Tramer, *Nuovo Chimento* **63B**, 228 (1981).

25. H. G. Kuttner, H. L. Selzle, and E. W. Schlag, *Chem. Phys. Lett.* **48**, 207 (1977).
26. M. Matsuzaki and S. Nagakura, *J. Luminescence* **12/13**, 787 (1976).
27. A. Matsuzaki and S. Nagakura, *Chem. Phys. Lett.* **37**, 204 (1976).
28. W. Goetz, A. J. McHugh, and D. A. Ramsay, *Can. J. Phys.* **48**, 1 (1970).
29. J. Nakamura, K. Hashimoto, and S. Nagakura, *J. Luminescence* **24/25**, 763 (1981).
30. C. Michel and C. Tric, *Chem. Phys.* **50**, 341 (1980).
31. H. G. Kuttner, H. L. Selzle, and E. W. Schlag, *Chem. Phys. Lett.* **28**, 1 (1978).
32. P. Dupre, R. Jost, and M. Lombardi, *Chem. Phys.* **91**, 355 (1984).
33. N. I. Sorokin, N. M. Bazhin, and G. G. Eremenchuk, *Chem. Phys. Lett.* **99**, 181 (1983).
34. P. Dupre, R. Jost, M. Lombardi, C. Michel, and A. Tramer, *Chem. Phys.* **82**, 25 (1983).
35. E. P. Peyroula, R. Jost, M. Lombardi, P. Dupre, and A. Tramer, *Chem. Phys.* **102**, 417 (1986).
36. C. Michel, M. Lombardi, and R. Jost, *Chem. Phys.* **109**, 357 (1986).
37. V. I. Makarov, Yu. N. Molin, S. A. Kochubei, and V. N. Ischenko, *Chem. Phys. Lett.* **266**, 303 (1997).
38. V. I. Makarov, Yu. N. Molin, S. A. Kochubei, and V. N. Ishchenko, *Dokl. RAS* **352**, 768 (1997).
39. V. I. Makarov, S. A. Kochubei, V. N. Ischenko, R. N. Musin, G. A. Bogdanchikov, and I. V. Khmelinskii, *Chem. Phys.* **242**, 37 (1999).
40. V. I. Makarov, S. A. Kochubei, V. N. Ischenko, and I. V. Khmelinskii, *Mol. Phys.* **96**, 1231 (1999).
41. V. I. Makarov, S. A. Kochubei, V. N. Ischenko, and I. V. Khmelinskii, *J. Chem. Phys.* **111**, 5783 (1999).
42. P. M. Felker, W. R. Lambert, and A. H. Zewail, *Chem. Phys. Lett.* **89**, 309 (1982).
43. Y. Matsumoto, L. H. Spangler, and D. W. Pratt, *Chem. Phys. Lett.* **98**, 333 (1983).
44. Y. Matsumoto, L. H. Spangler, and D. W. Pratt, *J. Chem. Phys.* **80**, 5539 (1984).
45. Y. Matsumoto, L. H. Spangler, and D. W. Pratt, *Laser Chem.* **2**, 91 (1983).
46. A. Amirov, *Chem. Phys.* **126**, 365 (1988).
47. N. Ohta and T. Takemura, *J. Chem. Phys.* **95**, 7133 (1991).
48. N. Ohta and T. Takemura, *J. Chem. Phys.* **91**, 4477 (1989).
49. N. Ohta and T. Takemura, *J. Chem. Phys.* **95**, 7120 (1991).
50. N. Ohta and T. Takemura, *J. Phys. Chem.* **94**, 3466 (1990).
51. G. Herzberg, *Molecular Spectra and Molecular Structure. I. Electronic Spectra and Electronic Structure of Diatomic Molecules*, New York, 1961.
52. G. Herzberg, *Molecular Spectra and Molecular Structure. III. Electronic Spectra and Electronic Structure of Polyatomic Molecules*, New York, 1966.
53. U. Fano, *Phys. Rev.* **24**, 1866 (1961).
54. R. A. Harris, *J. Chem. Phys.* **39**, 978 (1963).
55. M. Bixon and J. Jortner, *J. Chem. Phys.* **48**, 715 (1968).
56. J. Jortner and R. S. Berry, *J. Chem. Phys.* **48**, 2757 (1968).
57. D. P. Chock, J. Jortner, and S. A. Rice, *J. Chem. Phys.* **49**, 610 (1968).
58. G. Nicolis and S. A. Rice, *J. Chem. Phys.* **46**, 4445 (1967).
59. S. A. Rice, I. McLaughlin, and J. Jortner, *Chem. Phys.* **49**, 2756 (1968).

60. G. W. Robinson and R. P. Frosch, *J. Chem. Phys.* **37**, 1962 (1962).
61. M. P. Wright, R. P. Frosch, and G. W. Robinson, *J. Chem. Phys.* **33**, 934 (1960).
62. H. E. Radford and H. P. Broida, *J. Chem. Phys.* **38**, 644 (1963).
63. G. W. Robinson, *J. Chem. Phys.* **47**, 1967 (1967).
64. W. Siebrand, *J. Chem. Phys.* **47**, 2411 (1967).
65. W. Siebrand, *J. Chem. Phys.* **46**, 440 (1967).
66. W. Siebrand and D. F. Williams, *J. Chem. Phys.* **46**, 403 (1967).
67. M. Caner and R. Englman, *J. Chem. Phys.* **44**, 4054 (1966).
68. G. W. Robinson and R. P. Frosch, *J. Chem. Phys.* **38**, 1187 (1963).
69. E. C. Lim, *J. Chem. Phys.* **36**, 3497 (1962).
70. R. Srinivasan, *Chem. Phys.* **38**, 1039 (1962).
71. L. G. Van Uitert, E. F. Dearborn, and J. J. Rubin, *J. Chem. Phys.* **46**, 420 (1967).
72. B. J. Cohen and L. Goodman, *J. Chem. Phys.* **46**, 713 (1967).
73. R. Johnston and S. A. Rice, *J. Chem. Phys.* **49**, 2734 (1968).
74. P. R. Harrowell and K. F. Freed, *J. Chem. Phys.* **83**, 6288 (1985).
75. J. Jortner, S. A. Rice, and R. M. Hochstrasser, *Adv. Photochem.* **7**, 150 (1969).
76. L. Mower, *Phys. Rev.* **142**, 142 (1965).
77. F. Lahmani, A. Tramer, and C. Tric, *J. Chem. Phys.* **60**, 4431 (1974).
78. K. F. Freed, *Adv. Chem. Phys.* **42**, 207 (1980).
79. W. Dietz, *Ber. Bunsenges. Phys. Chem.* **92**, 403 (1988).
80. K. Takahashi, *J. Chem. Phys.* **73**, 309 (1980).
81. F. A. Novak and S. A. Rice, *J. Chem. Phys.* **73**, 858 (1980).
82. H. Kono, Y. Fujimura, and S. H. Lin, *J. Chem. Phys.* **75**, 2569 (1981).
83. H. K. Hong, *Chem. Phys.* **9**, 1 (1975).
84. I. H. Kuhn and F. Metz, *Chem. Phys.* **33**, 137 (1978).
85. K. F. Freed, *Adv. Chem. Phys.* **47**, 291 (1981).
86. K. F. Freed, *J. Chem. Phys.* **64**, 1604 (1976).
87. K. F. Freed and C. Tric, *Chem. Phys.* **33**, 249 (1978).
88. K. F. Freed, *J. Chem. Phys.* **52**, 1345 (1970).
89. M. Lombardi, *Excited States* **7**, 163 (1988).
90. J. L. Van Vleck, *Rev. Mod. Phys.* **23**, 213 (1951).
91. J. L. Van Vleck, *The Theory of Electronic and Magnetic Susceptibilities*, Oxford University Press, London, 1932.
92. J. L. Van Vleck, *Phys. Rev.* **94**, 1191 (1954).
93. V. I. Makarov and I. V. Khmelinskii, *Chem. Phys.* **207**, 115 (1996).
94. V. I. Makarov, *Mol. Phys.* **89**, 1803 (1996).
95. T. Imamura, N. Tamai, Y. Fukuda, I. Yamazaki, S. Nagakura, H. Abe, and H. Hayshi, *Chem. Phys. Lett.* **135**, 208 (1987).
96. V. I. Makarov, *Mol. Phys.* **89**, 867 (1996).
97. C. G. Stevens and J. C. D. Brand, *J. Chem. Phys.* **38**, 3324 (1973).
98. A.S.-C. Cheung and A. J. Merer, *Mol. Phys.* **46**, 111 (1982).
99. J. T. Hougen, *Can. J. Phys.* **42**, 433 (1964).

100. W. T. Raynes, *J. Chem. Phys.* **11**, 3020 (1964).
101. C. di Lauro, *J. Mol. Spectr.* **40**, 103 (1971).
102. J. C. D. Brand, C. di Lauro, and D. S. Liu, *Can. J. Phys.* **53**, 1853 (1975).
103. R. S. Henderson, *Phys. Rev.* **100**, 723 (1955).
104. G. Herzberg, *Infrared and Raman Spectra of Polyatomic Molecules*, New York, 1945.
105. J. T. Hougen, P. R. Bunker, and J. W. C. Johns, *J. Mol. Spectrosc.* **34**, 136 (1970).
106. P. A. Geldof, P. H. Rettschnick, and G. J. Hoyting, *Chem. Phys. Lett.* **10**, 549 (1971).
107. W. T. Raynes, *J. Chem. Phys.* **41**, 3020 (1964).
108. W. E. Howard and E. W. Schlag, *J. Chem. Phys.* **68**, 2679 (1978).
109. S. Golden, *J. Chem. Phys.* **16**, 78 (1948).
110. R. F. Curl and J. L. Kinsey, *J. Chem. Phys.* **35**, 1758 (1961).
111. C. H. Townes and A. L. Schawlov, *Microwave Spectroscopy*, Dover Publication, New York, 1975.
112. M. Goldman, *Quantum Description of High-Resolution NMR in Liquid*, Oxford Scientific Publications, Oxford University Press, New York, 1988.
113. K. Blum, *Density Matrix Theory and Applications*, Plenum Press, New York, 1981.
114. A. Abragam, *The Principles of Nuclear Magnetism*, Clarendon Press, Oxford, 1961.
115. B. R. Judd, *Angular Momentum Theory for Diatomic Molecules*, Academic Press, New York, 1975.
116. W. H. Huo, *J. Chem. Phys.* **52**, 3110 (1970).
117. S. H. Lin and Y. Fujimura, *Excited States* **4**, 237 (1979).
118. J. Jortner and R. D. Levine, in *Photoselective Chemistry, Advances in Chemical Physics*, Vol. 47, Wiley Interscience, New York, 1981, p. 1.
119. I. A. Godunov, A. V. Abramenkov, and V. I. Tyulin, *J. Struct. Chim.* **24**, 31 (1983).
120. M. G. Liverman, S. M. Beck, D. I. Monts, and R. E. Smalley, *J. Chem. Phys.* **70**, 192 (1979).
121. G. Moller and D. S. Tinti, *Mol. Phys.* **54**, 541 (1985).
122. J. L. Hencher and G. W. King, *J. Mol. Spectrosc.* **16**, 168 (1965).
123. W. J. Balfour and G. W. King, *J. Mol. Spectrosc.* **24**, 130 (1967).
124. J. R. Durig, S. C. Brown, and S. E. Hannum, *J. Chem. Phys.* **54**, 4428 (1971).
125. J. Tyrrell, *J. Am. Chem. Soc.* **98**, 5456 (1976).
126. A. P. Baronavski and J. R. McDonald, *J. Chem. Phys.* **67**, 4286 (1977).
127. J. M. Hollas and M. Z. bin Hussein, *J. Chem. Soc. Faraday Trans.* **86**, 2015 (1990).
128. V. I. Makarov, I. V. Khmelinskii, S. A. Kochubei, and V. N. Ishchenko, *J. Chem. Phys. Lett.*, submitted.
129. J. B. Con, *J. Chem. Phys.* **14**, 665 (1946).
130. G. W. King, *J. Chem. Phys.* **15**, 820 (1957).
131. J. H. Clark, C. B. Moore, and N. S. Nogar, *J. Chem. Phys.* **68**, 1264 (1978).
132. C. K. Ingold and G. W. King, *J. Chem. Soc.*, 2702 (1953).
133. K. Innes, *J. Chem. Phys.* **22**, 863 (1954).
134. J. T. Hougen and J. K. G. Watson, *Can. J. Phys.* **43**, 298 (1965).
135. J. K. G. Watson, M. Herman, J. C. Van Craen, and R. Colin, *J. Mol. Spectrosc.* **95**, 101 (1982).
136. N. Ochi and S. Tsuchiya, *Chem. Phys.* **52**, 319 (1991).

137. P. Dupre, R. Jost, M. Lombardi, P. G. Green, E. Abramson, and R. W. Field, *Chem. Phys.* **152**, 293 (1991).
138. P. G. Green, J. L. Kinsey, and R. W. Field, *J. Phys. Chem.* **91**, 5160 (1989).
139. A. Haijima, M. Fuji, and M. Ito, *J. Chem. Phys.* **92**, 959 (1990).
140. P. Dupre, P. G. Green, and R. W. Field, *Chem. Phys.* **196**, 211 (1995).
141. M. Fuji, A. Haijima, and M. Ito, *Chem. Phys. Lett.* **150**, 380 (1988).
142. T. Suzuki and N. Hashimoto, *J. Chem. Phys.* **110**, 2042 (1999).
143. V. I. Makarov, A. R. Cruz, and E. Quinones, *Chem. Phys.*, submitted.
144. J. C. Stephenson, J. A. Blazy, and D. S. King, *Chem. Phys.* **85**, 31 (1984).
145. V. I. Makarov and E. Quinones, *Chem. Phys.*, accepted for publication.
146. N. Ochi and S. Tsuchiya, *Chem. Phys.* **52**, 319 (1991).
147. J. A. Montgomery, Jr. and J. R. Peterson, *Chem. Phys. Lett.* **168**, 75 (1990).
148. C. W. Bausschlisher, Jr., S. R. Langhof, and P. R. Tayler, *Chem. Phys. Lett.* **171**, 42 (1990).
149. A. M. Wodtke and Y. T. Lee, *J. Phys. Chem.* **89**, 4744 (1985).
150. D. P. Baldwin, M. A. Buntine, and D. W. Chandler, *J. Chem. Phys.* **93**, 6578 (1990).
151. M. Drabbels, J. Heinze, and W. L. Meerts, *J. Chem. Phys.* **100**, 165 (1994).
152. P. Dupre and P. G. Green, *Chem. Phys. Lett.* **12**, 555 (1993).
153. P. Dupre, P. G. Green, and R. W. Field, *Chem. Phys.* **196**, 211 (1995).
154. Q. Cui, K. Morokuma, and J. F. Stanton, *Chem. Phys. Lett.* **263**, 46 (1996).
155. Q. Cui and K. Morokuma, *Chem. Phys. Lett.* **272**, 319 (1997).
156. D. G. Prichard, J. S. Muenter, and B. J. Howard, *Chem. Phys. Lett.* **135**, 9 (1987).
157. D. G. Prichard, R. N. Nandi, and J. S. Muenter, *J. Chem. Phys.* **89**, 115 (1988).
158. A. J. Colussi, S. P. Sander, and R. R. Friendl, *Chem. Phys. Lett.* **178**, 497 (1991).
159. G. W. Bryant, D. F. Eggers, and R. O. Watts, *Chem. Phys. Lett.* **151**, 309 (1988).
160. A. Amirov, *Chem. Phys.* **108**, 403 (1986).
161. K. E. Drable and J. Kommandeur, in *Excited State*, Academic New York, 1987.
162. J. Kommandeur, B. J. van der Meer, and H. Th. Jonkman, in *Intramolecular Dynamics*, Reidel, Dordrecht, 1982, p. 259.
163. J. van der Meer, H. Th. Jonkman, and J. Kommandeur, *Laser Chem.* **2**, 77 (1983).
164. J. van der Meer, H. Th. Jonkman, J. Kommandeur, W. L. Meerts, and W. A. Majewski, *Chem. Phys. Lett.* **92**, 565 (1982).
165. A. Frad, F. Lahmani, A. Trammer, and C. Tric, *J. Chem. Phys.* **60**, 4419 (1974).
166. A. Amirov and J. Jortner, *J. Chem. Phys.* **84**, 1500 (1986).
167. H. Saigusa and E. S. Lim, *J. Chem. Phys.* **78**, 91 (1983).
168. H. Saigusa and E. S. Lim, *Chem. Phys. Lett.* **88**, 455 (1982).
169. Y. Matsumoto, H. Spangler, and D. W. Pratt, *J. Chem. Phys.* **80**, 5542 (1984).
170. E. Riedle, H. J. Neusser, E. W. Schlag, and S. H. Lin, *J. Phys. Chem.* **88**, 198 (1984).
171. E. Riedle and H. J. Neusser, *J. Chem. Phys.* **80**, 4686 (1984).
172. A. Amirov, J. Jortner, M. Terazima, and E. C. Lim, *Chem. Phys. Lett.* **133**, 179 (1987).
173. A. Amirov, *J. Chem. Phys.* **86**, 4706 (1987).
174. B. F. Forch and E. C. Lim, *Chem. Phys. Lett.* **110**, 593 (1984).

175. H. Th. Jonkman, K. E. Drabe, and J. Kommandeur, *Chem. Phys. Lett.* **357**, 455 (1982).
176. N. Ohta and T. Takemura, *J. Chem. Phys.* **95**, 7119 (1991).
177. N. Ohta and T. Takemura, *Chem. Phys. Lett.* **84**, 308 (1981).
178. N. Ohta and T. Takemura, *Chem. Phys.* **82**, 41 (1983).
179. N. Ohta and T. Takemura, *Chem. Phys. Lett.* **106**, 308 (1984).
180. N. Ohta, *Laser Chem.* **10**, 109 (1989).

CLASSICAL AND QUANTUM MAGNETIZATION REVERSAL STUDIED IN NANOMETER-SIZED PARTICLES AND CLUSTERS

WOLFGANG WERNSDORFER

Laboratoire Louis Néel, Grenoble, France

CONTENTS

Advances in Chemical Physics, Volume 118, Edited by I. Prigogine and Stuart A. Rice.
ISBN 0-471-43816-2

I. INTRODUCTION

Since the late 1940s, nanometer-sized magnetic particles have generated continuous interest because the study of their properties has proved to be scientifically and technologically very challenging. In particular, it was recognized that the ferromagnetic state, with a given orientation of the particle moment, has a remanent magnetization if the particle is small enough. This was the starting point of huge permanent magnets and magnetic recording industries. However, despite intense activity during the last few decades, the difficulties in making nanoparticles of good enough quality has slowed the advancement of this field. As a consequence, for 50 years, these applications concentrated above and then near the micrometer scale. In the last decade, this has no longer been the case because of the emergence of new fabrication techniques that have led to the possibility of making small objects with the required structural and chemical qualities. In order to study these objects, new techniques were developed such as magnetic force microscopy, magnetometry based on micro-Hall probes, or micro-SQUIDs. This led to a new understanding of the magnetic behavior of nanoparticles, which is now very important for the development of new fundamental theories of magnetism and in modeling new magnetic materials for permanent magnets or high density recording.

In order to put this review into perspective, let us consider Fig. 1, which presents a scale of size ranging from macroscopic down to nanoscopic sizes. The unit of this scale is the number of magnetic moments in a magnetic system. At macroscopic sizes, a magnetic system is described by magnetic domains (Weiss 1907) [1] that are separated by domain walls. Magnetization reversal occurs via nucleation, propagation and annihilation of domain walls (see the hysteresis loop on the left in Fig. 1 which was measured on an individual elliptic CoZr particle of $1\,\mu m \times 0.8\,\mu m$ and a thickness of 50 nm). Shape and width of domain walls depend on the material of the magnetic system, on its size, shape and surface, and on its temperature [2]. The material dependence of the domain walls has motivated the definition of two length scales: (i) the domain wall width δ defined by $\delta = \sqrt{A/K}$ and (ii) the exchange length λ defined by $\lambda = \sqrt{A}/M_S$ where A is the exchange energy, K is the crystalline anisotropy constant, and M_S is the spontaneous magnetization. Qualitatively, the first definition shows that anisotropy energy favors a thin wall, while the exchange energy favors a thick wall. For very small crystalline anisotropy, the first definition suggests an infinite domain wall width which has a large total energy. This is due to the magnetostatic energy term that can be reduced by subdividing the ferromagnetic crystal into domains. Therefore, for very small crystalline anisotropy, the domain wall width is of the order of magnitude of the exchange length λ. Both length scales can range from submicrometer scales in alloys to atomic scales in rare earth systems. When the system size is of the order of magnitude of δ or λ, the formation of domain walls requires too much energy. Therefore, the magnetization remains in the so-called single-domain state.* Hence, the magnetization might reverse by uniform rotation, curling, or other nonuniform modes (see hysteresis loop in the middle of Fig. 1). In this review we discuss mainly this size range where the physics is rather simple (Sections II and III).

For system sizes well below δ and λ, one must take into account explicitly the magnetic moments (spins) and their couplings. The theoretical description is complicated by the particle's boundaries [3–5].

At the smallest size (below which one must consider individual atoms and spins) there are either free clusters made of several atoms [6, 7] or molecular clusters which are macromolecules with a central complex containing magnetic atoms. In the last case, measurements on the Mn_{12} acetate and Fe_8 molecular clusters showed that the physics can be described by a collective moment of spin $S = 10$ (Section V.A). By means of simple hysteresis loop measurements, the quantum character of these molecules showed up in well-defined steps which are due to resonance quantum tunneling between energy levels (see hysteresis loop on the right in Fig. 1).

*In the theory of micromagnetism the single-domain state describes the state where the magnetization is perfectly aligned [2], whereas experimentalists mean often a state without domain wall.

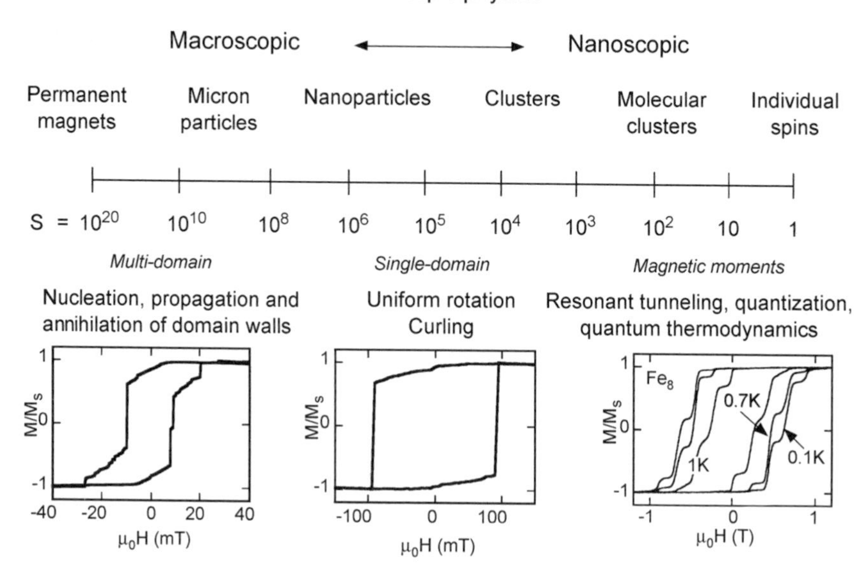

Figure 1. Scale of size that goes from macroscopic down to nanoscopic sizes. The unit of this scale is the number of magnetic moments in a magnetic system (roughly corresponding to the number of atoms). The hysteresis loops are typical examples of magnetization reversal via nucleation, propagation, and annihilation of domain walls (*left*), via uniform rotation (*middle*), and via quantum tunneling (*right*).

In the following sections, we review the most important theories and experimental results concerning the magnetization reversal of single-domain particles and clusters. Special emphasis is laid on single-particle measurements avoiding complications due to distributions of particle size, shape, and so on. Measurements on particle assemblies has been reviewed in Ref. 8. We mainly discuss the low-temperature regime in order to avoid spin excitations.

In Section II, we briefly review the commonly used measuring techniques. Among them, electrical transport measurements, Hall probes, and micro-SQUID techniques seem to be the most convenient techniques for low-temperature measurements. Section III discusses the mechanisms of magnetization reversal in single domain particles at zero kelvin. The influence of temperature on the magnetization reversal is reported in Section IV. Finally, Section V shows that for very small systems or very low temperature, magnetization can reverse via tunneling.

II. SINGLE-PARTICLE MEASUREMENT TECHNIQUES

The following sections review commonly used single-particle measuring techniques avoiding complications due to distributions of particle size, shape, and so on, which are always present in particle assemblies [8]. Special emphasis is laid on the micro-SQUID technique and the developed methods which allowed the most detailed studies at low temperatures.

A. Overview of Single-Particle Measurement Techniques

The dream of measuring the magnetization reversal of an individual magnetic particle goes back to the pioneering work of Néel [9, 10]. The first realization was published by Morrish and Yu in 1956 [11]. These authors employed a quartz-fiber torsion balance to perform magnetic measurements on individual micrometer-sized γ-Fe_2O_3 particles. With their technique, they wanted to avoid the complication of particle assemblies which are due to different orientations of the particle's easy axis of magnetization and particle–particle dipolar interaction. They aimed to show the existence of a single-domain state in a magnetic particle. Later on, other groups tried to study single particles, but the experimental precision did not allow a detailed study. A first breakthrough came via the work of Knowles [12] who developed a simple optical method for measuring the switching field, defined as the minimum applied field required to reverse the magnetization of a particle. However, the work of Knowles failed to provide quantitative information on well defined particles. More recently, insights into the magnetic properties of individual and isolated particles were obtained with the help of electron holography [13], vibrating reed magnetometry [14], Lorentz microscopy [15, 16], magneto-optical Kerr effect [17], and magnetic force microscopy [18, 19]. Recently, magnetic nanostructures have been studied by the technique of magnetic linear dichroism in the angular distribution of photoelectrons or by photoemission electron microscopy [20, 21]. In addition to magnetic domain observations, element-specific information is available via the characteristic absorption levels or threshold photoemission.* Among all mentioned techniques, most of the studies have been carried out using magnetic force microscopy at room temperature. This technique has an excellent spatial resolution, but time-dependent measurements are difficult due to the sample–tip interaction.

Only a few groups were able to study the magnetization reversal of individual nanoparticles or nanowires at low temperatures. The first magnetization measurements of individual single-domain nanoparticles and nanowires at very low temperatures were presented by Wernsdorfer et al. [22]. The detector (an Nb micro-bridge-DC-SQUID) and the studied particles were fabricated using electron-beam lithography. Coppinger et al. [23] investigated the magnetic properties of nanoparticles by resistance measurements. They observed the two-level fluctuations in the conductance of a sample containing self-organizing ErAs

*We refer to the literature concerning other domain observation techniques [1].

quantum wires and dots in a semi-insulating GaAs matrix. By measuring the electrical resistance of isolated Ni wires with diameters between 20 and 40 nm, Giordano and Hong studied the motion of magnetic domain walls [24, 25]. Other low-temperature techniques that may be adapted to single-particle measurements are Hall probe magnetometry [26–28], magnetometry based on magnetoresistance [29–31], or spin-dependent tunneling with Coulomb blockade [32, 33]. At the time of writing, the micro-SQUID technique allows the most detailed study of the magnetization reversal of nanometer-sized particles [34–39]. The following section reviews the basic ideas of the micro-SQUID technique.

B. Micro-SQUID Magnetometry

The superconducting quantum interference device (SQUID) has been used very successfully for magnetometry and voltage or current measurements in the fields of medicine, metrology, and science [40, 41]. SQUIDs are mostly fabricated from an Nb–AlO_x–Nb trilayer, several hundreds of nanometers thick. The two Josephson junctions are planar tunnel junctions with an area of at least 0.5 μm^2. In order to avoid flux pinning in the superconducting film the SQUID is placed in a magnetically shielded environment. The sample's flux is transferred via a superconducting pickup coil to the input coil of the SQUID. Such a device is widely used because the signal can be measured by simple lock-in techniques. However, this kind of SQUID is not well-suited for measuring the magnetization of single submicron-sized samples because the separation of SQUID and pickup coil leads to a relatively small coupling factor. A much better coupling factor can be achieved by coupling the sample directly with the SQUID loop. In this arrangement, the main difficulty arises from the fact that the magnetic field applied to the sample is also applied to the SQUID. The lack of sensitivity to a high field applied in the SQUID plane led us to the development of the microbridge-DC-SQUID technique [22] which allows us to apply several Tesla in the plane of the SQUID without dramatically reducing the SQUID's sensitivity.

1. *Choice of SQUID Configuration*

The main criteria for the choice of the micro-SQUID configuration were an easy coupling to a mesoscopic sample, a simple fabrication, a simple mode of operation, robustness and stability, the desired temperature range, and operation in high magnetic fields (in particular for fields applied in the SQUID plane). These criteria led to the use of microbridge junctions instead of the commonly used tunnel junctions.

The Josephson effect in microbridge junctions has first been suggested in 1964 by Anderson and Dayem [42]. These superconducting weak links seemed to be very promising in order to design planar DC-SQUIDs with a one-step thin-film technology. However, Dayem bridges exhibit a Josephson current-phase relation only when their dimensions are small compared to the coherence length ξ. Such dimensions were difficult to reach at those days. Nowadays, electron beam lithography allows one to directly fabricate reliable microbridge Josephson junctions made of materials like Al, Nb, and Pb.

2. *Fabrication of Micro-SQUIDs*

By using electron beam lithography, planar Al or Nb microbridge-DC-SQUIDs (of 0.5 to 4 μm in diameter) can be constructed (Fig. 2) [43–45]. Al SQUIDs can be obtained by evaporating a 20- to 50-nm thin film onto a PMMA mask, followed by standard lift-off techniques. The fabrication of Nb SQUIDs revealed to be more difficult. The direct evaporation of Nb onto a PMMA mask led to Nb SQUIDs of poor quality which manifested themselves in a very low critical temperature of the superconductivity. Better Nb qualities were achieved by using UHV facilities and substrate temperatures of about 800 K, or higher, in order to grow 20-nm thin Nb films on Si substrates.* The Nb films were covered by 5 nm of Si. The Nb SQUIDs were then patterned with reactive ion etching (RIE) using an AlO mask made by an electron beam lithography via a standard PMMA technique.

An alternative method to fabricate microbridge-DC-SQUIDs has recently been proposed by Bouchiat et al. [46]. The new method is based on local anodization of 3- to 6.5-nm-thick Nb strip lines under the voltage-biased tip of an atomic force microscope (AFM). Microbridge junctions and SQUID loops were obtained by either partial or total oxidation of the Nb layer. The first fabricated devices had about the same performance as micro-SQUIDs fabricated by electron beam lithography. The AFM-made SQUIDs should offer new features such as the fabrication at a chosen position allowing an optimized coupling to magnetic signals. In addition, we expect an increased intrinsic sensitivity: In the case of small magnetic clusters which are placed very close to the microbridge junctions, an improvement of the sensitivity of one to two orders of magnitude might be achieved due to the reduction of the microbridge size. It might allow us to detect the spin flips of about 100 magnetic moments.

3. *Magnetization Measurements Via Critical Current Measurements*

The microbridge-DC-SQUID have a hysteretic V–I curve (Fig. 3): Ramping the current up from zero, the SQUID transits from the superconducting to the normal state at the critical current I_c. Due to Joule heating in the normal state, the SQUID stays resistive down to currents much smaller than I_c. This hysteretic

*In order to have a SQUID that can be exposed to a high field applied in the SQUID plane, the SQUID was made out of a very thin layer preventing flux trapping. In most cases, we used 20-nm-thick Nb layers allowing measurements of hysteresis loops in magnetic fields of up to 2 T. Such SQUIDs might work at fields higher than 10 T when using very thin (<10 nm) Nb layers of high quality.

Figure 2. Scanning electron micrograph of an Nb microbridge-DC-SQUID fabricated by electron beam lithography. An Ni wire of diameter of about 90 nm was deposited on the SQUID (Section III.B.2).

V–I curve made it impossible to use standard SQUID electronics or lock-in amplifier to read out the SQUID. Therefore, Benoit et al. developed a method consisting in measuring the critical current of the micro-SQUID [43, 44]. A computer-controlled circuit triggers simultaneously a current ramp (Fig. 4) and a 100-MHz quartz clock. As soon as a dV/dt pulse is detected at the SQUID due to the transition from the superconducting to the normal state,* the clock stops and the current is set to zero. The clock reading is transferred to the computer, and the cycle begins again. The critical current is proportional to the duration of the current ramp. The repetition rate is about 10 kHz, limited by the time needed to settle the current. Because the critical current I_c is a periodic function of the flux going through the SQUID loop (Fig. 5), one can easily deduce the flux change in the SQUID loop by measuring the critical current. The sensitivity achieved by the critical current measurement technique was about $10^{-5}\Phi_0/\sqrt{\text{Hz}}$ for Al SQUIDs and $10^{-4}\Phi_0/\sqrt{\text{Hz}}$ for Nb SQUIDs ($\Phi_0 = h/2e = 2 \times 10^{-15}$ Wb).

*It is important to mention that the dV/dt pulse can be detected directly on the current biasing lead of the SQUID; that is, for most cases it is sufficient to connect the SQUID with a single wire and the mass of the cryostat.

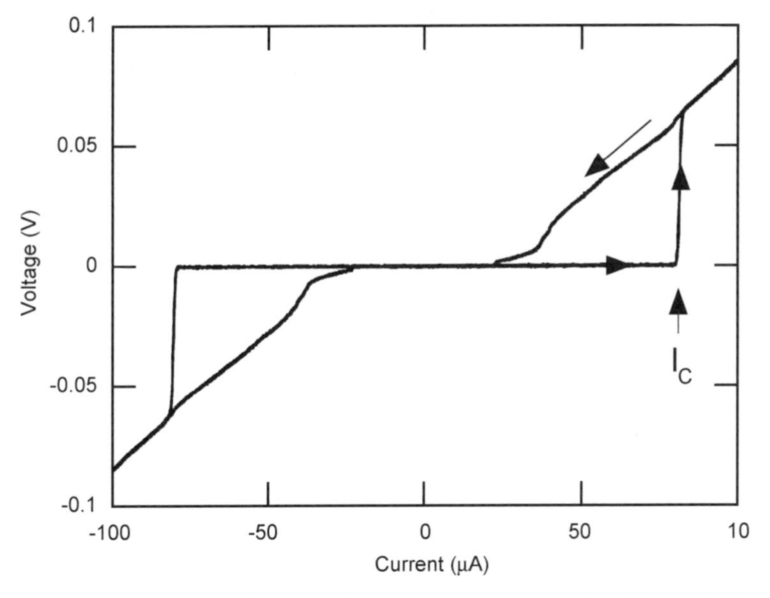

Figure 3. Oscilloscope reading of a voltage versus current curve of an Nb microbridge-DC-SQUID. The SQUID transits from the superconducting to the normal state at the critical current I_c.

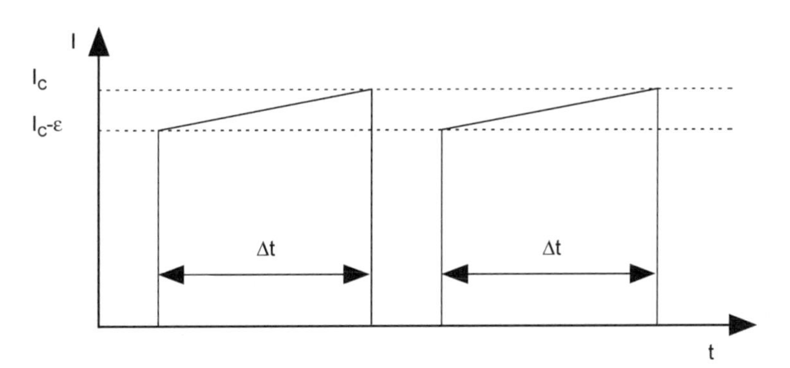

Figure 4. Current injected into a SQUID loop. First, the current is increased quickly up to a current $I_c - \varepsilon$ which is close to the critical current I_c. Then, the current is ramped up to I_c. As soon as a dV/dt pulse is detected at the SQUID due to the transition from the superconducting to the normal state, the current is set to zero. I_c is proportional to the duration of the current ramp Δt. The repetition rate of the cycle is up to 10 kHz.

In order to have good magnetic flux coupling, the mesoscopic systems, for instance the magnetic particles, is directly placed on the SQUID loop (Fig. 6) [45]. The SQUID detects the flux through its loop produced by the sample's

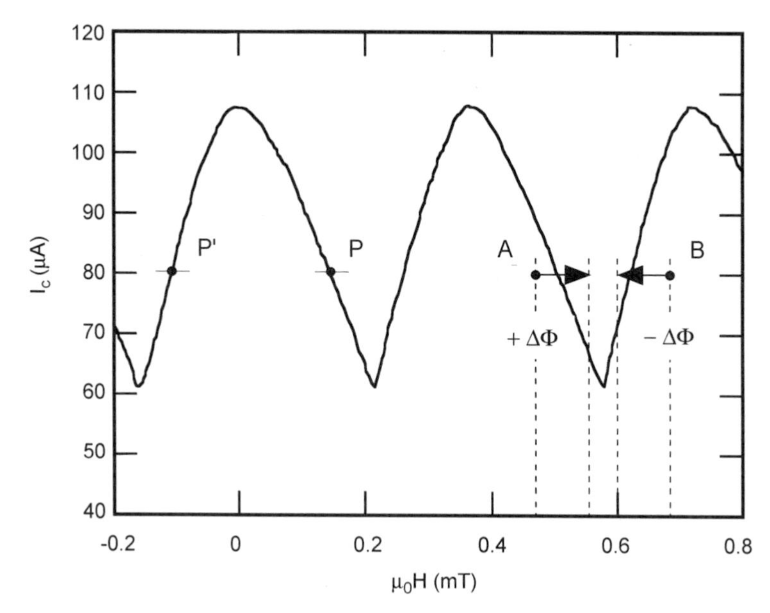

Figure 5. Critical current I_c of an Al micro-SQUID as a function of a magnetic field applied perpendicular to the SQUID plane. For the feedback mode an external field perpendicular to the SQUID loop is applied which keeps the critical current constant at a working point—for example, point **P** or **P**$'$. For the cold mode method the SQUID is biased close to the critical current so that it is in state **A** or **B**, respectively, for positive or negative flux jumps induced by the magnetization reversal.

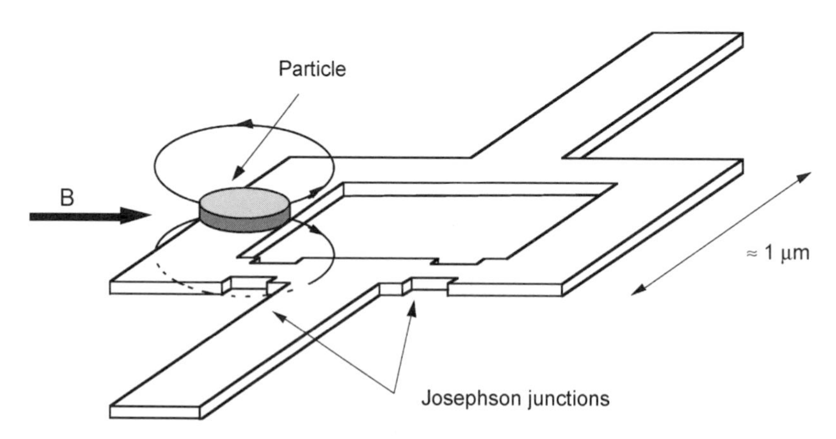

Figure 6. Schematic drawing of a planar microbridge-DC-SQUID on which a ferromagnetic particle is placed. The SQUID detects the flux through its loop produced by the sample magnetization. Due to the close proximity between sample and SQUID, a very efficient and direct flux coupling is achieved.

magnetization. For hysteresis loop measurements, the external field is applied in the plane of the SQUID (Fig. 6); thus the SQUID is only sensitive to the flux induced by the stray field of the sample's magnetization. The flux sensitivity of the critical current mode allowed us to detect the magnetization reversals corresponding to $10^4 \mu_B/\sqrt{Hz}$ (10^{-16}emu$/\sqrt{Hz}$)—that is, the magnetic moment of a Co nanoparticle with a diameter of 5 nm. The time resolution was given by the time between two measurements of the critical current. In this case, the achieved time resolution was 100 μs.

4. *Feedback Mode for Hysteresis Loop Measurements*

For the detection of flux variations larger than $\Phi_0/2$, the direct critical current method (Section II.B.3) is complicated by the fact that I_c is a nonmonotonous function of the flux (Fig. 5) which goes through the SQUID loop. In this cases, we used a feedback mode consisting in applying an external field perpendicular to the SQUID loop which compensates the stray field variations keeping the critical current constant at a working point as shown in Fig. 5. After each I_c measurement, the computer calculates the difference between a working point and I_c, multiplies it with a feedback factor, and adds it to the feedback (or compensation) field. The feedback field gives then the stray field variation of the sample's magnetization. It can easily be given in units of Φ_0. Knowing the coupling factor between particle and SQUID loop, absolute magnetic moment measurements are possible.

The time resolution of the feedback mode depends on the rate of the I_c measurements. For rates of about 10 kHz, we are able to follow flux variations in the millisecond range.

5. *Cold Mode Method for Magnetization Switching Measurements*

For studying the magnetization switching of nanoparticles, we developed a special mode, called the *cold mode method* [36–39], which are much faster and more sensitive than the previously presented modes. The achieved flux sensitivity allowed us to detect the magnetization switching corresponding to $10^3 \mu_B$ (10^{-17} emu)—that is, the switching of a Co nanoparticle with a diameter of 2 to 3 nm. The time resolution of the switching detection reached the nanosecond range.

The cold mode method is also specially adapted for studying the temperature dependence of the magnetization reversal and macroscopic quantum tunneling of magnetization. Indeed, the main difficulty associated with the SQUID detection technique lies in the Joule heating when the critical current is reached. After the normal state transition at the critical current, the SQUID dissipates for about 100 ns which slightly heats the magnetic particle coupled to the SQUID. This problem can be solved by the cold mode method which uses the SQUID only as a trigger [36–39].

The cold mode consists in biasing the SQUID close to the critical current while a field is applied perpendicular to the SQUID plane so that the SQUID is in state **A** (**B**) for a positive (negative) flux jump (Fig. 5) that can be induced by the magnetization reversal of a particle coupled to the SQUID loop. The magnetization reversal then triggers a transition of the SQUID from the superconducting to the normal state; that is, a *dV/dt* pulse can be detected on the current lead biasing the SQUID. Our SQUID electronics allows us to detect the *dV/dt* pulse some nanoseconds after the magnetization reversal, allowing very precise switching field measurements. Because this method only heats the sample after the magnetization reversal, we called it the *cold mode method.*

Another advantage of this mode is that the sample does not interfere with the rf-noise which is induced in oxide layer Josephson junctions of conventional SQUIDs because the hysteretic micro-SQUID is in the superconducting state before the magnetization reversal.

Finally, the cold mode method is very important for studying macroscopic quantum tunneling of magnetization. Quantum theory predicts that the escape rate from a metastable potential well by quantum tunneling is strongly reduced by the coupling of the magnetic system with its environment. Therefore, the measuring device must be weakly coupled to the magnetic particle. However, in order to measure the magnetization reversal, the SQUID must be strongly coupled to the magnetic particle that hinders the possibility of quantum tunneling. This problem can be solved by using the cold mode method. In order to show this schematically, Fig. 7 represents two energy potentials: One is the double well potential of the particle, and the other is the periodic potential of the SQUID. Before the magnetization reversal, both systems are in a metastable state: the particle because of an applied field which is close to the switching field and the SQUID because of a current through the SQUID loop which is close to the critical current. When the particle overcomes the saddle point or tunnels through the energy barrier, its magnetization rotates by only few degrees. For this starting process of the magnetization reversal, the coupling between particle and SQUID can be arranged to be very small.* Afterwards, the particle falls into

*In order to illustrate the dipolar couplings, let us consider the energy scales involved. For most of the particles measured so far below 1 K, the energy barrier height from the metastable state up to the saddle point is of the order of a few Kelvins whereas that from the lower state up to the saddle point is between 10^3 and 10^6 K. These energy scales should be compared with the energy necessary to drive the SQUID out of its metastable superconducting state which is of the order of a few Kelvin. Therefore, only a small energy transfer is necessary to measure the magnetization reversal. In addition, a proper orientation of the easy axis of magnetization with respect to the SQUID loop can further reduce the coupling during the first stage of the magnetization reversal. In the case of an easy axis of magnetization perpendicular to the current direction in the SQUID wire (Fig. 6), the coupling factor between SQUID loop and particle is proportional to $(1-\cos\varphi)$, where φ is the angle between the direction of magnetization and its easy axis. Therefore, the coupling is very weak at the first stage of magnetization reversal where φ is small.

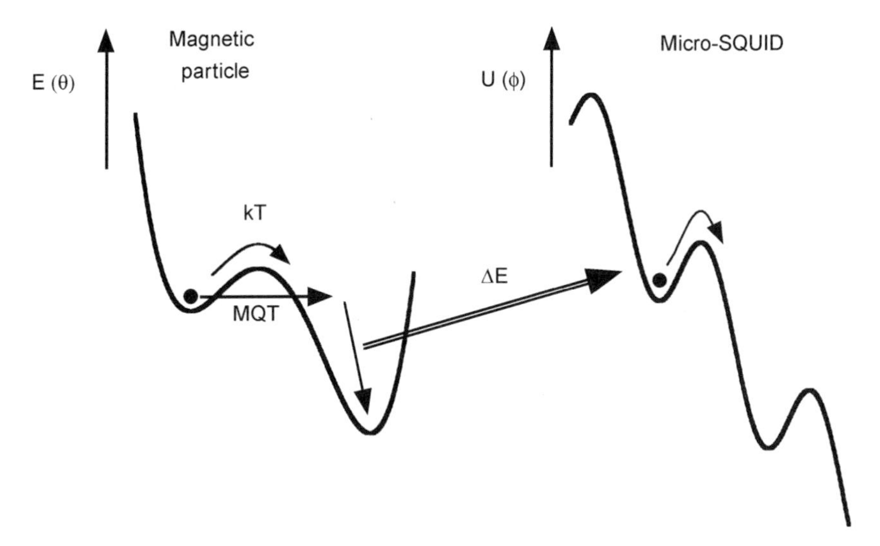

Figure 7. Energy scheme of the cold mode method. After the particle overcomes the saddle point or tunnels through the energy barrier, it falls into the lower potential well releasing energy. A very small fraction ΔE of this energy is transferred to the SQUID and drives the SQUID out of its metastable superconducting state.

the lower well which implies a rotation of magnetization of up to 180°. During this process, the coupling between particle and SQUID should be strong enough to drive the SQUID out of its metastable state. The corresponding transition from the superconducting into the normal state is easily measurable for a hysteretic SQUID. The main disadvantage of the cold mode is that only the switching field of magnetization reversal can be measured and not the magnetization before and during the magnetization reversal.

6. *Blind Mode Method for Three-Dimensional Switching Field Measurements*

A disadvantage of the micro-SQUID technique is that it does not function properly when a significant field is applied perpendicular to the SQUID loop. It works also only below the superconducting critical temperature of Nb ($T_c \approx 7$ K for our SQUIDs). These facts limited us to 2D measurements and to $T < T_c$ in the first experiments [35–37]. However, we showed recently that full-three-dimensional measurements can be done by using an indirect method [38]. In addition, this technique allows us to study the magnetization reversal for $T > T_c$.

Let us consider Fig. 8 showing switching field measurements (Section III) for in-the-SQUID-plane applied fields, detected using the *cold mode* (Section II.B.5). The three-dimensional switching field measurements and the studies as a function of temperature can be done using a three-step method which we call the *blind mode* (Fig. 8):

1. **Saturation.** The magnetization of the particles is saturated in a given direction (at $T = 35$ mK).
2. **Testing.** A test field is applied at a temperature between 35 mK and 30 K* which may or may not cause a magnetization switching.
3. **Probing.** After cooling to $T = 35$ mK, the SQUID is switched on and a field is swept in the plane of the SQUID to probe the resulting magnetization state using direct critical current measurements, the feedback mode, or the cold mode.

If the SQUID detects a magnetization jump in step 3, this means that the previously applied test field was weaker than the switching field for the direction being probed in step 2. The next interaction will then be done with a stronger test

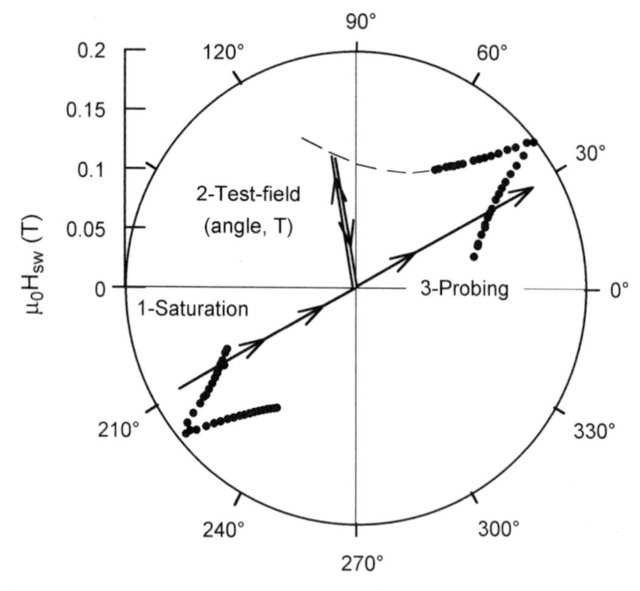

Figure 8. Schematics for the *blind mode* method presenting the angular dependence of the switching field near the easy axis of a 3-nm Co cluster (full circles) (Section III.A.2). First, the magnetization of the particle is saturated in a given direction (at $T = 35$ mK). Then, a test field is applied, at a temperature between 35 mK and 30 K, which may or may not cause a magnetization switching. Finally, the SQUID is switched on (at $T = 35$ mK) and a field is applied in the plane of the SQUID to probe the resulting magnetization state.

*Our highest temperatures of 30 K was only limited by the cooling time. Below 30 K, we achieved cooling rates of few Kelvins per second.

field. On the other hand, if the SQUID does not detect any magnetization jump in step 3, this means that the reversal occurred during step 2. The next interaction will then be done with a weaker field. When choosing the new test field with the help of a bisection algorithm, we needed about 8 repetitions of the three steps in order to get the switching field with good precision. This method allows us to scan the entire field space.*

7. *Micro-SQUID Arrays*

We also use arrays of micro-SQUIDs as a magnetometer for macroscopic samples. There are three applications that were particularly interesting: crystals of magnetic molecular clusters [47] (Section V.A), nucleation and depinning of magnetic domain walls in thin films [48], and arrays of magnetic dots [49]. The procedure consists in placing a sample on top of an array of micro-SQUIDs so that some SQUIDs are directly under the sample, some SQUIDs are at the border of the sample, and some SQUIDs are beside the sample (Fig. 9). When a SQUID is very close to the sample, it is sensing locally the magnetization reversal whereas when the SQUID is far away, it integrates over a bigger sample volume. Therefore, depending on the sample, one can obtain more insight in the magnetization reversal than with conventional techniques that measure only the total magnetization.

Our magnetometer works in the temperature range between 0.035 and 6 K and in fields up to 5 T with sweeping rates as high as 30 T/s, along with a field stability better than a microtesla. The time resolution is about 1 ms (Section II.B.4) allowing short-time measurements. The field can be applied in any direction of the micro-SQUID plane with a precision much better than 0.1° by separately driving three orthogonal coils [45].

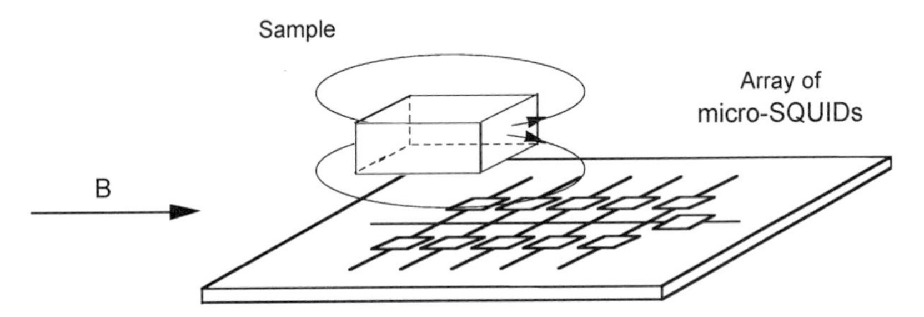

Figure 9. Schematic representation of our magnetometer which is an array of micro-SQUIDs. Its high sensitivity allows us to study single crystals of the order of 10 to 500 μm which are placed directly on the array.

*It is worth mentioning that special precautions are necessary for anisotropies that are more complex than uniaxial (Section III).

8. *Scanning SQUID Microscope*

The micro-SQUID technique has recently been used to build a scanning SQUID microscope [50]. The SQUID is designed by electron beam lithography at the apex of a silicon cantilever. The lever is attached to a force sensor, allowing to image magnetically, as well as topographically with a spatial resolution of 50 nm and a flux resolution of about $10^{-4}\,\phi_0$. The first application of this technique concerned the imaging of vortices in artificial networks.

9. *Outlook*

After the development of micro-SQUID technique in the early 1990s [43, 44], the study of magnetization reversal in magnetic nanostructures began in 1993 [45]. The first studied systems were micrometer-sized particles containing about 10^{10} magnetic moments. During the following years, the micro-SQUID technique has been improved to study smaller and smaller systems. In 2000, clusters containing about 10^3 magnetic moments could be studied. This achievement raises the question whether further improvements might be possible. The fundamental limit of a SQUID is the quantum limit which corresponds to a sensitivity of one magnetic moment for a SQUID with 1 μm^2. One might come close to this limit by using shunted SQUIDs [51]. Another possibility could be a reduction of the section of the microbridges [46]. Finally, the micro-SQUID technique could be improved by using superconducting materials with higher critical temperatures allowing measurements at higher temperatures.

III. MECHANISMS OF MAGNETIZATION REVERSAL AT ZERO KELVIN

As already briefly discussed in the introduction, for a sufficiently small magnetic sample it is energetically unfavorable to form a stable magnetic domain wall. The specimen then behaves as a single magnetic domain. For the smallest single-domain particles,* the magnetization is expected to reverse by uniform rotation of magnetization (Section III.A). For somewhat larger ones, nonuniform reversal modes are more likely—for example, the curling reversal mode (Section III.B.1). For larger particles, magnetization reversal occurs via a domain wall nucleation process starting in a rather small volume of the particle. For even larger particles, the nucleated domain wall can be stable for certain fields. The magnetization reversal happens then via nucleation and annihilation processes (Section III.B.3). In these sections we neglect temperature and quantum effects.

*In the theory of micromagnetism, the single-domain state describes the state where the magnetization is perfectly aligned [2], whereas experimentalists often mean a state without a domain wall.

The following section discusses in detail the uniform rotation mode that is used in many theories, in particular in Néel, Brown, and Coffey's theory of magnetization reversal by thermal activation (Section IV) and in the theory of macroscopic quantum tunneling of magnetization (Section V).

A. Magnetization Reversal by Uniform Rotation (Stoner–Wohlfarth Model)

The model of uniform rotation of magnetization, developed by Stoner and Wohlfarth [52], and Néel [53], is the simplest classical model describing magnetization reversal. One considers a particle of an ideal magnetic material where exchange energy holds all spins tightly parallel to each other, and the magnetization magnitude does not depend on space. In this case the exchange energy is constant, and it plays no role in the energy minimization. Consequently, there is competition only between the anisotropy energy of the particle and the effect of the applied field. The original study by Stoner and Wohlfarth assumed only uniaxial shape anisotropy, which is the anisotropy of the magnetostatic energy of the sample induced by its nonspherical shape. Thiaville has generalized this model for an arbitrary effective anisotropy which includes any magnetocrystalline anisotropy and even surface anisotropy [54]. In the simplest case of uniaxial anisotropy, the energy of a Stoner–Wohlfarth particle is given by

$$E = KV \sin^2\phi - \mu_0 M_S VH \cos(\theta - \phi) \tag{3.1}$$

where KV is the uniaxial anisotropy energy which depends on the shape of the particle, V is the volume of the particle, M_S is the spontaneous magnetization, H the magnitude of the applied field, and ϕ and θ are the angles of magnetization and applied field, respectively, with respect to the easy axis of magnetization. The potential energy of Eq. (3.1) has two minima separated by an energy barrier. For given values of θ and H, the magnetization lies at an angle ϕ which locally minimizes the energy. This position can by found by equating to zero the first derivative with respect to ϕ of Eq. (3.1): $\partial E/\partial\phi = 0$. The second derivative provides the criteria for maxima and minima. The magnetization reversal is defined by the minimal field value at which the energy barrier between the metastable minimum and the stable one vanishes—that is, at $\partial E/\partial\phi = \partial^2 E/\partial\phi^2 = 0$. A short analysis yields the angular dependence of this field, called the switching field H_{sw}^0 or in dimensionless units:

$$h_{sw}^0 = \frac{H_{sw}^0}{H_a} = \frac{1}{(\sin^{2/3}\theta + \cos^{2/3}\theta)^{3/2}} \tag{3.2}$$

where $H_a = 2K/(\mu_0 M_S)$ is the anisotropy field. The angular dependence of h^0_{sw} is plotted in Fig. 10.

Contrary to h^0_{sw}, the hysteresis loops cannot be expressed analytically and have to be calculated numerically. The result is seen in Fig. 11 showing the component of magnetization projected along the direction of the applied field; that is, $M_H = M_S \cos(\theta - \phi)$. Such loops are often called Stoner–Wohlfarth hysteresis loops.*

The main advantage of this classical theory is that it is sufficiently simple to add some extra features to it, as presented in the following.

1. *Generalization of the Stoner–Wohlfarth Model*

The original model of Stoner and Wohlfarth assumed only uniaxial shape anisotropy with one anisotropy constant [one second-order term, see Eq. (3.1)].

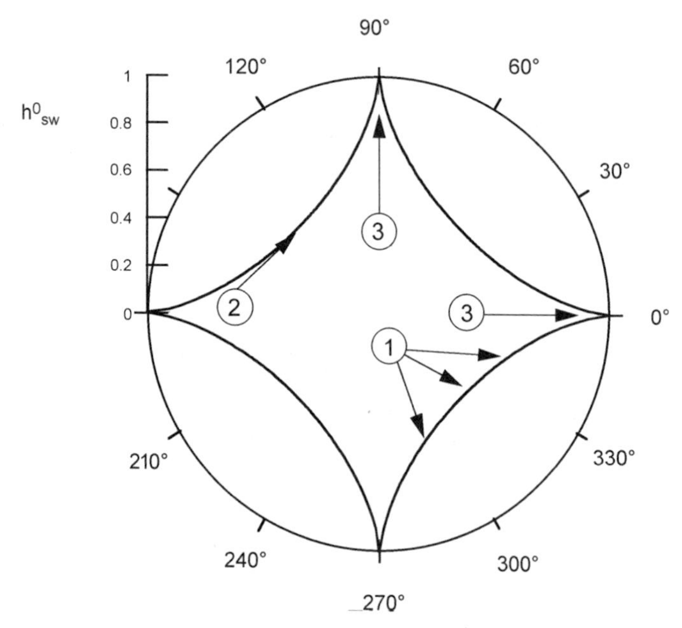

Figure 10. Angular dependence of the Stoner–Wohlfarth switching field $h^0_{sw} = H^0_{sw}/H_a$ [Eq. (3.2)]. This curve is often called the "Stoner–Wohlfarth astroid." Cases 1 to 3 correspond to Eqs. (3.5) to (3.9) concerning the field dependence of the anisotrophy barrier height.

*It is important to note that single-particle measurement techniques do not measure this component M_H. For example, for the micro-SQUID technique, with the easy axis of magnetization in the plane of the SQUID and perpendicular to the current direction in the SQUID wire (Fig. 6), one measures a magnetic flux that is proportional to $M_S \cos \phi$ (Fig. 12).

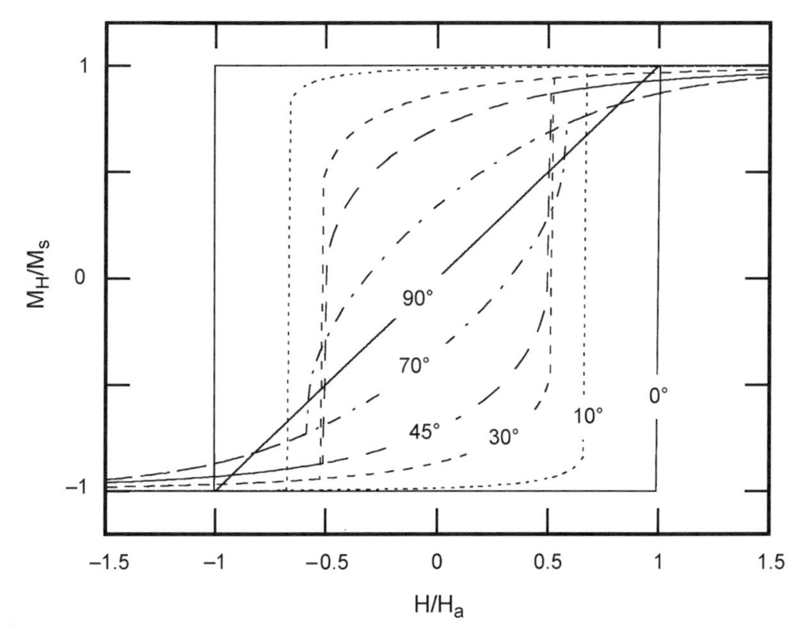

Figure 11. Hysteresis loops of a Stoner–Wohlfarth particle for different field angles θ. The component of magnetization along the direction of the applied field is plotted; that is, $M_{\mathrm{H}} = M_{\mathrm{S}} \cos(\phi - \theta)$.

This is sufficient to describe highly symmetric cases like a prolate spheroid of revolution or an infinite cylinder. However, real systems are often quite complex, and the anisotropy is a sum of mainly shape (magnetostatic), magnetocrystalline, magnetoelastic, and surface anisotropy. One additional complication arises because the different contributions of anisotropies are often aligned in an arbitrary way one with respect to each other. All these facts motivated a generalization of the Stoner–Wohlfarth model for an arbitrary effective anisotropy which was done by Thiaville [54, 55].

Similar to the Stoner–Wohlfarth model, one supposes that the exchange interaction in the cluster couples all the spins strongly together to form a giant spin whose direction is described by the unit vector $\vec{m}$. The only degrees of freedom of the particle's magnetization are the two angles of orientation of $\vec{m}$. The reversal of the magnetization is described by the potential energy:

$$E(\vec{m}, \vec{H}) = E_0(\vec{m}) - \mu_0 V M_s \vec{m}.\vec{H} \tag{3.3}$$

where V and M_s are the magnetic volume and the saturation magnetization of the particle respectively, $\vec{H}$ is the external magnetic field, and $E_0(\vec{m})$ is the magnetic anisotropy energy which is given by

$$E_0(\vec{m}) = E_{\text{shape}}(\vec{m}) + E_{\text{MC}}(\vec{m}) + E_{\text{surface}}(\vec{m}) + E_{\text{ME}}(\vec{m}) \tag{3.4}$$

E_{shape} is the magnetostatic energy related to the cluster shape. E_{MC} is the magnetocrystalline anisotropy (MC) arising from the coupling of the magnetization with the crystalline lattice, similar as in bulk. E_{surface} is due to the symmetry breaking and surface strains. In addition, if the particle experiences an external stress, the volumic relaxation inside the particle induces a magnetoelastic (ME) anisotropy energy E_{ME}. All these anisotropy energies can be developed in a power series of $m_x^a m_y^b m_z^c$ with $p = a + b + c = 2, 4, 6, \ldots$ giving the order of the anisotropy term. Shape anisotropy can be written as a biaxial anisotropy with two second-order terms. Magnetocrystalline anisotropy is in most cases either uniaxial (hexagonal systems) or cubic, yielding mainly second- and fourth-order terms. Finally, in the simplest case, surface and magnetoelastic anisotropies are of second-order.

Thiaville proposed a geometrical method to calculate the particle's energy and to determine the switching field for all angles of the applied magnetic field yielding the critical surface of switching fields which is analogous to the Stoner–Wohlfarth astroid (Fig. 10).

The main interest of Thiaville's calculation is that measuring the critical surface of the switching field allows one to find the effective anisotropy of the nanoparticle. The knowledge of the latter is important for temperature-dependent studies (Section IV) and quantum tunneling investigations (Section V). Knowing precisely the particle's shape and the crystallographic axis allows one to determine the different contributions to the effective anisotropy.

Thiaville's calculation predicts also the field dependence of the energy barrier height ΔE close to switching ($\varepsilon = (1 - H/H_{\text{sw}}^0) \ll 1$) which is important to know for temperature-dependent studies (Sections IV and V). Three cases have to be distinguished:

1. In the majority of cases except the two following cases (see case 1 in Fig. 10):

$$\Delta E = 4KV \frac{\sqrt{2}}{3} \frac{\cos\gamma}{\sqrt{\rho}} \varepsilon^{3/2} = E_0 \varepsilon^{3/2} \tag{3.5}$$

where KV is the anisotropy energy constant, γ is the angle of incidence between the local normal to the critical surface and the field sweeping direction, and ρ is the radius of curvature of the focal curve [55] at H_{sw}^0. It is important to emphasize that all these variables can be found experimentally by measuring the critical surface of the switching field. For uniaxial anisotropy (i.e., the 2D Stoner–Wohlfarth case), Eq. (3.5) becomes

$$\Delta E = 4KV\left(\frac{2}{3}\right)^{3/2} \frac{|\cos\theta|^{1/3}}{1+|\cos\theta|^{2/3}}\varepsilon^{3/2} \tag{3.6}$$

where θ is the angle of the applied field with respect to the easy axis of magnetization [Eq. (3.1)].

2. At glancing incidence (see case 2 in Fig. 10) with respect to the critical surface ($\gamma = \pi/2$), the power of ε is different, yielding

$$\Delta E = E_0' \varepsilon^3 \tag{3.7}$$

where E_0' has been calculated only in the two-dimensional case [54].

3. At a cusp point where $\gamma = \pi/2$ (see case 3 in Fig. 10),

$$\Delta E = E_0'' \varepsilon^2 \tag{3.8}$$

where E_0'' has been calculated only in the two-dimensional case [54]. In the case of uniaxial anisotropy i.e., the 2D Stoner–Wohlfarth case), Eq. (3.8) becomes

$$\Delta E = KV\varepsilon^2 = KV\left(1 - \frac{H}{H_a}\right)^2 \tag{3.9}$$

where $H_a = (2K/\mu_0 M_s)$ [Eq. (3.1)]. This equation is only valid for $\theta = 0$ and $\pi/2$, and it is valid for $0 \leq H \leq H_a$. This famous result of Néel [9, 10] has often been wrongly used for assemblies of nanoparticles where it is very difficult to achieve the conditions $\theta = 0$ and $\pi/2$ [56]. Up to now, only the power 3/2 and 2 (cases 1 and 3) has been found by single-particle measurements [36, 37] (see also Figs. 28 and 29).

Another simple analytical approximation for the field dependence of the energy barrier $\Delta E(H)$ was derived numerically by Pfeiffer [57, 58]

$$\Delta E = KV(1 - H/H_{sw}^0)^a = E_0 \varepsilon^a \tag{3.10}$$

where $a = 0.86 + 1.14h_{sw}^0$, and h_{sw}^0 is given by Eq. (3.2). This approximation is good for the intermediate field regime—that is, for fields H not too close to H_a and not too small.

2. *Experimental Evidence for Magnetization Reversal by Uniform Rotation*

In order to demonstrate experimentally the uniform rotation mode, the angular dependence of the magnetization reversal has often been studied (see references in Ref. 2). However, a comparison of theory with experiment is difficult because magnetic particles often have a nonuniform magnetization state that is due to rather complicated shapes and surfaces, crystalline defects, and surface anisotropy. In general, for many particle shapes the demagnetization fields inside the particles are nonuniform leading to nonuniform magnetization states [2]. An example is presented in Fig. 12 which compares typical hysteresis loop measurements of an elliptical Co particle, fabricated by electron beam lithography, with the prediction of the Stoner–Wohlfarth model (Fig. 12). Before magnetization reversal, the magnetization decreases more strongly than predicted because the magnetic configuration is not collinear as in the Stoner–Wohlfarth model, but instead presents deviations mainly near the particle surface. The angular dependence of the switching field agrees with the Stoner–Wohlfarth model only for angles $\theta \geq 30°$ where nonlinearities and defects play a less important role [22, 59].

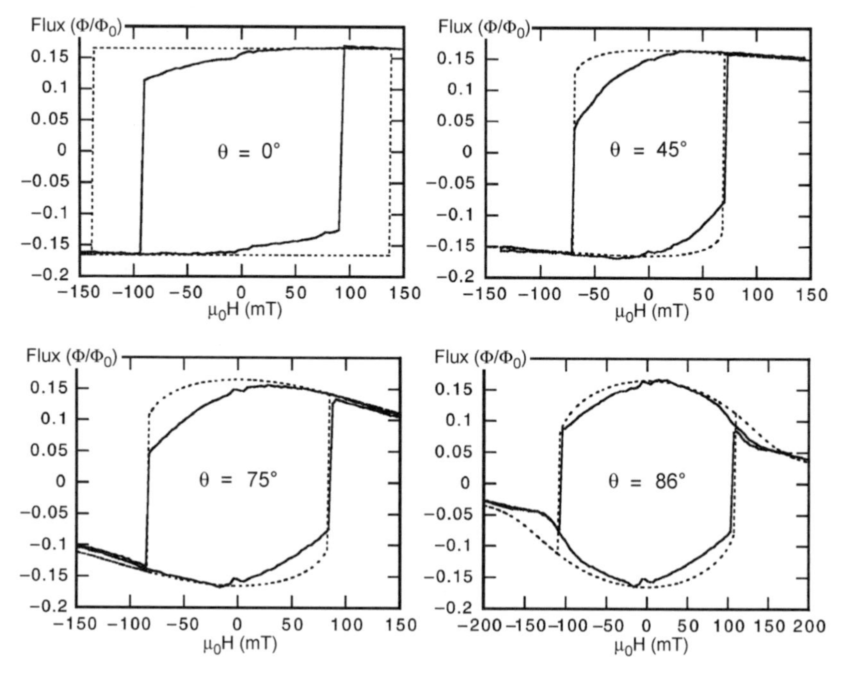

Figure 12. Hysteresis loops of a nanocrystalline elliptic Co particle of $70 \times 50 \times 25\,\text{nm}^3$. The *dashed line* is the prediction of the Stoner–Wohlfarth model of uniform rotation of magnetization. The deviations are due to nonuniform magnetization states.

Studies of magnetization reversal processes in ultrathin magnetic dots with in-plane uniaxial anisotropy showed also switching fields that are very close to the Stoner–Wohlfarth model, although magnetic relaxation experiments clearly showed that nucleation volumes are by far smaller than an individual dot volume [49]. These studies show clearly that switching field measurements as a function of the angles of the applied field cannot be taken unambiguously as a proof of a Stoner–Wohlfarth reversal.

The first clear demonstration of the uniform reversal mode has been found with Co nanoparticles [36], and BaFeO nanoparticles [37], the latter having a dominant uniaxial magnetocrystalline anisotropy. The three-dimensional angular dependence of the switching field measured on BaFeO particles of about 20 nm could be explained with the Stoner–Wohlfarth model taking into account the shape anisotropy and hexagonal crystalline anisotropy of BaFeO [38]. This explication is supported by temperature- and time-dependent measurements yielding activation volumes which are very close to the particle volume (Section IV).

We present here the first measurements on individual cobalt clusters of 3 nm in diameter containing about a thousand atoms (Figs. 13 and 14) [39]. In order to achieve the needed sensitivity, Co clusters preformed in the gas phase are directly embedded in a co-deposited thin Nb film that is subsequently used to pattern micro-SQUIDs. A laser vaporization and inert gas condensation source is used to produce an intense supersonic beam of nanosized Co clusters which can be deposited in various matrices under ultra-high-vacuum (UHV) conditions. Due to the low-energy deposition regime, clusters do not fragment upon impact on the substrate [60]. The niobium matrix is simultaneously deposited from a UHV electron gun evaporator leading to continuous films with a low concentration of embedded Co clusters [61]. These films are used to pattern planar microbridge-DC-SQUIDs by electron beam lithography. The later ones allow us to detect the magnetization reversal of a single Co cluster for an applied magnetic field in any direction and in the temperature range between 0.03 and 30 K (Section II. B). However, the desired sensitivity is only achieved for Co clusters embedded into the microbridges where the magnetic flux coupling is high enough. Due to the low concentration of embedded Co clusters, we have a maximum of 5 noninteracting particles in a microbridge which is 300 nm long and 50 nm wide. We can separately detect the magnetization switching for each cluster. Indeed they are clearly different in intensity and orientation because of the random distribution of the easy magnetization directions. The *cold mode* method (Section II.B.5) in combination with the *blind* method (Section II.B.6) allows us to detect separately the magnetic signal for each cluster.

High-resolution transmission electron microscopy observations showed that the Co clusters are well-crystallized in a *f.c.c.* structure (Fig. 13) with a sharp size distribution [61]. They mainly form truncated octahedrons (Fig. 14) [39].

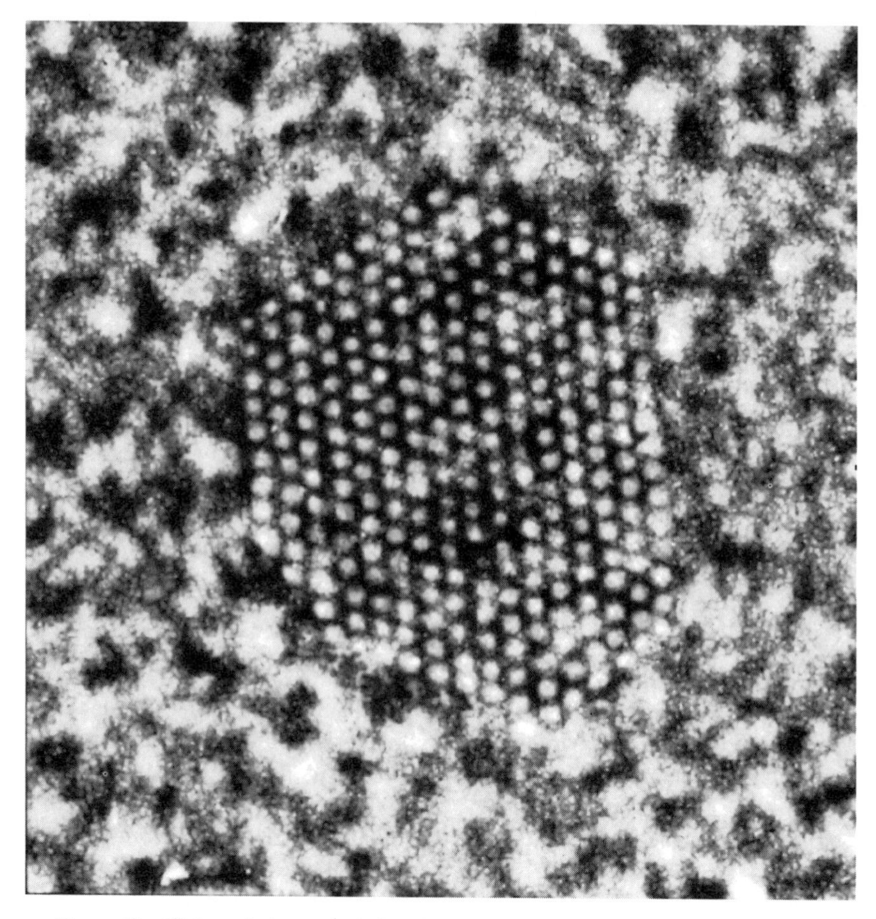

Figure 13. High-resolution transmission electron microscopy observation along a [110] direction of 3-nm cobalt cluster exhibiting an *f.c.c.* structure.

Figure 15 displays a typical measurement of switching fields in three dimensions of a 3-nm Co cluster at $T = 35\,\text{mK}$. This surface is a three-dimensional picture directly related to the anisotropy involved in the magnetization reversal of the particle (Section III. A). It can be reasonably fitted with the generalized Stoner and Wohlfarth model [54] (Section III.A.1). We obtain the following anisotropy energy:

$$E_0(\vec{m})/v = -K_1 m_z^2 + K_2 m_x^2 - K_4(m_{x'}^2 m_{y'}^2 + m_{x'}^2 m_{z'}^2 + m_{y'}^2 m_{z'}^2) \qquad (3.11)$$

where K_1 and K_2 are the anisotropy constants along z and x, the easy and hard magnetization axis, respectively. K_4 is the forth order anisotropy constant and

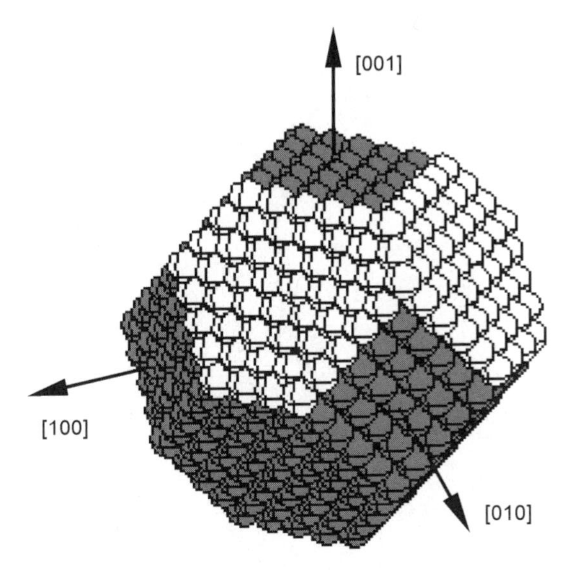

Figure 14. Scheme of a typical cluster shape with light-gray atoms belonging to the 1289 atoms truncated octahedron basis and dark-gray atoms belonging to the (111) and (001) added facets.

the $(x'y'z')$ coordinate system is deduced from (xyz) by a 45° rotation around the z axis. We obtained $K_1 = 2.2 \times 10^5 \text{J/m}^3$, $K_2 = 0.9 \times 10^5 \text{J/m}^3$, and $K_4 = 0.1 \times 10^5 \text{J/m}^3$. The corresponding theoretical surface is showed in Fig. 16. Furthermore, we measured the temperature dependence of the switching field distribution (Section IV.C.2). We deduced the blocking temperature of the particle $T_B \approx 14\,\text{K}$, and the number of magnetic atoms in this particle: $N \approx 1500$ atoms (Section IV.C.2). Detailed measurements on about 20 different particles showed similar three-dimensional switching field distributions with comparable anisotropy $(K_1 = (2.0 \pm 0.3) \times 10^5 \text{J/m}^3$, $(K_2 = (0.8 \pm 0.3) \times 10^5$ J/m^3, and $(K_4 = (0.1 \pm 0.05) \times 10^5 \text{J/m}^3)$ and size $(N = 1500 \pm 200$ atoms).

In the following, we analyze various contributions to the anisotropy energy of the Co clusters. Fine structural studies using EXAFS measurements [61] were performed on 500-nm-thick niobium films containing a very low concentration of cobalt clusters. They showed that niobium atoms penetrate the cluster surface to almost two atomic monolayers because cobalt and niobium are miscible elements. Further magnetic measurements [61] on the same samples showed that these two atomic monolayers are magnetically dead. For this reason, we estimated the shape anisotropy of the typical nearly spherical deposited cluster in Fig. 14 after removing two atomic monolayers from the surface. By calculating all the dipolar interactions inside the particle assuming a bulk magnetic moment of $\mu_{a:} = 1.7\mu_B$, we estimated the shape anisotropy constants: $K_1 \approx 0.3 \times 10^5 \text{J/m}^3$ along the easy magnetization axis and $K_2 \approx 0.1 \times 10^5 \text{J/m}^3$

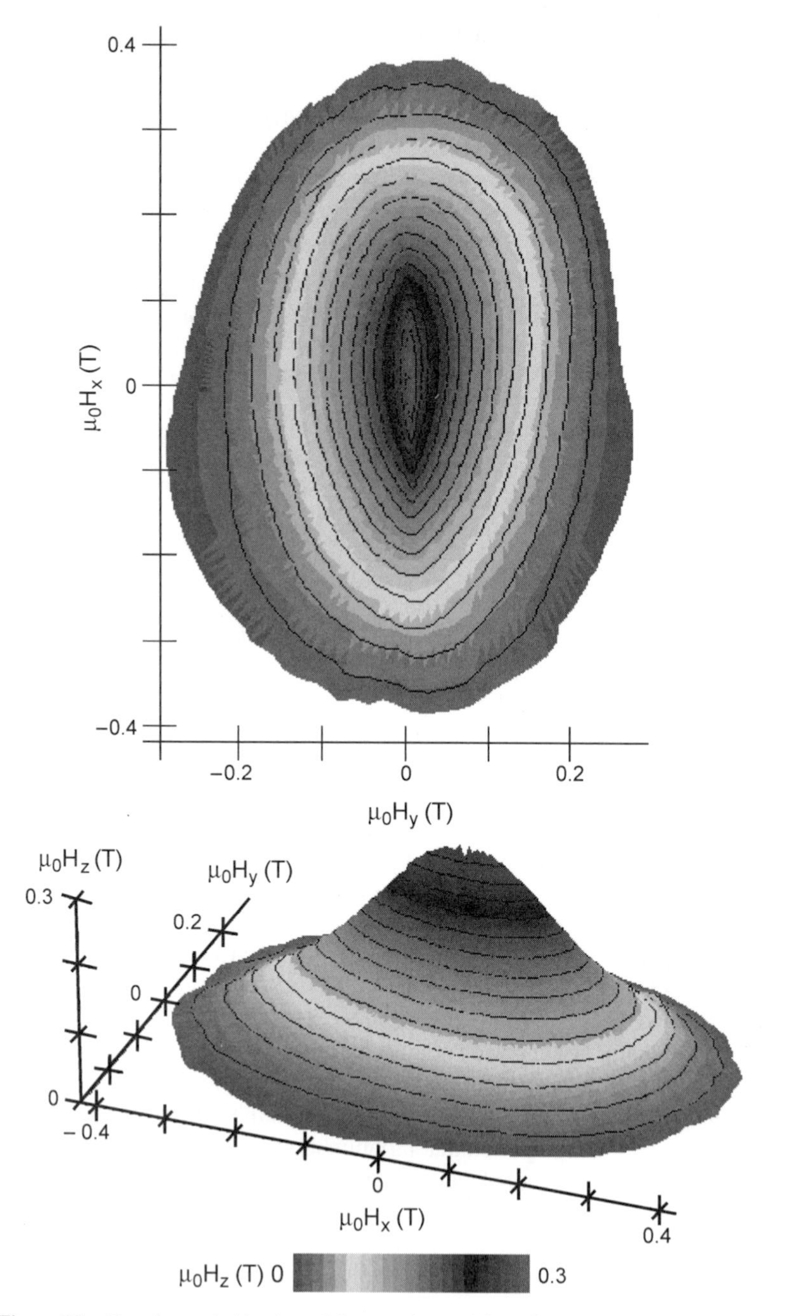

Figure 15. Top view and side view of the experimental three-dimensional angular dependence of the switching field of a 3-nm Co cluster at 35 mK. This surface is symmetrical with respect to the H_x–H_y plane, and only the upper part ($\mu_0\ H_z > 0$ T) is shown. Continuous lines on the surface are contour lines on which $\mu_0\ H_z$ is constant.

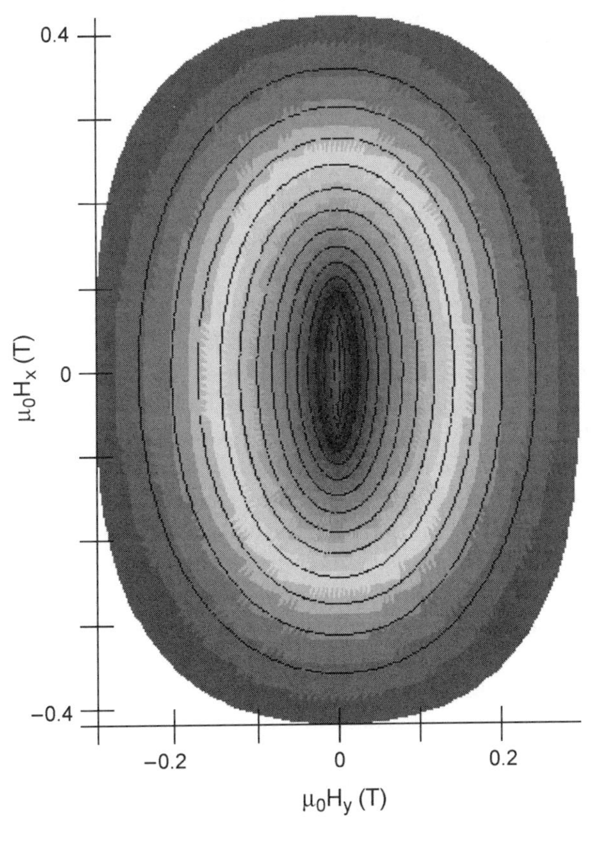

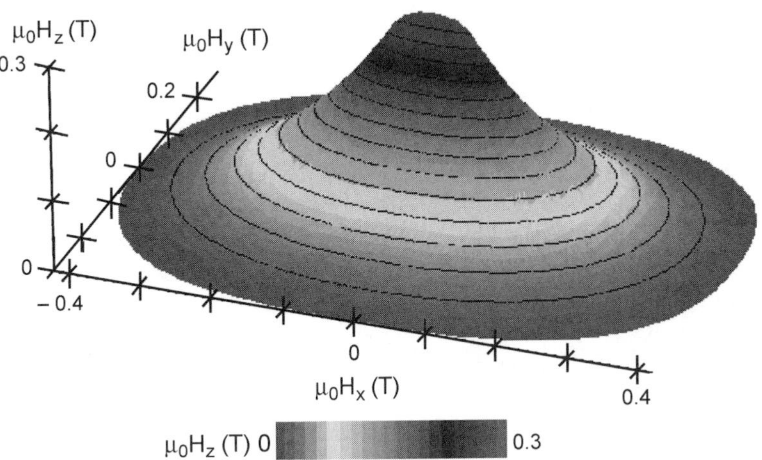

Figure 16. Top view and side view of the theoretical switching field surface considering second- and fourth-order terms in the anisotropy energy.

along the hard magnetization axis. These values are much smaller than the measured ones which means that E_{shape} is not the main cause of the second-order anisotropy in the cluster.

The fourth-order term $K_4 = 0.1 \times 10^5$ J/m^3 should arise from the cubic magnetocrystalline anisotropy in the *f.c.c.* cobalt clusters. However, this value is smaller than the values reported in previous works [62, 63]. This might by due to the different atomic environment of the surface atoms with respect to that of bulk *f.c.c.* Co. Taking the value of the bulk [62, 63], ($K_{bulk} = 1.2 \times 10^5$ J/m^3 only for the core atoms in the cluster, we find $K_{MC} \approx 0.2 \times 10^5$ J/m^3, which is in reasonable agreement with our measurements.

We expect that the contribution of the magnetoelastic anisotropy energy K_{MC} coming from the matrix-induced stress on the particle is also small. Indeed, using the co-deposition technique, niobium atoms cover uniformly the cobalt cluster creating an isotropic distribution of stresses. In addition, they can relax preferably inside the matrix and not in the particle volume because niobium is less rigid than cobalt. We believe therefore that only interface anisotropy $K_{surface}$ can account for the experimentally observed second-order anisotropy terms. Niobium atoms at the cluster surface might enhance this interface anisotropy through surface strains and magnetoelastic coupling. This emphasizes the dominant role of the surface in nanosized systems.

In conclusion, the three-dimensional switching field measurements of individual clusters give access to their magnetic anisotropy energy. A quantitative understanding of the latter is still difficult, but it seems that the cluster–matrix interface provides the main contribution to the magnetic anisotropy. Such interfacial effects could be promising to control the magnetic anisotropy in small particles in order to increase their blocking temperature up to the required range for applications.

3. *Uniform Rotation With Cubic Anisotropy*

We have seen in the previous section that the magnetic anisotropy is often dominated by second-order anisotropy terms. However, for nearly symmetric shapes, fourth-order terms* can be comparable or even dominant with respect to the second-order terms. Therefore, it is interesting to discuss further the features of fourth-order terms. We restrict the discussion to the 2D problem [54, 64] (see Ref. 55 for 3D).

The reversal of the magnetization is described by Eq. (3.3) which can be rewritten in 2D:

$$E(\theta) = E_0(\theta) - \mu_0 v M_s (H_x \cos(\theta) + H_y \sin(\theta)) \tag{3.12}$$

*For example, the fourth-order terms of *f.c.c.* magnetocrystalline anisotropy.

where v and M_s are the magnetic volume and the saturation magnetization of the particle, respectively, θ is the angle between the magnetization direction and x, H_x and H_y are the components of the external magnetic field along x and y, and $E_0(\theta)$ is the magnetic anisotropy energy. The conditions of critical fields ($\partial E/\partial\theta = 0$ and $\partial E^2/\partial\theta^2 = 0$) yield a parametric form of the locus of switching fields:

$$H_x = -\frac{1}{2\mu_0 v M_s}\left(\sin(\theta)\frac{dE}{d\theta} + \cos(\theta)\frac{d^2E}{d\theta^2}\right) \tag{3.13}$$

$$H_y = +\frac{1}{2\mu_0 v M_s}\left(\cos(\theta)\frac{dE}{d\theta} - \sin(\theta)\frac{d^2E}{d\theta^2}\right) \tag{3.14}$$

As an example we study a system with uniaxial shape anisotropy and cubic anisotropy. The total magnetic anisotropy energy can be described by

$$E_0(\theta) = vK_1 \sin^2(\theta + \theta_0) + vK_2 \sin^2(\theta)\cos^2(\theta) \tag{3.15}$$

where K_1 and K_2 are anisotropy constants (K_1 could be a shape anisotropy and K_2 the cubic crystalline anisotropy of a *f.c.c.* crystal.) θ_0 is a constant which allows to turn one anisotropy contribution with respect to the other one. Figure 17 displays an example of a critical curve which can easily be calculated from Eqs. (3.13) to (3.15). When comparing the standard Stoner–Wohlfarth astroid (Fig. 10) with Fig. 17, we can realize that the critical curve can cross itself several times. In this case, the switching field of magnetization depends on the path followed by the applied field. In order to understand this point, let us follow the energy potential [Eq. (3.15)], when sweeping the applied field as indicated in Fig. 18. When the field is in **A**, the energy E has two minima and the magnetization is in the metastable potential well. As the field increases, the metastable well becomes less and less stable. Let us compare two paths, one going along $\mathbf{A} \rightarrow \mathbf{B}_1 \rightarrow \mathbf{C} \rightarrow \mathbf{D} \rightarrow \mathbf{E}$, the other over $\mathbf{B}_2$ instead of $\mathbf{B}_1$. Figure 19 shows E in the vicinity of the metastable well for different field values along the considered paths (the stable potential well is not presented). One can realize that the state of the magnetization in **C** depends on the path followed by the field: Going over $\mathbf{B}_1$ leads to the magnetization state in the left metastable well (1), whereas going over $\mathbf{B}_2$ leads to the right metastable well (2). The latter path leads to magnetization switching in **D**, and the former one leads to a switching in **E**. Note that a small magnetization switch happens when reaching $\mathbf{B}_1$ or $\mathbf{B}_2$. Point **I** is a supercritical bifurcation.

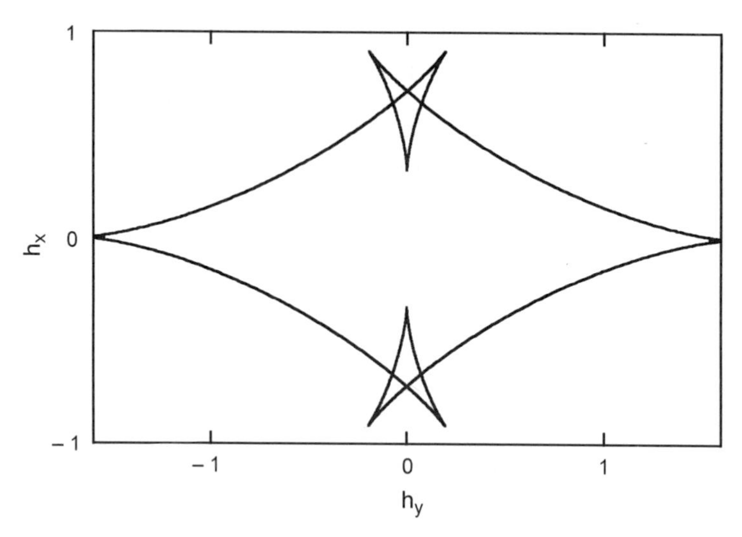

Figure 17. Angular dependence of the switching field obtained from Eqs. (3.13) to (3.15) with $K_1 > 0$ and $K_2 = -2/3K_1$. The field is normalized by the factor $2K_1/(\mu_0 M_s)$.

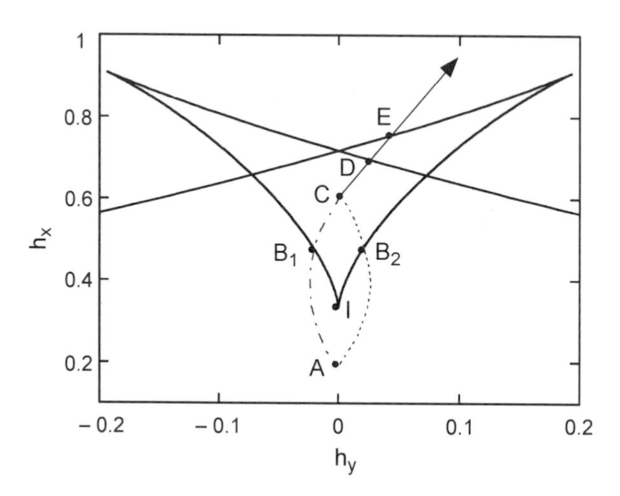

Figure 18. Enlargement of angular dependence of the switching field of Fig. 17. Two possible paths of the applied field are indicated: Starting from point **A** and going over the point $\mathbf{B}_1$ leads to magnetization reversal in **E**, whereas going over the point $\mathbf{B}_2$ leads to reversal in **D**.

The first measurement of such a field path dependence of the switching field were performed on single-domain FeCu nanoparticles of about 15 nm with a cubic crystalline anisotropy and a small arbitrarily oriented shape anisotropy [65]. Figure 20 presents switching field measurements of a 15-nm cobalt nano-

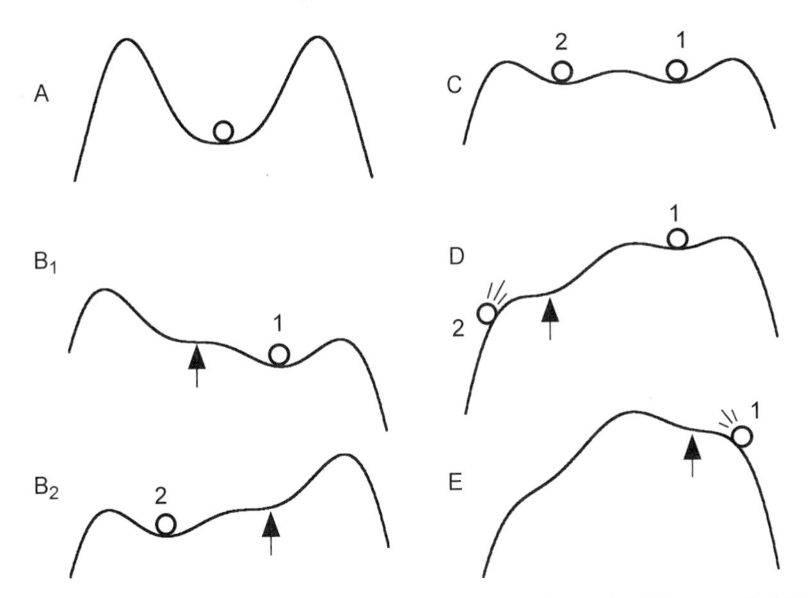

Figure 19. Scheme of the potential energy near the metastable state for different applied fields as indicated in Fig. 18. The balls represent the state of the magnetization, and the arrows locate the appearing or disappearing well.

particle showing clearly a contribution of cubic crystalline anisotropy and the field path dependence of the switching field.

B. Nonuniform Magnetization Reversal

We have seen in the previous sections that for extremely small particles, magnetization should reverse by uniform rotation. For somewhat larger single-domain particles, nonuniform reversal modes are more likely. The simplest one is the curling mode that is discussed in the following section.

1. Magnetization Reversal by Curling

The simplest nonuniform reversal mode is the curling reversal mode [2, 66]. The critical parameter is the exchange length $\lambda = \sqrt{A}/M_s$, delimiting the region of uniform rotation and curling (A is the exchange constant). Therefore in the case of the size $R > \lambda$ (R is, for example, the radius of a cylinder), magnetization reversal via curling is more favorable. In the following, we review the analytical result of an ellipsoid of rotation, which can be applied approximately to most of the shapes of nanoparticles or nanowires [67].

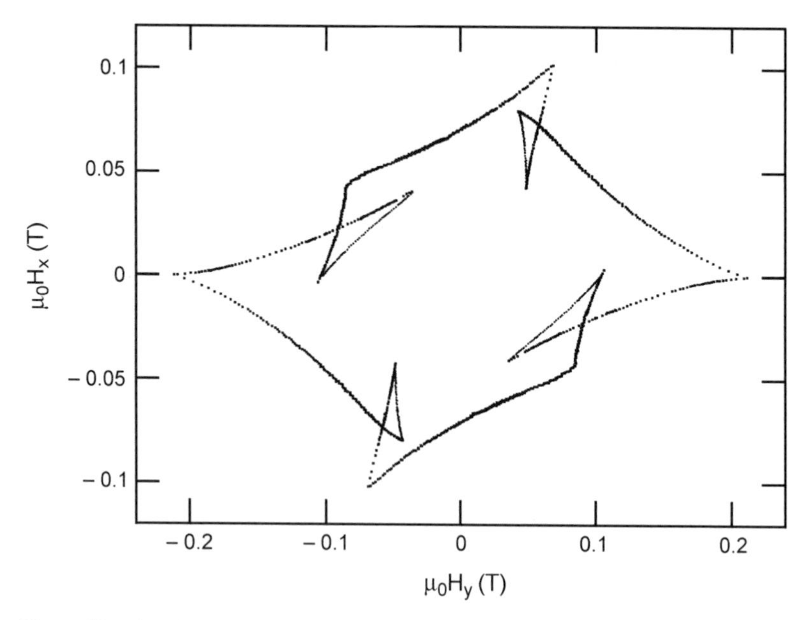

Figure 20. Angular dependence of the switching field of a 15-nm Co nanoparticle showing a strong influence of cubic crystalline anisotropy.

The variation of the switching field with the angle θ (defined between the applied field and the long axis of the ellipsoid) is given by [68]

$$H_{\mathrm{sw}}^{0} = \frac{M_{\mathrm{s}}}{2} \frac{a_x a_z}{\sqrt{a_z^2 \sin^2\theta + a_x^2 \cos^2\theta}} \tag{3.16}$$

where $a_{x,z} = 2\,N_{x,z} - k/S^2$, $N_{x,z}$ are the demagnetization factors, $S = R/\lambda$, and R is the minor semiaxis of the ellipsoid. The parameter k is a monotonically decreasing function of the aspect ratio of the ellipsoid. This function is plotted in Fig. 1 of Ref. 69. The smallest and highest value of k is that for an infinite cylinder ($k = 1.079$) and a sphere ($k = 1.379$), respectively.

For a long ellipsoid of rotation, the demagnetization factors are given by

$$N_z = \frac{1}{n^2 - 1} \frac{n}{\sqrt{n^2 - 1}} \ln\left(n - 1 + \sqrt{n^2 - 1}\right), \qquad N_x = \frac{1 - N_z}{2} \tag{3.17}$$

where n is the ratio of the length to the diameter.

For an infinite cylinder, Eq. (3.16) becomes [66, 70]

$$H_{\rm sw}^{0} = \frac{M_{\rm s}}{2} \frac{h_t(1+h_t)}{\sqrt{h_t^2 + (1+2h_t)\cos^2\theta}} \tag{3.18}$$

where $h_t = -1.079/S^2$. Equation (3.18) is a good approximation for a very long ellipsoid of rotation. It is plotted in Fig. 21 for several radii of an infinite cylinder.

The case of uniform rotation of magnetization was generalized by Thiaville to an arbitrary anisotropy energy function and to three dimensions (Section III.A.1). For the curling mode, this generalization is not possible. However, approximated calculations were proposed [67, 71, 72] and micromagnetic simulations were performed [73].

2. *Experimental Evidence for Magnetization Reversal by Curling*

We report here the first studies of isolated nanoscale wires with diameters smaller than 100 nm, for which single-domain states could be expected [35, 74]. The cylindrical geometry, with its large shape anisotropy, is well-suited

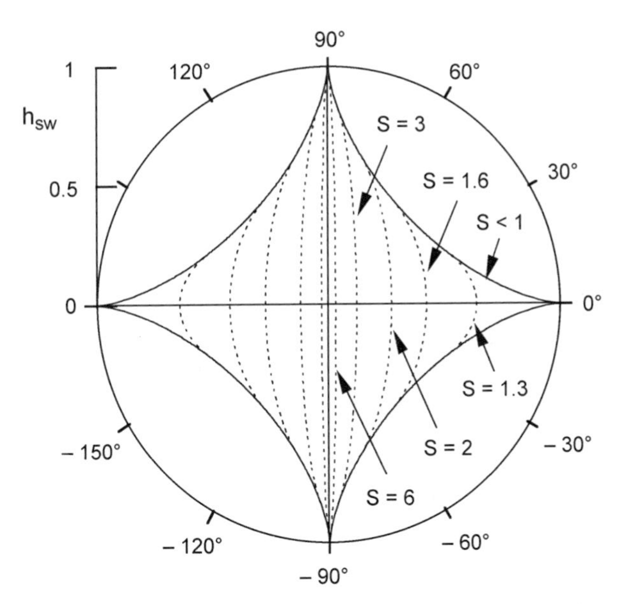

Figure 21. Angular dependence of the switching field of an infinite cylinder for several reduced cylinder radii S. For $S<1$, the switching field is given by the uniform rotation mode (Section III.A).

for comparison with theory. Ni wires were produced by filling electrochemically the pores of commercially available nanoporous track-etched polycarbonate membranes of thicknesses of 10 μm. The pore size was chosen in the range of 30 to 100 nm [75, 76]. In order to place one wire on the SQUID detector, we dissolved the membrane in chloroform and put a drop on a chip of some hundreds of SQUIDs. Magnetization measurements were performed on SQUIDs with a single isolated wire (Fig. 1).

The angular dependence of the switching field of wires with 45- and 92-nm diameter are shown in Fig. 22. These measurements are in quantitative agreement with the curling mode [Eq. (3.16)]. Nevertheless, dynamical measurements showed a nucleation volume that is much smaller than the wire volume [35, 74] (Section IV.C.3). Therefore, we believe that the magnetization reversal starts close to curling instability, but the nucleation happens in a small fraction of the wire only, then rapidly propagating along the whole sample. This picture is also in good agreement with micromagnetic simulations [73] and the micromagnetic model of Braun [77].*

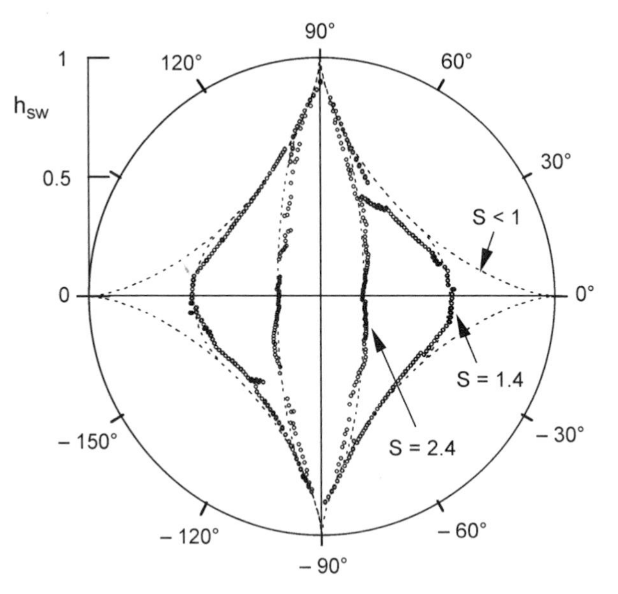

Figure 22. Angular dependence of the switching field of two Ni wires with a diameter of 45 nm and 92 nm, that is, $S = 1.4$ and 2.4. The switching fields are normalized by 125 mT and 280 mT, respectively.

*Note that the curling model only predicts an instability. There has never been a claim to describe what happens afterwards [2].

The angular dependence of the switching field of Ni wires with larger diameters (270–450 nm) were measured at room temperature by Lederman et al. [78]. Their results could roughly be explained by the curling mode.

3. Magnetization Reversal by Nucleation and Annihilation of Domain Walls

For magnetic particles that have at least two dimensions much larger than the domain wall width, the magnetization reversal may occur via nucleation/propagation and annihilation of one or several domain walls happening at two or more applied fields. We focus here on a 30-nm-thick elliptic Co particle defined by electron beam lithography and lift-off techniques out of sputtered thin films (inset of Fig. 23) [79]. The Co film has a nanocrystalline structure leading to a magnetically soft material with a coercive field value of 3 mT at 4 K. Therefore we neglected the magnetocrystalline anisotropy. The nanofabricated particles have an elliptic contour with in-plane dimensions of 300 nm $\times$ 200 nm and a thickness of 30 nm.

In order to study the domain structure of our particles, we measure the angular dependence of hysteresis loops. Figure 23 shows a typical hysteresis loop of an individual Co particle. The magnetic field is applied in the plane of the particle. The hysteresis loop is mainly characterized by two magnetization jumps. Starting from a saturated state, the first jump can be associated with domain wall nucleation and the second jump can be associated with domain wall annihilation. During these jumps, the magnetization switches in less than 100 μs (our time resolution, see Section II.B.4). The reversible central region of the hysteresis loops is evidence for the motion of the domain wall through the particle.

The simplest domain structure, showing such a hysteresis loop, has been proposed by van den Berg [80] in zero field and calculated for fields smaller than the saturation field by Bryant and Suhl [81].* This domain structure has been observed experimentally on low anisotropy circular thin film disks (100 μm in diameter) using high-resolution Kerr techniques [82]. In zero field, the particle has a vortex-like domain wall as shown in Fig. 23. When a magnetic field is applied, this domain wall is pushed to the border of the particle. For higher fields the domain wall is annihilated and the particle becomes single domain. The main conditions of the van den Berg model are (i) that the magnetic material is very soft and (ii) that the system is two-dimensional. The first condition is satisfied by our particles because they are made of a randomly oriented nanocrystallized Co, being very soft. The second condition is not quite well satisfied however, MFM measurements confirmed this domain structure [83]. Furthermore, we obtained

*Note that these models are 2D and neglect the domain wall width. Therefore, they can give only a qualitative description.

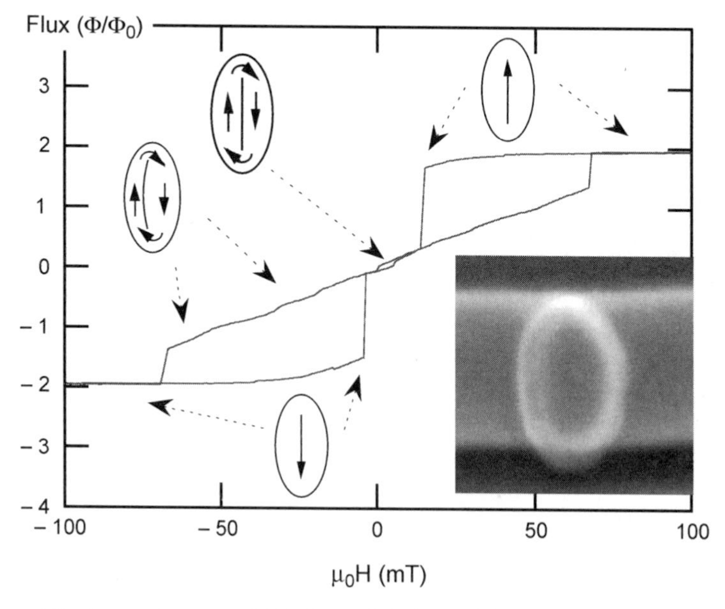

Figure 23. Hysteresis loops at 0.1 K of the elliptic Co particle seen in the inset which shows the electron micrograph of the wire of the SQUID loop with a Co particle (300 nm × 200 nm × 30 nm). The in-plane field is applied along the long axis of the particles. The domain wall structure in an elliptical particle is also presented schematically as proposed by van den Berg. Arrows indicate the spin direction. The two small magnetization jumps near $\approx \pm 5$ mT might be due to the reversal of the center vortex of the domain structure.

similar results for thinner particles (10 and 20 nm). More complicated domain structures as proposed by van den Berg [79, 80] may be excluded by the fact that similar Co particles of length smaller than 200 nm are single domain [22, 59].

After studying the magnetization reversal of individual particles, the question arises as to how the properties of a macroscopic sample are based on one-particle properties. In order to answer this question, we fabricated a sample consisting of 1.8×10^7 nearly identical elliptic Co particles of about the same dimensions and material as the individual particle studied above. These particles are placed on an Si substrate with a spacing of 2 μm. Because of this large spacing, dipole interactions between particles are negligible. Figure 24 shows the hysteresis loop of the array of Co particles when the field is applied parallel to the long axis of the particle. This hysteresis loop shows the same characteristics as the hysteresis loop of one particle (Fig. 23) that is, nucleation and annihilation of domain walls. Because of switching field distributions mainly due to surface defects and a slight distribution of particle sizes, shapes, and so on, the domain wall nucleation and annihilation are no longer discontinuous

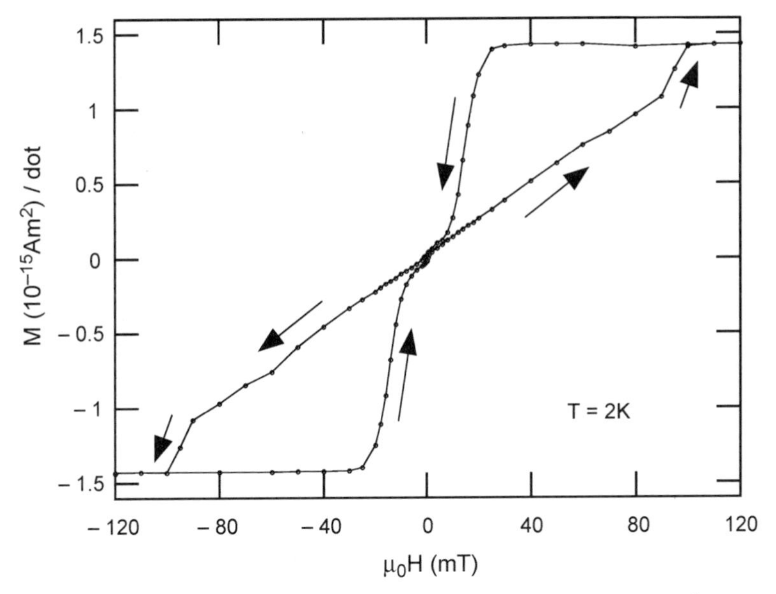

Figure 24. Hysteresis loops of the magnetic moment of the array of 1.8×10^7 Co particles (300 nm $\times$ 200 nm $\times$ 30 nm). The in-plane field is applied along the long axis of the particles.

although still irreversible. They take place along continuous curves with a width of about 10 mT.

IV. INFLUENCE OF TEMPERATURE ON THE MAGNETIZATION REVERSAL

The thermal fluctuations of the magnetic moment of a single-domain ferromagnetic particle and its decay towards thermal equilibrium were introduced by Néel [9, 10] and further developed by Bean and Livingston [84, 85] and Brown [86–88]. The simplest case is an assembly of independent particles having no magnetic anisotropy. In the absence of an applied magnetic field, the magnetic moments are randomly oriented. The situation is similar to paramagnetic atoms where the temperature dependence of the magnetic susceptibility follows a Curie behavior, and the field dependence of magnetization is described by a Brillouin function. The only difference is that the magnetic moments of the particles are much larger than those of the paramagnetic atoms. Therefore, the quantum mechanical Brillouin function can be replaced by the classical limit for larger magnetic moments, namely the Langevin function. This theory is called superparamagnetism. The situation changes, however, as soon as magnetic anisotropy is present which establishes one or more preferred orientations of the particle's magnetization (Section III). In the following, we present an overview over the simplest model describing thermally activated magnetization reversal of single isolated nanoparticles which is called the Néel–Brown model. After a brief review of the model (Section IV. A), we present experimental methods to study the thermally activated magnetization reversal (Section IV. B). Finally, we discuss some applications of the Néel–Brown model (Section IV.C).

A. Néel–Brown Model of Thermally Activated Magnetization Reversal

In Néel and Brown's model of thermally activated magnetization reversal, a single-domain magnetic particle has two equivalent ground states of opposite magnetization separated by an energy barrier which is due to shape and crystalline anisotropy. The system can escape from one state to the other by thermal activation over the barrier. Just as in the Stoner–Wohlfarth model, they assumed uniform magnetization and uniaxial anisotropy in order to derive a single relaxation time. Néel supposed further that the energy barrier between the two equilibrium states is large in comparison to the thermal energy $k_B T$ which justified a discrete orientation approximation [9, 10]. Brown criticized Néel's model because the system is not explicitly treated as a gyromagnetic one [86–88]. Brown considered the magnetization vector in a particle to wiggle around an energy minimum, then jump to the vicinity of the other minimum, then wiggle around there before jumping again. He supposed that the orientation of the magnetic moment may be described by a Gilbert equation with a random field term that is assumed to be white noise. On the basis of these assumptions, Brown was able to derive a Fokker–Planck equation for the distribution of magnetization orientations. Brown did not solve his differential equation. Instead he tried some analytic approximations and an asymptotic expansion for the case of the field parallel or perpendicular to the easy axis of magnetization. More recently, Coffey et al. [89, 90] found by numerical methods an exact solution of Brown's differential equation for uniaxial anisotropy and an arbitrary applied field direction. They also derived an asymptotic general solution for the case of large energy barriers in comparison to the thermal energy $k_B T$. This asymptotic solution is of particular interest for single-particle measurements and is reviewed in the following.

For a general asymmetric bistable energy potential $E = E(\vec{m}, \vec{H})$ [Eq. (3.3)] with the orientation of magnetization $\vec{m} = \vec{M}/M_s$ (M_s is the spontaneous magnetization), $\vec{H}$ is the applied field, and with minima at $\vec{n}_1$ and $\vec{n}_2$ separated by a potential barrier containing a saddle point at $\vec{n}_0$ (with the $\vec{n}_i$ coplanar), and in the case of $\beta(E_0 - E_i) \gg 1$ where $\beta = 1/k_B T$, and $E_i = E(\vec{n}_i, \vec{H})$, Coffey et al. showed that the longest relaxation time* is given by the following equation which is valid in the intermediate to high damping limit (IHD) defined by $\alpha\,\beta\,(E_0-E_i)>1$ [91]:

$$\tau^{-1}=\frac{\Omega_0}{2\pi\omega_0}\left[\omega_1 e^{-\beta(E_0-E_1)}+\omega_2 e^{-\beta(E_0-E_2)}\right] \tag{4.1}$$

where ω_0 and Ω_0 are the saddle and damped saddle angular frequencies:

$$\omega_0=\frac{\gamma}{M_s}\sqrt{-c_1^{(0)}c_2^{(0)}} \tag{4.2}$$

$$\Omega_0=\frac{\gamma}{M_s}\frac{\alpha}{1+\alpha^2}\left[-c_1^{(0)}-c_2^{(0)}+\sqrt{(c_2^{(0)}-c_1^{(0)})^2-4\alpha^{-2}c_1^{(0)}c_2^{(0)}}\right] \tag{4.3}$$

ω_1 and ω_2 are the well angular frequencies:

$$\omega_i=\frac{\gamma}{M_s}\sqrt{c_1^{(i)}c_2^{(i)}} \tag{4.4}$$

with $i=1$ and 2. $c_1^{(j)}$ and $c_2^{(j)}$ ($j=0$, 1, 2) are the coefficients in the truncated Taylor series of the potential at well and saddle points—that is, the curvatures of the potential at well and saddle points. γ is the gyromagnetic ratio, $\alpha=\nu\,\gamma M_s$ is the dimensionless damping factor and ν is the friction in Gilbert's equation (ohmic damping).

Whereas in the low damping limit (LD), defined by $\alpha\,\beta(E_0-E_i)<1$, the longest relaxation time is given by [92, 93]

$$\tau^{-1}=\frac{\alpha}{2\pi}\left[\omega_1\beta(E_0-E_1)e^{-\beta(E_0-E_1)}+\omega_2\beta(E_0-E_2)e^{-\beta(E_0-E_2)}\right] \tag{4.5}$$

In this case, the energy dissipated in one cycle of motion in the well is very small in comparison to the thermal energy k_BT.

Experimentally, relaxation is observed only if τ is of the order of magnitude of the measuring time of the experiment. This implies for all known single-particle measurement techniques that $\beta(E_0-E_i)\gg 1$; that is, the asymptotic solutions (4.1) and (4.5) are always a very good approximation to the exact

*The inverse of the longest relaxation time is determined by the smallest nonvanishing eigenvalue of the appropriate Fokker–Planck equation [89, 90]. All other eigenvalues can be neglected in the considered asymptotic limit of $\beta(E_0-E_i)\gg 1$.

solution of Brown's Fokker–Planck equation [94]. Due to an applied field, $\beta(E_0 - E_1) \gg \beta(E_0 - E_2)$ (taking E_2 as the metastable minimum) might be true. Then the first exponential in Eq. (4.1) and (4.5) can be neglected.

Concerning the possible values of α, we remark that little information is available. Typical values should be between 0.01 and 5 [8], meaning that in practice $\alpha\beta(E_0 - E_i)$ can be $\gg 1$, $\ll 1$, or ≈ 1. Thus the distinction between Eqs. (4.1) and (4.5) becomes important.

Finally, we note that $c_1^{(j)}$ and $c_2^{(j)}$ ($j = 0, 1, 2$) can be found experimentally by measuring the critical surface of the switching field and applying the calculation of Thiaville (Section III.A.1) [55].

B. Experimental Methods for the Study of the Néel–Brown Model

As discussed in the previous section, in the Néel–Brown model of thermally activated magnetization reversal a single-domain magnetic particle has two equivalent ground states of opposite magnetization separated by an energy barrier due to, for instance, shape and crystalline anisotropy. The system can escape from one state to the other either by thermal activation over the barrier at high temperatures or by quantum tunneling at low temperatures (Section V). At sufficiently low temperatures and at zero field, the energy barrier between the two states of opposite magnetization is much too high to observe an escape process. However, the barrier can be lowered by applying a magnetic field in the opposite direction to that of the particle's magnetization. When the applied field is close enough to the switching field at zero temperature H_{sw}^0, thermal fluctuations are sufficient to allow the system to overcome the barrier, and the magnetization is reversed.

In the following, we discuss three different experimental methods for studying this stochastic escape process which are called waiting time, switching field, and telegraph noise measurements.

1. *Waiting Time Measurements*

The waiting time method consists in measuring the probability that the magnetization has not switched after a certain time. In the case of an assembly of identical and isolated particles, it corresponds to measurements of the relaxation of magnetization. However, in most particle assemblies, broad distributions of switching fields lead to logarithmic decay of magnetization, and the switching probability is hidden behind the unknown distributions functions [8]. For individual particle studies, waiting time measurements give direct access to the switching probability (Fig. 25). At a given temperature, the magnetic field H is increased to a waiting field H_w near the switching field H_{sw}^0. Next, the elapsed time until the magnetization switches is measured. This process is repeated several hundred times, yielding a waiting time histogram. The integral of this histogram and proper normalization yields the probability that the magnetization

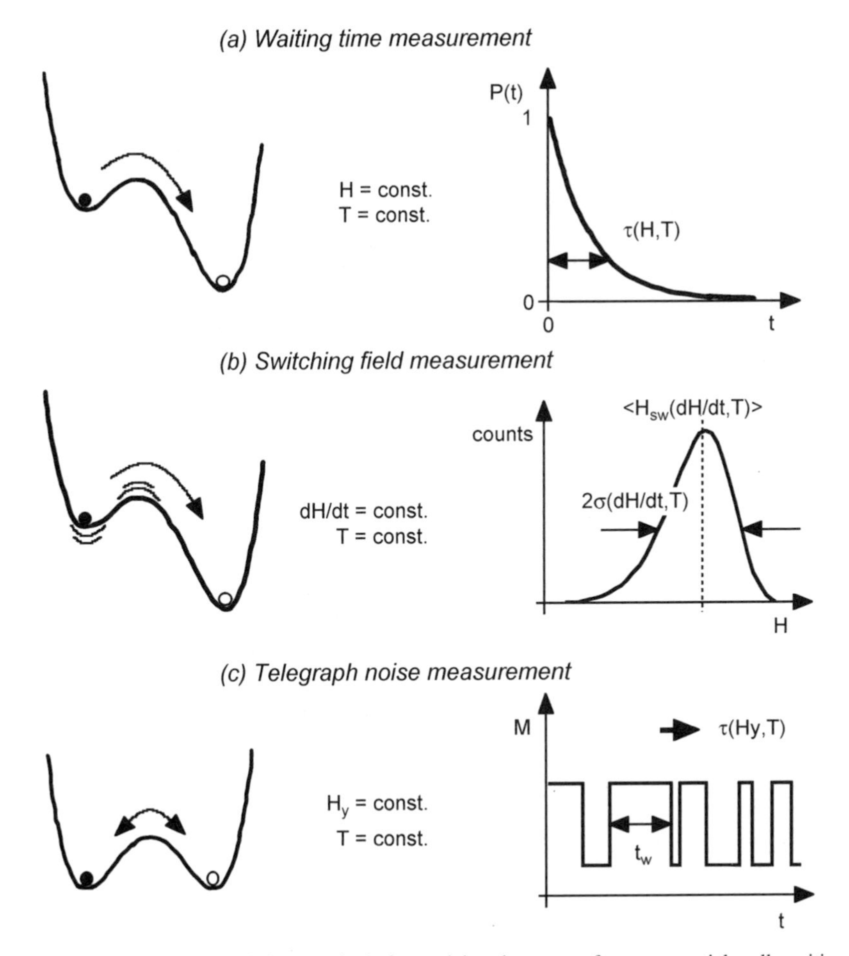

Figure 25. Schema of three methods for studying the escape from a potential well: waiting time and telegraph noise measurements give direct access to the switching time probability $P(t)$, whereas switching field measurements yield histograms of switching fields.

has not switched after a time t. This probability is measured at different waiting fields H_w and temperatures in order to explore several barrier heights and thermal activation energies.

According to the Néel–Brown model, the probability that the magnetization has not switched after a time t is given by

$$P(t) = e^{-t/\tau} \tag{4.6}$$

and τ (inverse of the switching rate) can be expressed by an Arrhenius law of the form

$$\tau^{-1}(\varepsilon) = B\varepsilon^{a+b-1}e^{-A\varepsilon^a} \tag{4.7}$$

where $\varepsilon = (1 - H/H_{sw}^0)$ and A, B, a, and b depend on damping, temperature, energy barrier height [Eqs. (3.5)–(3.10) and (4.1)–(4.5)], curvatures at well and saddle points, and reversal mechanism (thermal or quantum) (cf. Table 1 of Ref. 95). For simplicity, experimentalists have often supposed a constant pre-exponential factor τ_0^{-1} instead of $B\varepsilon^{a+b-1}$.

The adjustment of Eq. (4.6) to the measured switching probabilities yields a set of mean waiting times $\tau^{-1}(H_w, T)$. In order to adjust the Néel–Brown model to this set of data, we propose the following relation that can be found by inserting $\varepsilon = (1 - H_w/H_{sw}^0)$ into Eq. (4.7):

$$H_w = H_{sw}^0\left(1 - \left[\frac{1}{A}\ln(\tau B\varepsilon^{a+b-1})\right]^{1/a}\right) \tag{4.8}$$

When plotting the H_w values as a function of $[T \ln (\tau\, B\, \varepsilon^{a+b-1})]^{1/a}$, all points should gather on a straight line (master curve) by choosing the proper value for the constants B, a, and b [a and b should be given by Eqs. (3.5)–(3.10) and (4.1)–(4.5)]. A can be obtained from the slope of the master.

The number of exploitable decades for τ values is limited for waiting time measurements: Short-time (milliseconds) experiments are limited by the inductance of the field coils* and long-time (minutes) studies by the stability of the experimental setup. Furthermore, the total acquisition time for a set of $\tau^{-1}(H_w, T)$ is rather long (weeks). Thus a more convenient method is needed for single-particle measurements—namely, the switching field method.

2. *Switching Field Measurements*

For single-particle studies, it is often more convenient to study magnetization reversal by ramping the applied field at a given rate and measuring the field value as soon as the particle magnetization switches. Next, the field ramp is reversed and the process repeated. After several hundred cycles, switching field histograms are established, yielding the mean switching field $\langle H_{sw}\rangle$ and the width σ_{sw} (rms deviation). Both mean values are measured as a function of the field sweeping rate and temperature (Fig. 25).

From the point of view of thermally activated magnetization reversal, switching field measurements are equivalent to waiting time measurements because the

*A solution to this problem might be a superposition of a constant applied field and a small pulse field.

time scale for the sweeping rate is typically more than 8 orders of magnitude greater than the time scale of the exponential prefactor, which is in general around 10^{-10} s. We can therefore apply the Néel–Brown model described above. The mathematical transformation from a switching time probability [Eqs.(4.6)–(4.7)] to a switching field probability was first given by Kurkijärvi [96] for the critical current in SQUIDs. A more general calculation was evaluated by Garg [95]. In many cases, the mean switching field $\langle H_{\mathrm{sw}} \rangle$ can be approximated by the first two terms of the development of Garg [95]:

$$\langle H_{\mathrm{sw}}(T, v) \rangle \approx H_{\mathrm{sw}}^{0} \left(1 - \left[\frac{1}{A} \ln \left(\frac{H_{\mathrm{sw}}^{0} B}{vaA^{1-b/a}} \right) \right]^{1/a} \right) \tag{4.9}$$

where the field sweeping rate is given by $v = dH/dt$. The width of the switching field distribution σ_{sw} can be approximated by the first term of Garg's development:

$$\sigma_{\mathrm{sw}} \approx H_{\mathrm{sw}}^{0} \frac{\pi}{\sqrt{6}a} \left(\frac{1}{A} \right)^{1/a} \left[\ln \left(\frac{H_{\mathrm{sw}}^{0} B}{vaA^{1-b/a}} \right) \right]^{(1-a)/a} \tag{4.10}$$

In the case of a constant preexponential factor τ_0^{-1}, the calculation of $\langle H_{\mathrm{sw}} \rangle$ and σ_{sw} is more simple and is given by Eqs. (4) and (5) in Ref. 36, respectively.

Similar to the waiting time measurements, a scaling of the model to a set of $\langle H_{\mathrm{sw}}(T, v) \rangle$ values can be done by plotting the $\langle H_{\mathrm{sw}}(T, v) \rangle$ values as a function of $[T \ln((H_{\mathrm{sw}}^{0} B)/(vaA^{1-b/a}))]^{1/a}$. All points should gather on a straight line by choosing the proper value for the constants [a and b should be given by Eqs. (3.5)–(3.10) and (4.1)–(4.5)].

The entire switching field distribution $P(H)$ can be calculated iteratively by the following equation [96]:

$$P(H) = \tau^{-1}(H) v^{-1} \left[1 - \int_{0}^{H} P(H') dH' \right] \tag{4.11}$$

3. *Telegraph Noise Measurements*

In order to study the superparamagnetic state* of a single particle, it is simply necessary to measure the particle's magnetization as a function of time. We call

*At zero applied field, a single-domain magnetic particle has two equivalent ground states of opposite magnetization separated by an energy barrier. When the thermal energy $k_{\mathrm{B}}T$ is sufficiently high, the total magnetic moment of the particle can fluctuate thermally, like a single spin in a paramagnetic material. Such magnetic behavior of an assembly of independent single-domain particles is called superparamagnetism [8, 84, 85].

this telegraph noise measurement as stochastic fluctuations between two states are expected. According to the Néel–Brown model, the mean time τ spent in one state of magnetization is given by an Arrhenius law of the form of Eq. (4.7). As τ increases exponentially with decreasing temperature, it is very unlikely that an escape process will be observed at low temperature. However, applying a constant field in direction of a hard axis (hard plane) of magnetization reduces the height of the energy barrier (Fig. 25). When the energy barrier is sufficiently small, the particle's magnetization can fluctuate between two orientations which are close to a hard axis (hard plane) of magnetization. The time spent in each state follows an exponential switching probability law as given by Eqs. (4.6) and (4.7) with $a = 2$ [Eq. (3.8)].*

C. Experimental Evidence for the Néel–Brown Model

The Néel–Brown model is widely used in magnetism, particularly in order to describe the time dependence of the magnetization of collections of particles, thin films, and bulk materials. However until recently, all the reported measurements, performed on individual particles, were not consistent with the Néel–Brown theory. This disagreement was attributed to the fact that real samples contain defects, ends, and surfaces that could play an important, if not dominant, role in the physics of magnetization reversal. It was suggested that the dynamics of reversal occurs via a complex path in configuration space, and that a new theoretical approach is required to provide a correct description of thermally activated magnetization reversal even in single-domain ferromagnetic particles [19, 59]. Similar conclusions were drawn from numerical simulations of the magnetization reversal [97–101].

A few years later, micro-SQUID measurements on individual Co nanoparticles showed for the first time a very good agreement with the Néel–Brown model by using waiting time, switching field, and telegraph noise measurements [36–39]. It was also found that sample defects, especially sample oxidation, play a crucial role in the physics of magnetization reversal.

In the following subsections, we review some typical results concerning nanoparticles (Section IV.C.1), clusters (Section IV.C.2), and wires (Section IV.C.3). In Section IV.C.4, we point out the main deviations from the Néel–Brown model which are due to defects.

1. Application to Nanoparticles

One of the important predictions of the Néel–Brown model concerns the exponential not-switching probability $P(t)$ [Eq. (4.6)] which can be measured directly via waiting time measurements (Section IV.B.1): At a given temperature, the

*Note that for a slightly asymmetric energy potential, one switching probability can be so long that two-level fluctuation becomes practically unobservable.

magnetic field is increased to a waiting field H_w which is close to the switching field. Then, the elapsed time is measured until the magnetization switches. This process is repeated several hundred times, in order to obtain a waiting time histogram. The integral of this histogram gives the not-switching probability $P(t)$ which is measured at several temperatures T and waiting fields H_w. The inset of Fig. 26 displays typical measurements of $P(t)$ performed on a Co nanoparticle. All measurements show that $P(t)$ is given by an exponential function described by a single relaxation time τ.

The validity of Eqs. (3.5) and (4.7) is tested by plotting the waiting field H_w as a function of $[T \ln(\tau/\tau_0)]^{2/3}$.* If the Néel–Brown model applies, all points should collapse onto one straight line (master curve) by choosing the proper values for τ_0. Figure 26 shows that the data set $\tau(H_w,T)$ falls on a master curve provided that $\tau_0 \approx 3 \times 10^{-9}$ s. The slope and intercept yield the values $E_0 = 214{,}000$ K and $H^0_{sw} = 143.05$ mT. The energy barrier E_0 can be approximately converted to a

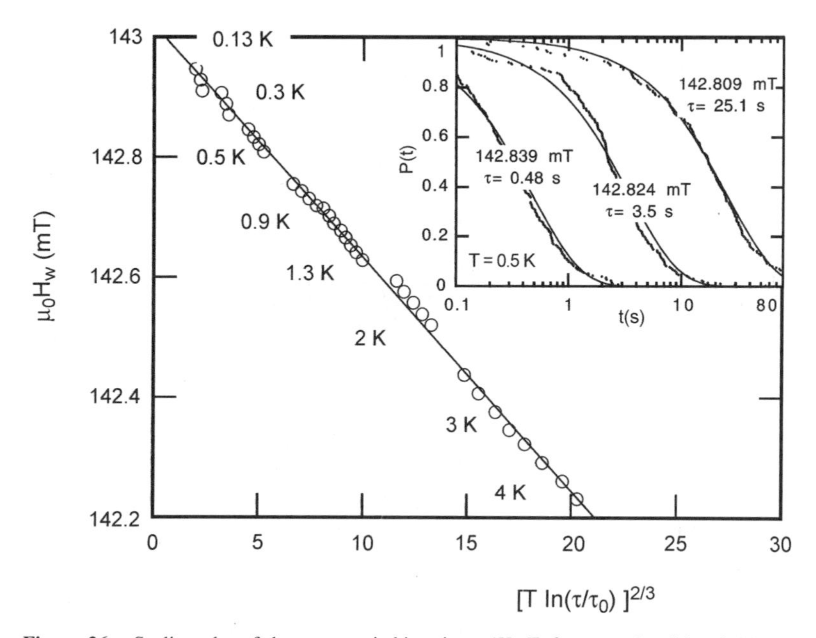

Figure 26. Scaling plot of the mean switching time $\tau(H_w,T)$ for several waiting fields H_w and temperatures (0.1 s < $\tau(H_w,T)$ < 60 s) for a Co nanoparticle. The scaling yields $\tau_0 \approx 3 \times 10^{-9}$ s. *Inset*: Examples of the probability of not-switching of magnetization as a function of time for different applied fields and at 0.5 K. *Full lines* are data fits with an exponential function: $P(t) = e^{-t/\tau}$.

*$a = 3/2$ because the field was applied at about 20° from the easy axis of magnetization [Eq. (3.5)].

thermally "activated volume" by using $V = E_0/(\mu_0 M_S H_{sw}^0) \approx (25\,\mathrm{nm})^3$ which is very close to the particle volume estimated by SEM. This agreement is another confirmation of a magnetization reversal by uniform rotation. The result of the waiting time measurements are confirmed by switching field and telegraph noise measurements [36, 37]. The field and temperature dependence of the exponential prefactor τ_0 is taken into account in Ref. 90.

2. *Application to Co Clusters*

Figure 27 presents the angular dependence of the switching field of a 3-nm Co cluster measured at different temperatures. At 0.03 K, the measurement is very close to the standard Stoner–Wohlfarth astroid (Fig. 10). For higher temperatures the switching field becomes smaller and smaller. It reaches the origin at about 14 K, yielding the blocking temperature $T_B = 14\,\mathrm{K}$ of the cluster

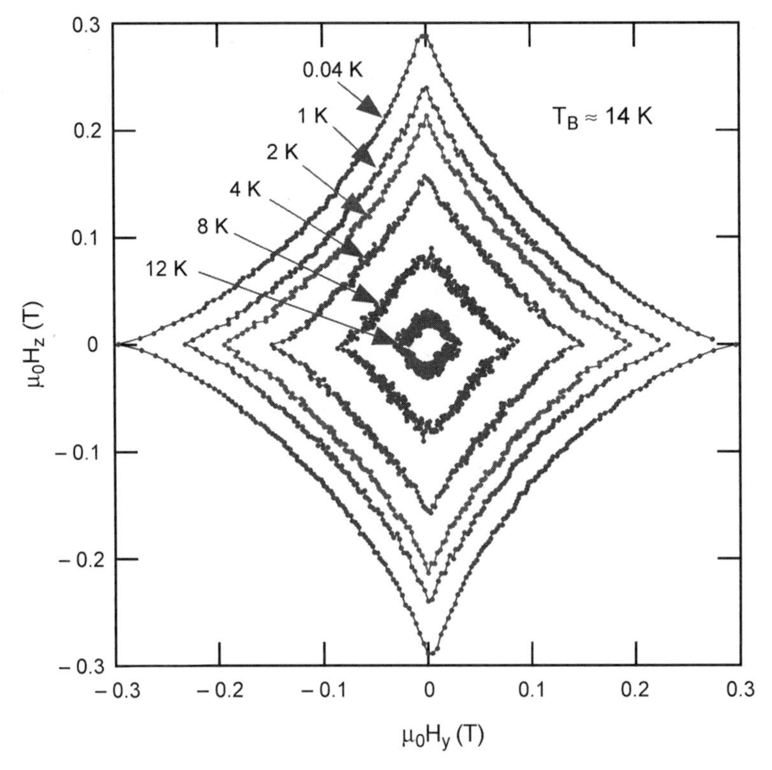

Figure 27. Temperature dependence of the switching field of a 3-nm Co cluster, measured in the plane defined by the easy and medium hard axes (H_y–H_z plane in Fig. 15). The data were recorded using the blind mode method (Section II.B.6) with a waiting time of the applied field of $\Delta t = 0.1$ s. The scattering of the data is due to stochastic and in good agreement with Eq. (4.10).

magnetization. T_B is defined as the temperature for which the waiting time Δt becomes equal to the relaxation time τ of the particle's magnetization at $\vec{H} = \vec{0}$. T_B can be used to estimate the total number N_{tot} of magnetic Co atoms in the cluster. Using an Arrhenius-like law [Eq. (4.7)] which can be written as $\Delta t = \tau = \tau_0 \exp(K_{at} N_{tot}/k_B T_B)$, where τ_0^{-1} is the attempt frequency typically between 10^{10} to 10^{11} Hz [102], K_{at} is an effective anisotropy energy per atom and k_B is the Boltzmann constant. Using the expression of the switching field at $T = 0$ K and for $\theta = 0$: $\mu_0 H_{sw} = 2K_{at}/\mu_{at} = 0.3\,T$ (Fig. 27), the atomic moment $\mu_{at} = 1.7\,\mu_B$, $\Delta t = 0.01$ s, $\tau_0 = 10^{-10}$ s, and $T_B = 14$ K, we deduce $N_{tot} \approx 1500$, which corresponds very well to a 3-nm Co cluster (Fig. 13).

Figure 28 presents a detailed measurement of the temperature dependence of the switching field at 0° and 45° and for three waiting times Δt. This measurement allows us to check the predictions of the field dependence of the barrier height. Equation (4.9) predicts that the mean switching field should be proportional to $T^{1/a}$ where a depends on the direction of the applied field [Eqs. (3.5) to (3.10)]: $a = 2$ for $\theta = 0°$ and 90°, and $a = 3/2$ for all other angles which are not too close to $\theta = 0°$ and 90°. We found a good agreement with this model (Fig. 29).

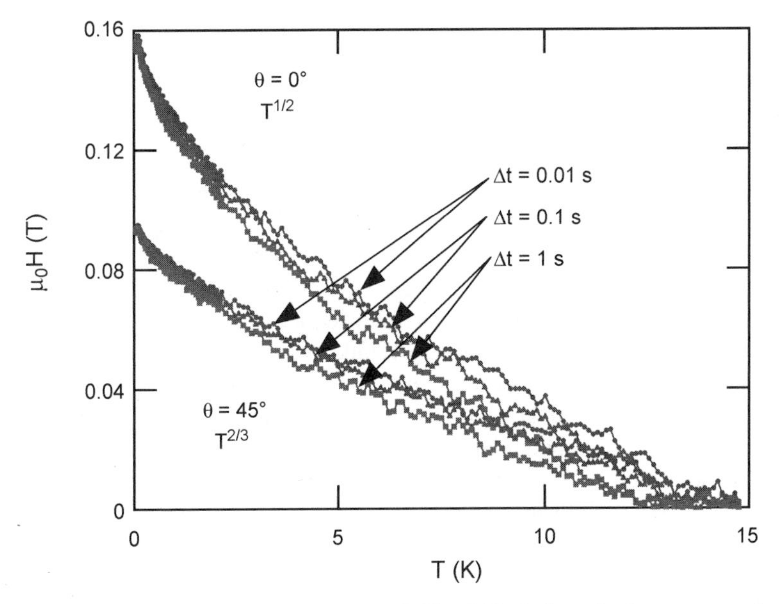

Figure 28. Temperature dependence of the switching field of a 3-nm Co cluster, measured at 0° and 45°. The data were recorded using the blind mode method (Section II.B.6) with different waiting time Δt of the applied field. The scattering of the data is due to stochastics and is in good agreement with Eq. (4.10).

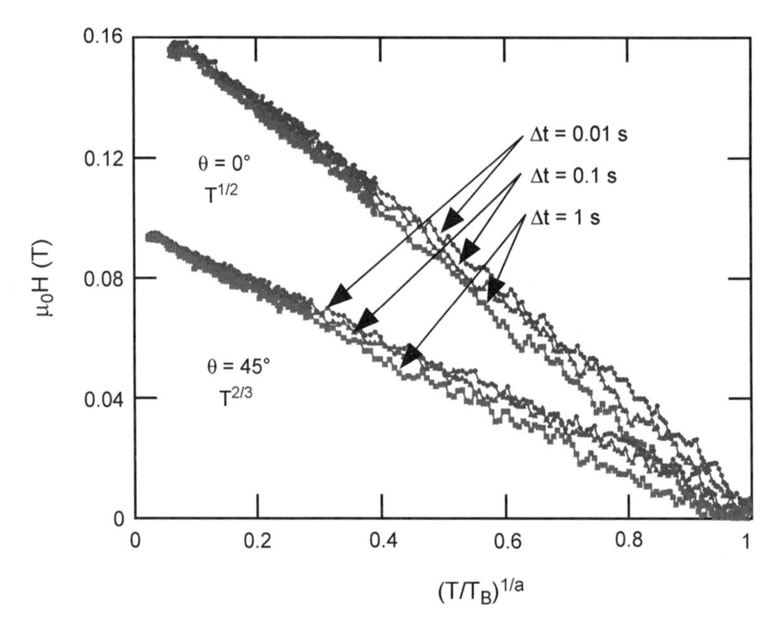

Figure 29. Temperature dependence of the switching field of a 3-nm Co cluster as in Fig. 28 but plotted as a function of $(T/T_B)^{1/a}$ with $T_B = 14$ K and $a = 2$ or 3/2 for $\theta = 0°$ and 45°, respectively, and for three waiting times Δt.

3. Application to Ni Wires

Electrodeposited wires (with diameters ranging from 40 to 100 nm and lengths up to 5000 nm; see Fig.1) were studied [35, 74] using the micro-SQUID technique (Section III.B.2). For diameter values under 50 nm, the switching probability as a function of time could be described by a single exponential function [Eq. (4.6)]. The mean waiting time τ followed an Arrhenius law [Eq. (4.7)] as proposed by the Néel–Brown model. Temperature and field sweeping rate dependence of the mean switching field could be described by the model of Kurkijärvi (Section IV.B.2) which is based on thermally assisted magnetization reversal over a simple potential barrier. These measurement allowed us to estimate an activation volume which was two orders of magnitude smaller than the wire volume. This confirmed the idea of the reversal of the magnetization caused by a nucleation of a reversed fraction of the cylinder, rapidly propagating along the whole sample. This result was also in good agreement with a micromagnetic model of Braun [77].

A pinning of the propagation of the magnetization reversal occurred for a few samples, where several jumps were observed in the hysteresis curves. The pinning of a domain wall was probably due to structure defects. The dynamic

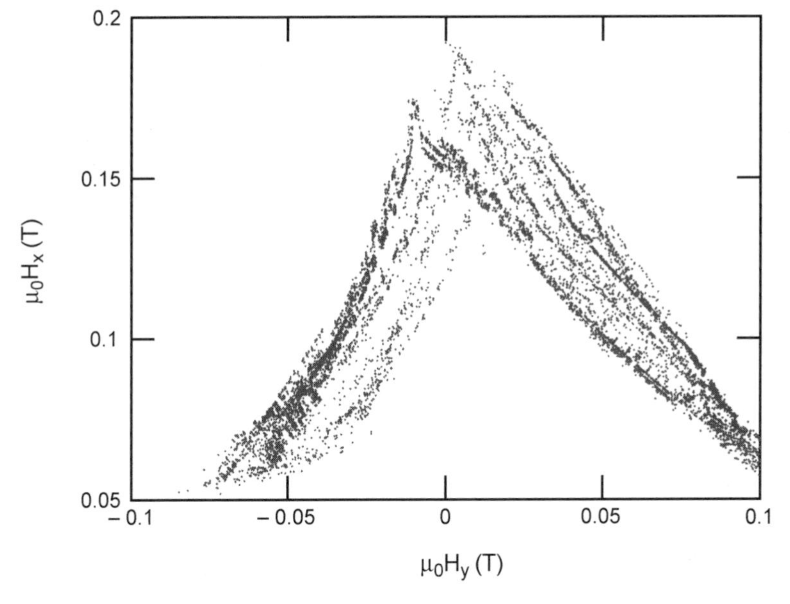

Figure 30. Angular dependence of switching fields of a 3-nm Fe cluster having (probably) a slightly oxidized surface. Each point corresponds to one of the 10,000 switching field measurements. The huge variations of the switching field might be due to exchange bias of frustrated spin configurations. However, quantum effects like those described in Section V are not completely excluded.

reversal properties of depinning were quite different from those of nucleation of a domain wall. For example, the probability of depinning as a function of time did not follow a single exponential law. A similar effect was also observed in single submicron Co particles having one domain wall [79], showing a domain wall annihilation process (Section III.B.3).

4. Deviations From the Néel–Brown Model

Anomalous magnetic properties of oxidized or ferrimagnetic nanoparticles have been reported previously by several authors [103, 104]. These properties are, for example, the lack of saturation in high fields and shifted hysteresis loops after cooling in the presence of a magnetic field. These behaviors have been attributed to uncompensated surface spins of the particles and surface spin disorder [105, 106].

Concerning our single-particle studies, we systematically observed aging effects which we attribute to an oxidation of the surface of the sample, forming antiferromagnetic CoO or NiO [59, 74]. We found that the antiferromagnetic coupling between the core of the particle or wire and its oxidized surface changed the dynamic reversal properties. For instance, we repeated the measurements of the magnetization reversal of a Ni wire two days after fabrication, six weeks after, and finally after three months [74]. Between these measurements, the wire stayed in a dry box. The quasi-static micro-SQUID measurements did reveal only small changes. The saturation magnetization measured after six weeks was unchanged and was reduced by one to two percent after three months. The angular dependence of the switching field changed also only slightly. The dynamic measurements showed the aging effects more clearly, as evidenced by

- A nonexponential probability of not switching
- An increase of the width of the switching field distributions
- A decrease of the activation energy

We measured a similar behavior on lithographic fabricated Co particles with an oxidized border [59].

Figure 30 presents the angular dependence of switching fields of a 3-nm Fe cluster having a slightly oxidized surface. Huge variation of the switching fields can be observed which might be due to exchange bias of frustrated spin configurations at the surface of the cluster (Fig. 31).

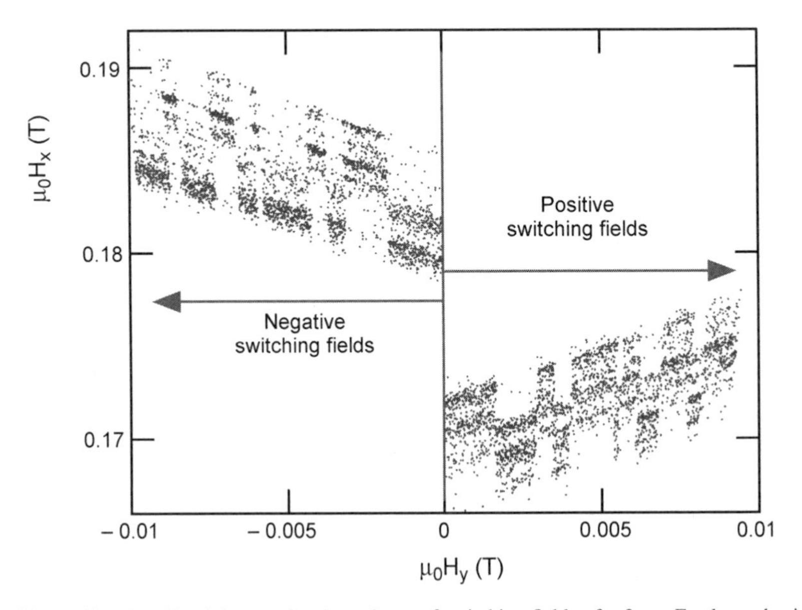

Figure 31. Details of the angular dependence of switching fields of a 3-nm Fe cluster having (probably) a slightly oxidized surface. Each point corresponds to one of the 3000 switching field measurements. Stochastic fluctuation between different switching field distributions are observed. The "mean" hysteresis loop is shifted to negative fields.

We propose that the magnetization reversal of a ferromagnetic particle with an antiferromagnetic surface layer is mainly governed by two mechanisms which are both due to spin frustration at the interface between the ferromagnetic core and the antiferromagnetic surface layer(s). The first mechanism may come from the spin frustration differing slightly from one cycle to another, thus producing a varying energy landscape. These energy variations are less important at high temperatures when the thermal energy ($k_B T$) is much larger. However, at lower temperature the magnetization reversal becomes sensitive to the energy variations. During the hysteresis loop the system chooses randomly a path through the energy landscape which leads to broad switching field distributions. A second mechanism may become dominant at high temperatures: the magnetization reversal may be governed by a relaxation of the spin frustration, hence by a relaxation of the energy barrier. This relaxation is thermally activated—that is, slower at lower temperatures.

V. MAGNETIZATION REVERSAL BY QUANTUM TUNNELING

Studying the boundary between classical and quantum physics has become a very attractive field of research which is known as "mesoscopic" physics (Fig. 1). New and fascinating mesoscopic effects can occur when characteristic system dimensions are smaller than the length over which the quantum wave function of a physical quantity remains sensitive to phase changes. Quantum interference effects in mesoscopic systems have, until now, involved phase interference between paths of particles moving in real space as in SQUIDs or mesoscopic rings. For magnetic systems, similar effects have been proposed for spins moving in spin space, such as magnetization tunneling out of a metastable potential well or coherent tunneling between classically degenerate directions of magnetization [107, 108].

We have seen in the previous sections that the intrinsic quantum character of the magnetic moment can be neglected for nanoparticles with dimensions of the order of the domain wall width δ and the exchange length λ—that is, particles with a collective spin of $S = 10^5$ or larger. However, recent measurements on molecular clusters with a collective spin of $S = 10$ suggest that quantum phenomena might be observed at larger system sizes with $S \gg 1$. Indeed, it has been predicted that macroscopic quantum tunneling of magnetization can be observed in magnetic systems with low dissipation. In this case, it is the tunneling of the magnetization vector of a single-domain particle through its anisotropy energy barrier or the tunneling of a domain wall through its pinning energy. These phenomena have been studied theoretically and experimentally [108].

The following sections review the most important results concerning the observed quantum phenomena in molecular clusters which are mesoscopic model systems to test quantum tunneling theories and the effects of the environmental decoherence. Their understanding requires a knowledge of many physical phenomena, and they are therefore particularly interesting for fundamental studies. We then focus on magnetic quantum tunneling (MQT) studied in individual nanoparticles or nanowires. We concentrate on the necessary experimental conditions for MQT and review some experimental results which suggest that quantum effects might even be important in nanoparticles with $S = 10^5$ or larger.

A. Quantum Tunneling of Magnetization in Molecular Clusters

Magnetic molecular clusters are the final point in the series of smaller and smaller units from bulk matter to atoms. Up to now, they have been the most promising candidates for observing quantum phenomena because they have a well-defined structure with well-characterized spin ground state and magnetic anisotropy. These molecules can be regularly assembled in large crystals where all molecules often have the same orientation. Hence, macroscopic measurements can give direct access to single molecule properties. The most prominent examples are a dodecanuclear mixed-valence manganese-oxo cluster with acetate ligands, short Mn_{12} acetate [111], and an octanuclear iron(III) oxo-hydroxo cluster of formula $[Fe_8O_2(OH)_{12}(tacn)_6]^{8+}$ where tacn is a macrocyclic ligand, short Fe_8 (Fig. 32) [112]. Both systems have a spin ground state of $S = 10$ and an Ising-type magnetocrystalline anisotropy, which stabilizes the spin states with $m = \pm 10$ and generates an energy barrier for the reversal of the magnetization of about 67 K for Mn_{12} acetate and 25 K for Fe_8.

Thermally activated quantum tunneling of the magnetization has first been evidenced in both systems [113–116]. Theoretical discussion of this assumes that thermal processes (principally phonons) promote the molecules up to high levels with small quantum numbers $|m|$, not far below the top of the energy barrier, and the molecules then tunnel inelastically to the other side. Thus the transition is almost entirely accomplished via thermal transitions, and the characteristic relaxation time is strongly temperature-dependent. An alternative explication was also presented [117]. For Fe_8, however, the relaxation time becomes temperature-independent below 0.36 K [116, 118], showing that a pure tunneling mechanism between the only populated ground states $m = \pm S = \pm 10$ is responsible for the relaxation of the magnetization. On the other hand, in the Mn_{12} acetate system one sees temperature-independent relaxation only for strong applied fields and below about 0.6 K [119, 120]. During the last years, several new molecular magnets were presented (see, for instance, Refs. 121–124) which show also tunneling at low temperatures.

The following subsections review the most appealing results concerning the Fe_8 system which can be seen as an *ideal* "model molecule" to study quantum phenomena in magnetic nanostructures. We stress that the tunneling in large

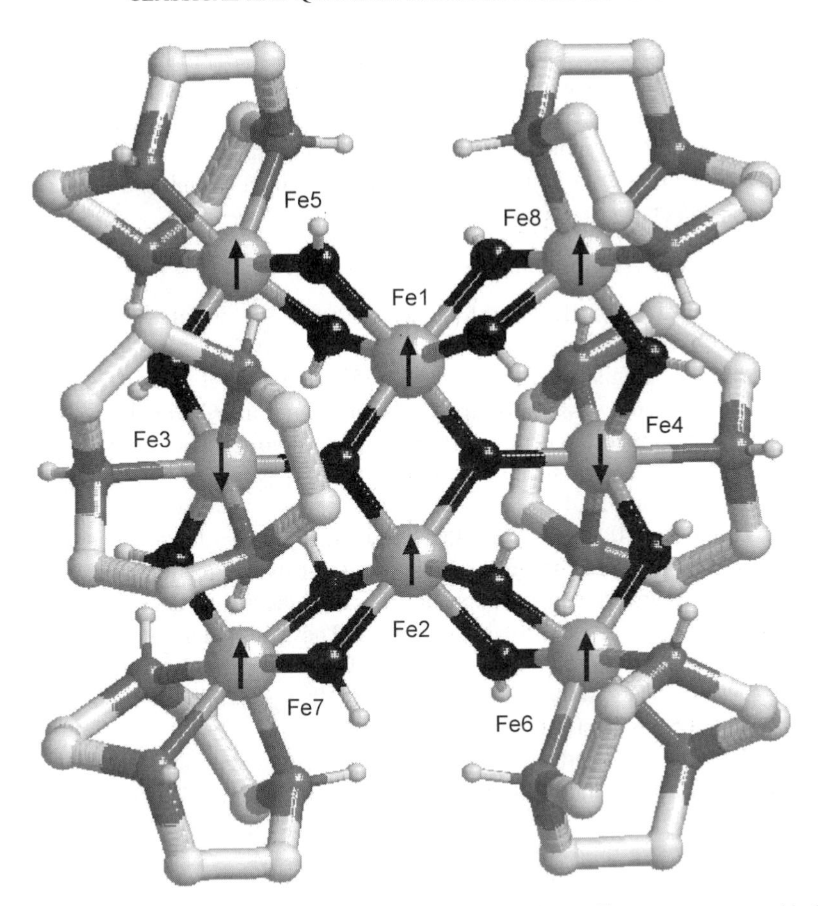

Figure 32. Schematic view of the magnetic core of the Fe_8 cluster. The oxygen atoms are black, the nitrogen atoms are gray, and carbon atoms are white. For the sake of clarity, only the hydrogen atoms that are exchanged with deuterium are shown as small spheres (Section V.B.4). The arrows represent the spin structure of the ground state $S = 10$ as experimentally determined through polarized neutron diffraction experiments [109]. The exact orientation of easy, medium, and hard axis of magnetization (Fig. 33) can be found in Ref. 110.

spins is remarkable because it does not show up at the lowest orders of perturbation theory.

All measurements on Fe_8 were performed using an array of micro-SQUIDs (Section II.B.7). The high sensitivity of this magnetometer allows us to study single Fe_8 crystals [125] of sizes of the order of 10 to 500 μm. For ac-susceptibility measurements and magnetization measurements at $T > 6\,K$, we used a home-built Hall probe magnetometer [26, 126]. It works in the temperature range between 0.03 K and 100 K, for frequencies between 1 Hz and 100 kHz.

After discussing the magnetic anisotropy of Fe_8, we present the observed quantum phenomena. The discussions of the following sections neglect environmental decoherence effects for the sake of simplicity. In Section V.2, we focus on effects of the environment (dipolar coupling, nuclear spins, and temperature) onto the tunneling. This review should help to set up a complete theory which describes *real* magnetic quantum systems.

1. *Magnetic Anisotropy in Fe_8*

The octanuclear iron(III) oxo-hydroxo cluster of formula $[Fe_8O_2(OH)_{12}(tacn)_6]^{8+}$ where tacn is a macrocyclic ligand, short Fe_8 (Fig. 32), was first synthesized by Wieghardt et al. in 1984 [125]. Four central iron (III) ions with $S = 5/2$ are bridged by two oxo groups. The other four iron (III) ions are bridged by hydroxo groups to the central iron ions in an almost planar arrangement, as shown in Fig. 32. The clusters have approximate D_2 symmetry but crystallize in the triclinic system [110].

Fe_8 has an $S = 10$ ground state that originates from antiferromagnetic interactions that do not give complete compensation of the magnetic moment [127]. Spin–orbital moments can be neglected because the magnetic ions are in an "orbital singlet" as a result of Hund's rules. The spin structure of the ground state schematized by the arrows in Fig. 32 has been recently confirmed by a single-crystal polarized neutron investigation that provided a magnetization density map of the cluster [109].

The simplest model describing the spin system of Fe_8 molecular clusters (called the giant spin model) has the following Hamiltonian [112]:

$$H = -DS_z^2 + E(S_x^2 - S_y^2) + g\mu_B\mu_0\vec{S}\cdot\vec{H} \tag{5.1}$$

S_x, S_y, and S_z are the three components of the spin operator, D and E are the anisotropy constants which were determined via HF-EPR ($D/k_B \approx 0.275$ K and $E/k_B \approx 0.046$ K [112]), and the last term of the Hamiltonian describes the Zeeman energy associated with an applied field $\vec{H}$. This Hamiltonian defines hard, medium, and easy axes of magnetization in x, y, and z directions, respectively (Fig. 33). It has an energy level spectrum with $(2S+1) = 21$ values which, to a first approximation, can be labeled by the quantum numbers $m = -10, -9, \ldots, 10$ choosing the z-axis as quantization axis. The energy spectrum, shown in Fig. 34, can be obtained by using standard diagonalisation techniques of the $[21 \times 21]$ matrix describing the spin Hamiltonian $S = 10$. At $\vec{H} = 0$, the levels $m = \pm 10$ have the lowest energy. When a field H_z is applied, the energy levels with $m < -2$ increase, while those with $m > 2$ decrease (Fig. 34). Therefore, energy levels of positive and negative quantum numbers cross at certain fields H_z. It turns out that for Fe_8 the levels cross at fields given by $\mu_0 H_z \approx$

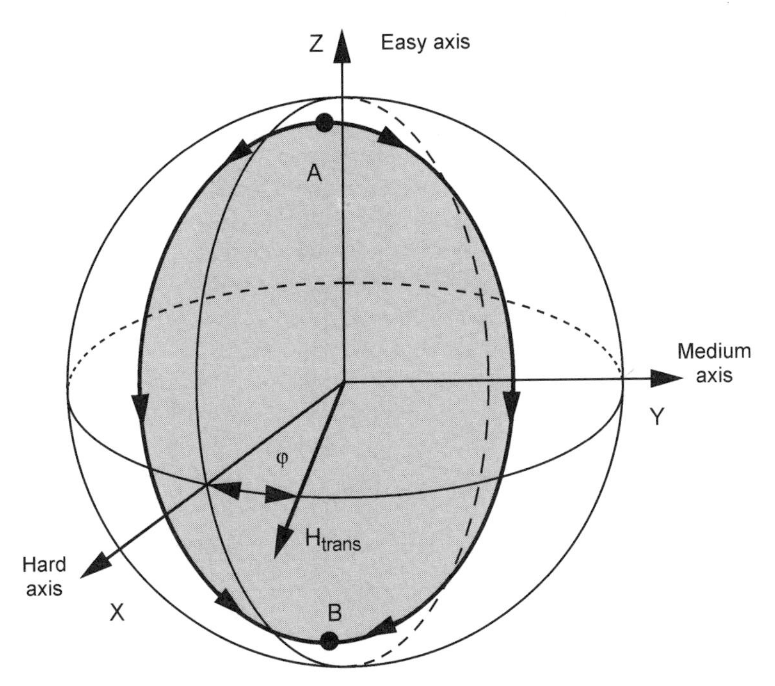

Figure 33. Unit sphere showing degenerate minima **A** and **B** which are joined by two tunnel paths (heavy lines). The hard, medium, and easy axes are taken in *x*-, *y*-, and *z*-direction, respectively. The constant transverse field H_{trans} for tunnel splitting measurements is applied in the *xy*-plane at an azimuth angle φ. At zero applied field $\vec{H} = 0$, the giant spin reversal results from the interference of two quantum spin paths of opposite direction in the easy anisotropy *yz*-plane. For transverse fields in direction of the hard axis, the two quantum spin paths are in a plane which is parallel to the *yz*-plane, as indicated in the figure. By using Stokes' theorem it has been shown [128] that the path integrals can be converted in an area integral, yielding–that destructive interference—that is, a quench of the tunneling rate—occurs whenever the shaded area is $k\,\pi/S$, where k is an odd integer. The interference effects disappear quickly when the transverse field has a component in the *y*-direction because the tunneling is then dominated by only one quantum spin path.

$n \times 0.22\,\mathrm{T}$, with $n = 1, 2, 3\ldots$. The inset of Fig. 34 displays the details at a level crossing where transverse terms containing S_x or S_y spin operators turn the crossing into an "avoided level crossing." The spin S is "in resonance" between two states when the local longitudinal field is close to an avoided level crossing. The energy gap, the so-called "tunnel spitting" Δ, can be tuned by an applied field in the *xy*-plane (Fig. 33) via the S_xH_x and S_yH_y Zeeman terms (Section V.A.3).

The effect of these avoided level crossings can be seen in hysteresis loop measurements (Fig. 35). When the applied field is near an avoided level crossing, the magnetization relaxes faster, yielding steps separated by plateaus. As the

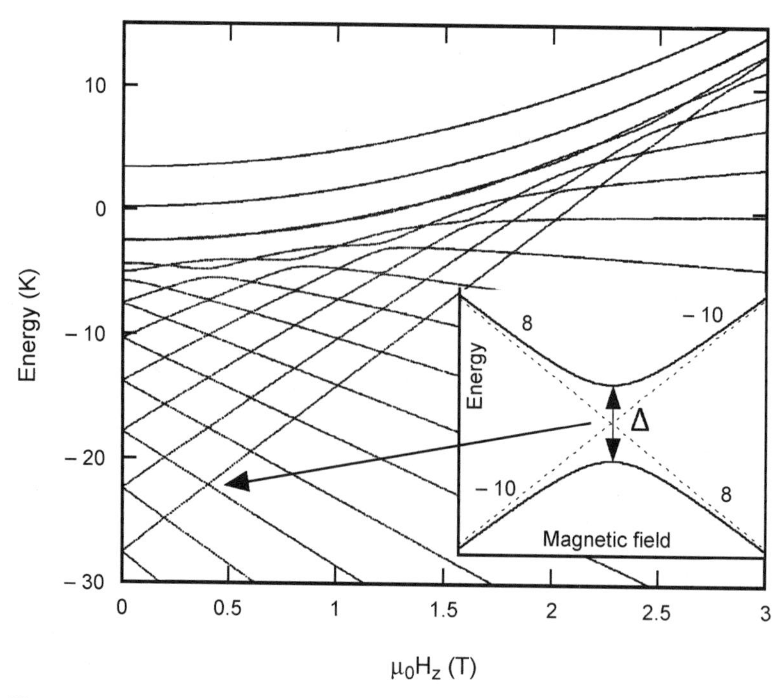

Figure 34. Zeeman diagram of the 21 levels of the $S = 10$ manifold of Fe_8 as a function of the field applied along the easy axis [Eq. (5.1)]. From bottom to top, the levels are labeled with quantum numbers $m = \pm 10, \pm 9, \ldots, 0$. The levels cross at fields given by $\mu_0 H_z \approx n \times 0.22\,\mathrm{T}$, with $n = 1, 2, 3, \ldots$. The *inset* displays the detail at a level crossing where the transverse terms (terms containing S_x or/and S_y spin operators) turn the crossing into an avoided level crossing. The greater the tunnel splitting Δ, the higher the tunnel rate.

temperature is lowered, there is a decrease in the transition rate due to reduced thermal-assisted tunneling. A similar behavior was observed in Mn_{12} acetate clusters [113–115] where equally separated steps were observed at $H_z \approx n \times 0.45\,\mathrm{T}$. The main difference between both clusters is that the hysteresis loops of Fe_8 become temperature-independent below 0.36 K whereas measurements on Mn_{12} acetate indicate a temperature independence only for strong applied fields and below 0.6 K [119–120].

Another important difference is that the step heights (i.e., the relaxation rates) change periodically when a constant transverse field is applied (Fig. 36). It is the purpose of the next subsections to present a detailed study of this behavior.

2. *Landau–Zener Tunneling in* Fe_8

The nonadiabatic transition between the two states in a two-level system has first been discussed by Landau, Zener, and Stückelberg [130–132]. The original work

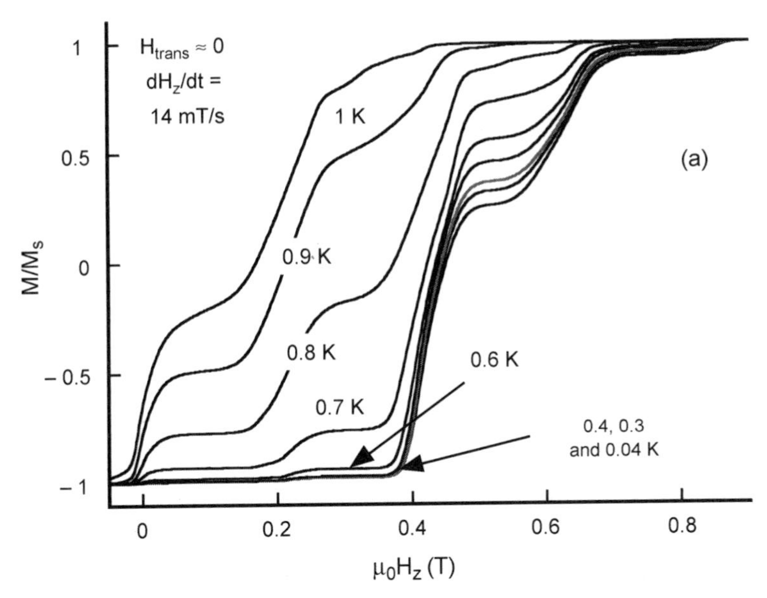

Figure 35. Hysteresis loops of a single crystal of Fe_8 molecular clusters: (a) at different temperatures and a constant sweeping rate $dH_z/dt = 0.014$ T/s and (b) at 0.04 K and different field sweeping rates. The loops display a series of steps, separated by plateaus. As the temperature is lowered, there is a decrease in the transition rate due to reduced thermal assisted tunneling. The hysteresis loops become temperature-independent below 0.35 K, demonstrating quantum tunneling at the lowest energy levels. The resonance widths at small fields H_z of about 0.05 T are mainly due to dipolar fields between the molecular clusters [118, 129].

by Zener concentrates on the electronic states of a biatomic molecule, while Landau and Stückelberg considered two atoms that undergo a scattering process. Their solution of the time-dependent Schrödinger equation of a two-level system could be applied to many physical systems and it became an important tool for studying tunneling transitions. The Landau–Zener model has also been applied to spin tunneling in nanoparticles and clusters [133–138]. The tunneling probability P when sweeping the longitudinal field H_z at a constant rate over an avoided energy level crossing (Fig. 37) is given by

$$P_{m,m'} = 1 - \exp\left[-\frac{\pi \Delta_{m,m'}^2}{2\hbar g \mu_B |m - m'| \mu_0 dH_z/dt} \right] \tag{5.2}$$

Here, m and m' are the quantum numbers of the avoided level crossing, dH_z/dt is the constant field sweeping rates, $g \approx 2$, μ_B the Bohr magneton, and $\hbar$ is Planck's constant.

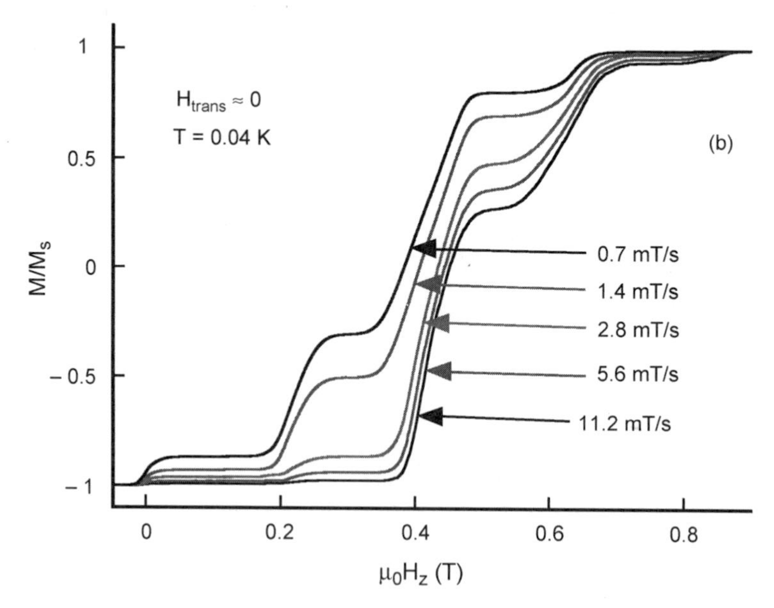

Figure 35. *(Continued).*

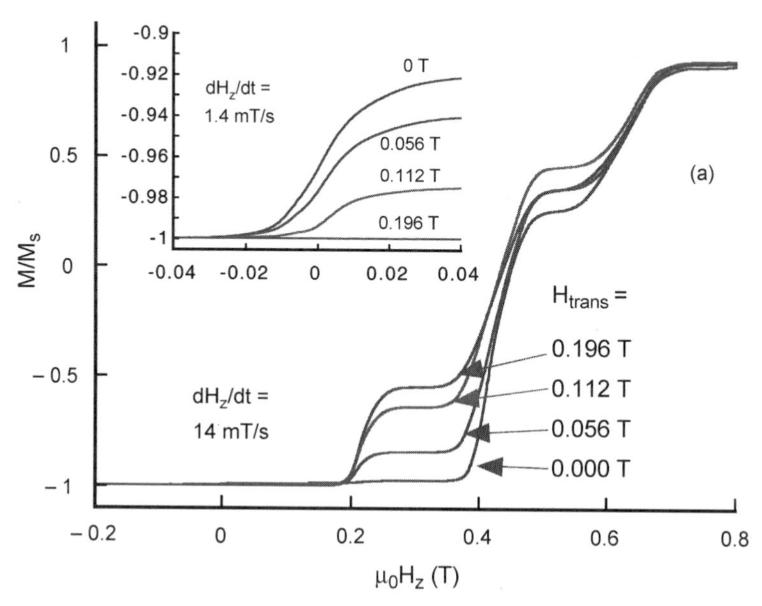

Figure 36. Hysteresis loops measured along H_z in the presence of a constant transverse field at 0.04 K. *Insets*: Enlargement around the field $H_z = 0$. Notice that the sweeping rate is ten times slower for the measurements in the insets than that of the main figures.

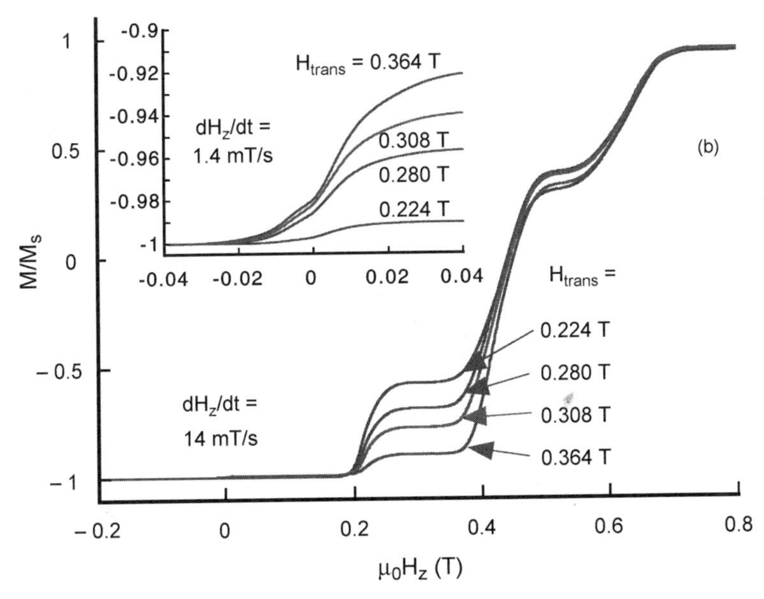

Figure 36. *(Continued).*

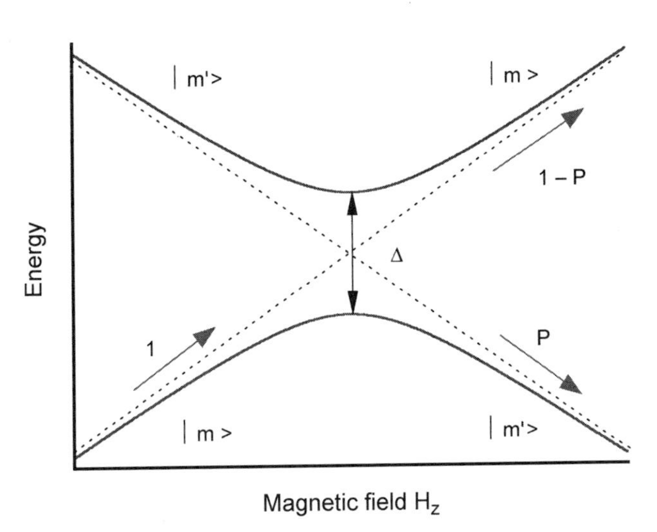

Figure 37. Detail of the energy level diagram near an avoided level crosssing. m and m' are the quantum numbers of the energy level. $P_{m,m'}$ is the Landau–Zener tunnel probability when sweeping the applied field from the left to the right over the anticrossing. The greater the gap Δ and the slower the sweeping rate, the higher is the tunnel rate [Eq. (5.2)].

With the Landau–Zener model in mind, we can now start to understand qualitatively the hysteresis loops (Fig. 35). Let us start at a large negative magnetic field H_z. At very low temperature, all molecules are in the $m=-10$ ground state. When the applied field H_z is ramped down to zero, all molecules will stay in the $m=-10$ ground state. When ramping the field over the $\Delta_{-10,10}$-region at $H_z \approx 0$, there is a Landau–Zener tunnel probability $P_{-10,10}$ to tunnel from the $m=-10$ to the $m=10$ state. $P_{-10,10}$ depends on the sweeping rate [Eq. (5.2)]; that is, the slower the sweeping rate, the larger the value of $P_{-10,10}$. This is clearly demonstrated in the hysteresis loop measurements showing larger steps for slower sweeping rates (Fig. 35). When the field H_z is now further increased, there is a remaining fraction of molecules in the $m=-10$ state which became a metastable state. The next chance to escape from this state is when the field reaches the $\Delta_{-10,9}$ region. There is a Landau–Zener tunnel probability $P_{-10,9}$ to tunnel from the $m=-10$ to the $m=9$ state. As $m=9$ is an excited state, the molecules in this state desexcite quickly to the $m=10$ state by emitting a phonon. An analogous procedure happens when the applied field reaches the $\Delta_{-10,10-n}$-regions ($n=2, 3, \ldots$) until all molecules are in the $m=10$ ground state; that is, the magnetization of all molecules is reversed. As phonon emission can only change the molecule state by $\Delta m=1$ or 2, there is a phonon cascade for higher applied fields.*

In order to apply quantitatively the Landau–Zener formula [Eq. (5.2)], we first saturated the crystal of Fe_8 clusters in a field of $H_z=-1.4$ T, yielding an initial magnetization $M_{in}=-M_s$.† Then, we swept the applied field at a constant rate over one of the resonance transitions and measured the fraction of molecules which reversed their spin. This procedure yields the tunneling rate $P_{-10,10-n}$ and thus the tunnel splitting $\Delta_{-10,10-n}$ [Eq. (5.2)] with $n=0, 1, 2, \ldots$.

For very small tunneling probabilities $P_{-10,10-n}$, we did multiple sweeps over the resonance transition. The magnetization M after N sweeps is given by

$$M(N) \approx M_{eq} + (M_{in} - M_{eq})e^{-kP_{-10,10-n}N} = M_{eq} + (M_{in} - M_{eq})e^{-\Gamma t} \tag{5.3}$$

Here M_{in} is the initial magnetization, $M_{eq}(H_z)$ is the equilibrium magnetization, $N=(1/A)(dH_z/dt)t$ is the number of sweeps over the level crossing, $\Gamma=kP_{-10,10-n}\,(1/A)(dH_z/dt)$ is the overall Landau–Zener transition rate, $k=2$ for $n=0$ and $k=1$ for $n=1, 2, \ldots$, and A is the amplitude of the ramp-field.‡ We have therefore a simple tool to obtain the tunnel splitting by measuring $P_{-10,10-n}$,

*Phonon-induced transitions with $|\Delta m|>2$ are very small [139–141].

†In order to avoid heating problems for measurements of Δ for $n>1$, we started in a thermally annealed sample with $M_{in}=0.95\,M_s$ instead of $M_{in}=-M_s$ or $M_{in}=0$.

‡We supposed here that the forth and back sweeps give the same tunnel probability. This is a good approximation for $P \ll 1$ where next-nearest-neighbor (molecule) effects can be neglected.

or $M(N)$ for $P_{-10,10-n} \ll 1$. We first checked the predicted Landau–Zener sweeping field dependence of the tunneling rate. This can be done, for example, by plotting the relaxation of magnetization as a function of $t = N\ (A/dH_z/dt)$. The Landau–Zener model predicts that all measurements should fall on one curve which was indeed the case for sweeping rates between 1 and 0.001 T/s (Fig. 38) for the $m = \pm 10$ transition. The deviations at lower sweeping rates are mainly due to the *hole-digging mechanism* [129] which slows down the relaxation (see Section V.B.2).* In the ideal case, we should find an exponential curve [Eq. (5.3)]. However, we found clear deviations from the exponential curve (Fig. 39), which might be due to molecules with different amounts of nuclear spins. For example, two percent of natural iron has a nuclear spin; that is, about 10 percent of Fe_8 has at least one nuclear spin on the iron. This interpretation is supported by measurements on isotopically substituted Fe_8 samples (Fig. 39).

Another way of checking the Landau–Zener sweeping field dependence of the tunneling rate is presented in Fig. 40 showing a sweeping rate independent

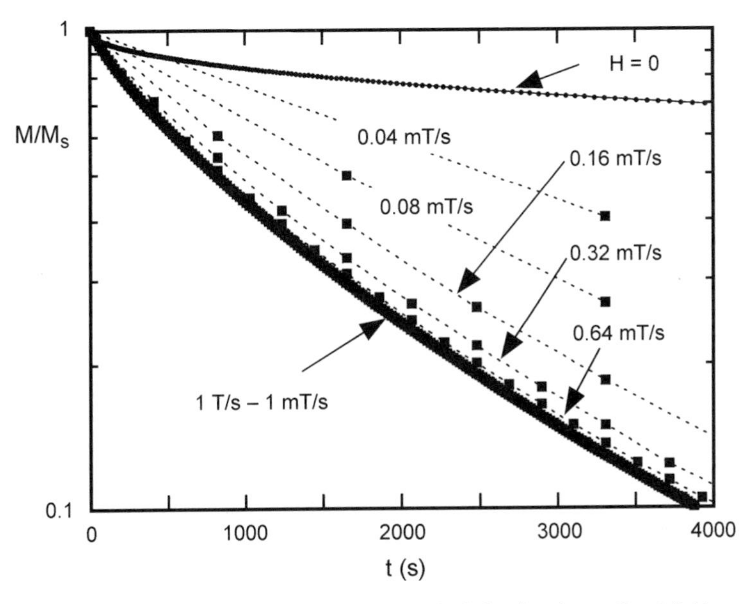

Figure 38. Scaling plot for the Landau–Zener method showing the predicted field sweeping rate dependence for 1 T/s to 1 mT/s. Each point indicates the magnetization after a field sweep over the $m = \pm 10$ resonance. The dotted lines are guides for the eyes. For comparison, the figure displays also a relaxation curve at a constant field $\vec{H} = 0$ (Fig. 36) which shows much slower relaxation [129].

*Roughly speaking, at very low field sweeping rates internal fields change faster than the external field.

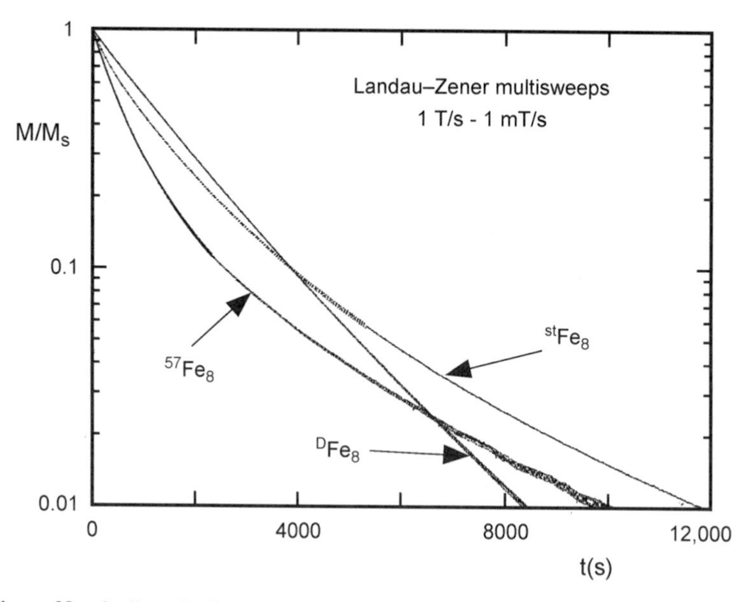

Figure 39. Scaling plot for the Landau–Zener method showing the predicted field sweeping rate dependence for 1 T/s to 1 mT/s similar to Fig. 38 but for three isotopically substituted Fe_8 samples. Further details are presented in Section V.B.4.

$\Delta_{-10,10}$ between 1 and 0.001 T/s.* The measurements on isotopically substituted Fe_8 samples show a small dependence of $\Delta_{m,m'}$ on the hyperfine coupling (Fig. 40). Such an effect has been predicted for a constant applied field by Tupitsyn et al. [142], and for a ramped field by Rose [136]. Further details are presented in Section V.B.4.

We also compared the tunneling rates found by the Landau–Zener method with those found using a square-root decay method that was proposed by Prokof'ev and Stamp [143], and we found a good agreement [129, 144] (Section V. B.1).

Our measurements showed for the first time that the Landau–Zener method is particularly adapted for molecular clusters because it works even in the presence of dipolar fields that spread the resonance transition provided that the field sweeping rate is not too small. Furthermore, our measurements show a small but clear influence of the hyperfine coupling which should be included in a generalized Landau–Zener model [136].

3. *Oscillations of Tunnel Splitting*

An applied field in the *xy*-plane can tune the tunnel splittings $\Delta_{m,m'}$ via the S_x and S_y spin operators of the Zeeman terms that do not commute with the spin

*Recent measurements confirmed the good agreement up to 30 T/s.

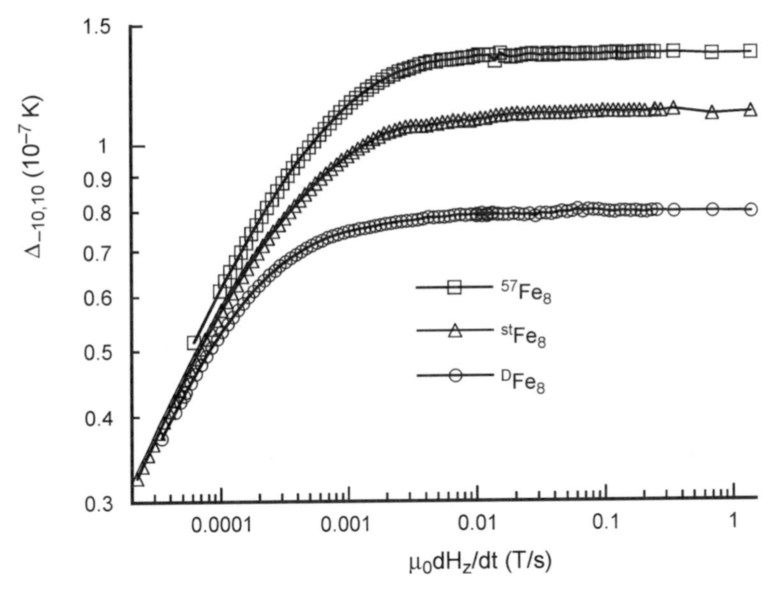

Figure 40. Field sweeping rate dependence of the tunnel splitting $\Delta_{-10,10}$ measured by a Landau–Zener method for three Fe_8 samples, for $H_x = 0$. The Landau–Zener method works in the region of high sweeping rates where $\Delta_{-10,10}$ is sweeping rate independent. Note that the differences of $\Delta_{-10,10}$ between the three isotopically substituted samples are rather small in comparison to the oscillations in Fig. 41.

Hamiltonian. This effect can be demonstrated by using the Landau–Zener method (Section V.A.2). Figure 41 presents a detailed study of the tunnel splitting $\Delta_{\pm 10}$ at the tunnel transition between $m = \pm 10$, as a function of transverse fields applied at different angles φ, defined as the azimuth angle between the anisotropy hard axis and the transverse field (Fig. 42). For small angles φ the tunneling rate oscillates with a period of $\sim 0.4\,\mathrm{T}$, whereas no oscillations showed up for large angles φ [47]. In the latter case, a much stronger increase of $\Delta_{\pm 10}$ with transverse field is observed. The transverse field dependence of the tunneling rate for different resonance conditions between the state $m = -10$ and $(10-n)$ can be observed by sweeping the longitudinal field around $\mu_0 H_z = n \times 0.22\,\mathrm{T}$ with $n = 0, 1, 2, \ldots$. The corresponding tunnel splittings $\Delta_{-10,10-n}$ oscillate with almost the same period of $\sim 0.4\,\mathrm{T}$ (Fig. 41). In addition, comparing quantum transitions between $m = -10$ and $(10-n)$, with n even or odd, revealed a parity (or symmetry) effect that is analogous to the Kramers' suppression of tunneling predicted for half-integer spins [145, 146]. This behavior has been observed for $n = 0$ to 4.* A similar strong dependence on the azimuth angle φ was observed for all the resonances.

*The tunneling rate were too fast for $n > 4$.

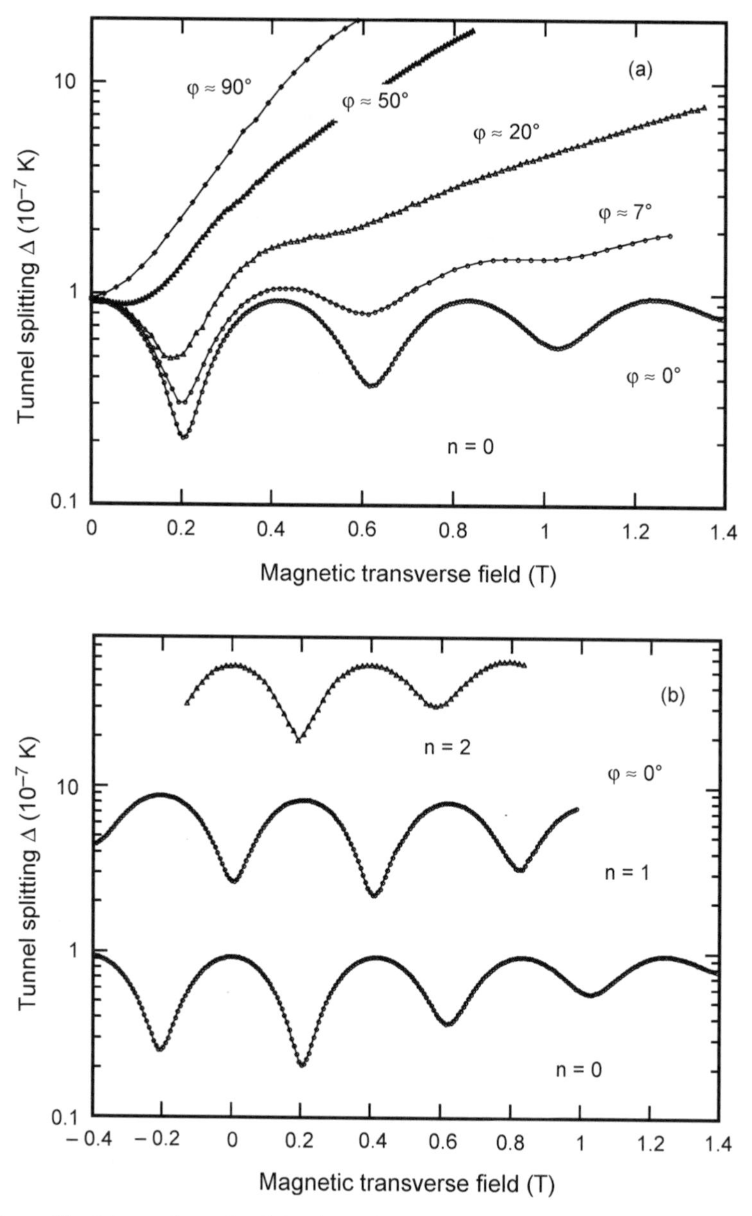

Figure 41. Measured tunnel splitting Δ as a function of transverse field for (a) several azimuth angles φ at $m = \pm 10$ and (b) $\varphi \approx 0°$, as well as for quantum transition between $m = -10$ and $(10-n)$. Note the parity effect that is analogous to the suppression of tunneling predicted for half-integer spins. It should also be mentioned that internal dipolar and hyperfine fields hinder a quench of Δ which is predicted for an isolated spin (Figs. 42 and 45).

a. Semiclassical Descriptions. Before showing that the above results can be derived by an exact numerical calculation using the quantum operator formalism, it is useful to discuss semiclassical models. The original prediction of oscillation of the tunnel splitting was done by using the path integral formalism [147]. Here [128], the oscillations are explained by constructive or destructive interference of quantum spin phases (Berry phases) of two tunnel paths (instanton trajectories) (Fig. 33). Since our experiments were reported, the Wentzel–Kramers–Brillouin theory has been used independently by Garg [148] and Villain and Fort [149]. The surprise is that although these models [128, 148, 149] are derived semiclassically, and should have higher-order corrections in $1/S$, they appear to be exact as written! This has first been noted in Refs. 148 and 149 and then proven in Ref. 150. Some extensions or alternative explications of Garg's result can be found in Refs. 151–154.
The period of oscillation is given by [128]

$$\Delta H = \frac{2k_B}{g\mu_B}\sqrt{2E(E+D)} \tag{5.4}$$

where D and E are defined in Eq. (5.1). We find a period of oscillation of $\Delta H = 0.26\,\mathrm{T}$ for $D = 0.275\,\mathrm{K}$ and $E = 0.046\,\mathrm{K}$ as in Ref. 112. This is somewhat smaller than the experimental value of $\sim 0.4\,\mathrm{T}$. We believe that this is due to higher-order terms of the spin Hamiltonian which are neglected in Garg's calculation. These terms can easily be included in the operator formalism as shown in the next subsection.

b. Exact Numerical Diagonalization. In order to quantitatively reproduce the observed periodicity we included fourth-order terms in the spin Hamiltonian [Eq. (5.1)] as recently employed in the simulation of inelastic neutron scattering measurements [155, 156] and performed a diagonalization of the $[21 \times 21]$ matrix describing the $S = 10$ system. For the calculation of the tunnel splitting we used $D = 0.289\,\mathrm{K}$, $E = 0.055\,\mathrm{K}$ [Eq. (5.1)] and the fourth-order terms as defined in [155] with $B_4^0 = 0.72 \times 10^{-6}\,\mathrm{K}$, $B_4^2 = 1.01 \times 10^{-5}\,\mathrm{K}$, $B_4^4 = -0.43 \times 10^{-4}\,\mathrm{K}$, which are close to the values obtained by EPR measurements [110] and neutron scattering measurements [156]. The calculated tunnel splittings for the states involved in the tunneling process at the resonances $n = 0$, 1, and 2 are reported in Figure 42, showing the oscillations as well as the parity effect for odd resonances. The calculated tunneling splitting is, however, $\sim$1.5 times smaller than the observed one. This small discrepancy could be reduced by introducing higher-order terms. We believe that this is not relevant because the above model neglects, for example, the influence of nuclear spins which seems to increase the measured (effective) tunnel splittings (Fig. 40 and Section V.B.4). Our choice of the fourth-order terms suppresses the oscillations of large

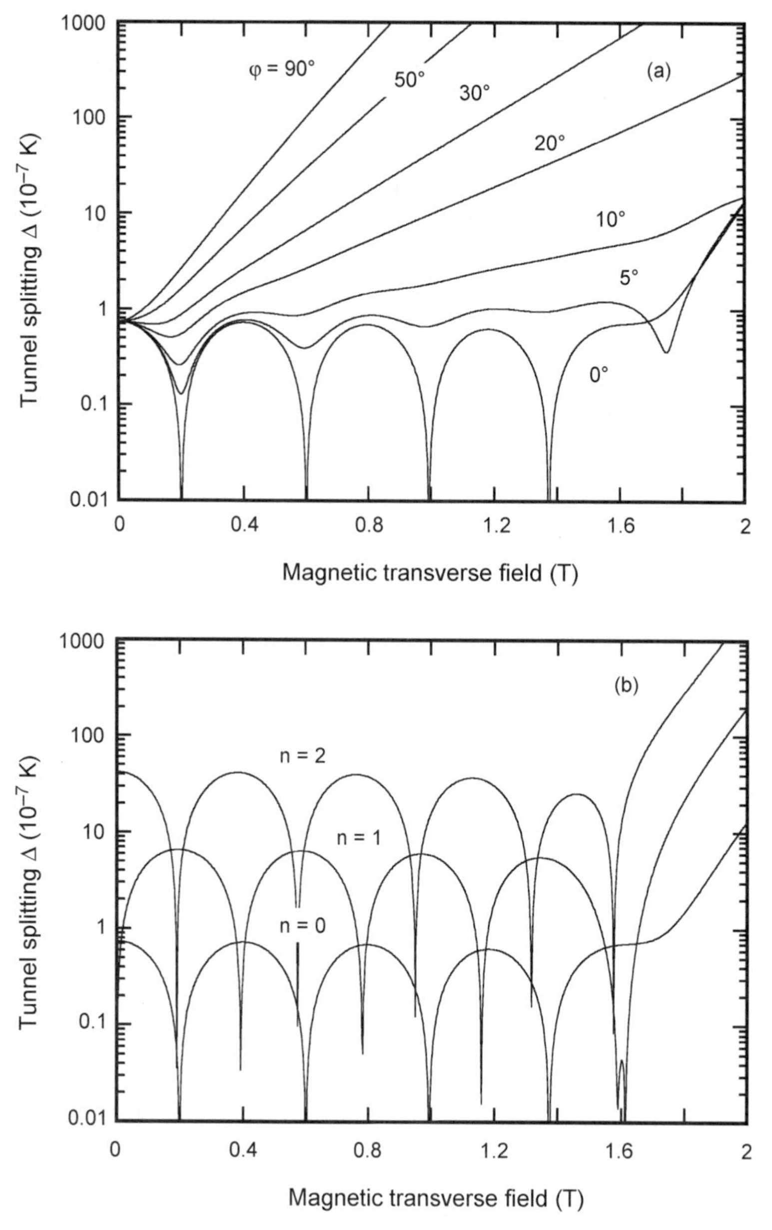

Figure 42. Calculated tunnel splitting Δ as a function of transverse field for (a) quantum transition between $m = \pm 10$ at several azimuth angles φ and (b) quantum transition between $m = -10$ and $(10-n)$ at $\varphi = 0°$ (Section V.A.3.b). The fourth-order terms suppress the oscillations of Δ for large transverse fields $|H_x|$.

transverse fields (Fig. 42). This region could not be studied in the current setup. Future measurements should focus on the higher-field region in order to find a better effective Hamiltonian.

B. Environmental Decoherence Effects in Molecular Clusters

At temperatures below 0.36 K, Fe_8 molecular clusters display a clear crossover from thermally activated relaxation to a temperature-independent quantum regime, with a pronounced resonance structure of the relaxation time as a function of the external field (Section V.A.1). It was surprising, however, that the observed relaxation of the magnetization in the quantum regime was found to be nonexponential and the resonance width orders of magnitude too large [116, 118]. The key to understand this seemingly anomalous behavior involves the hyperfine fields as well as the evolving distribution of the weak dipole fields of the nanomagnets themselves [143]. Both effects were shown to be the main source of decoherence at very low temperature. At higher temperatures, phonons are another source of decoherence.

In the following sections, we focus on the low temperature and low field limits, where phonon-mediated relaxation is astronomically long and can be neglected. In this limit, the $m = \pm S$ spin states are coupled due to the tunneling splitting $\Delta_{\pm S}$ which is about 10^{-7} K for Fe_8 (Section V.A.3) and 10^{-11} K for Mn_{12} [157] with $S = 10$. In order to tunnel between these states, the longitudinal magnetic energy bias $\xi = g\mu_B S H_{local}$ due to the local magnetic field H_{local} on a molecule must be smaller than $\Delta_{\pm S}$, implying a local field smaller than 10^{-8} T for Fe_8 clusters. Since the typical intermolecular dipole fields are of the order of 0.05 T, it seems at first that almost all molecules should be blocked from tunneling by a very large energy bias. Prokof'ev and Stamp have proposed a solution to this dilemma by proposing that fast dynamic nuclear fluctuations broaden the resonance, and the gradual adjustment of the dipole fields in the sample caused by the tunneling brings other molecules into resonance and allows continuous relaxation [143]. Some interesting predictions are briefly reviewed in the following section.

1. Prokof'ev–Stamp Theory

Prokof'ev and Stamp were the first who realized that there are localized couplings of environmental modes with mesoscopic systems which cannot be modeled with an "oscillator bath" model [158] describing delocalized environmental modes such as electrons, phonons, photons, and so on. They found that these localized modes such as nuclear and paramagnetic spins are often strong and described them with a spin bath model [159]. We do not review this theory* but focus on one

*For a review, see Ref. 160.

particular application which is interesting for molecular clusters [143]. Prokof'ev and Stamp showed that at a given longitudinal applied field H_z, the magnetization of a crystal of molecular clusters should relax at short times with a square-root time dependence which is due to a gradual modification of the dipole fields in the sample caused by the tunneling:

$$M(H_z, t) = M_{\text{in}} + (M_{\text{eq}}(H_z) - M_{\text{in}})\sqrt{\Gamma_{\text{sqrt}}(H_z)t} \tag{5.5}$$

Here M_{in} is the initial magnetization at time $t = 0$ (after a rapid field change), and $M_{\text{eq}}(H_z)$ is the equilibrium magnetization at H_z. The rate function $\Gamma_{\text{sqrt}}(H_z)$ is proportional to the normalized distribution $P(H_z)$ of molecules which are in resonance at H_z:

$$\Gamma_{\text{sqrt}}(H_z) = c\frac{\xi_0}{E_D}\frac{\Delta^2_{\pm S}}{4\hbar}P(H_z) \tag{5.6}$$

where ξ_0 is the line width coming from the nuclear spins, E_D is the Gaussian half-width of $P(H_z)$, and c is a constant of the order of unity which depends on the sample shape. If these simple relations are exact, then measurements of the short time relaxation as a function of the applied field H_z give directly the distribution $P(H_z)$, and they allows one to measure the tunnel splitting $\Delta_{\pm S}$ which is described in the next section.

2. *Hole Digging Method to Study Dipolar Distributions and Hyperfine Couplings*

Motivated by the Prokof'ev–Stamp theory [143], we developed a new technique—which we call the *hole digging method*—that can be used to observe the time evolution of molecular states in crystals of molecular clusters. It allowed us to measure the statistical distribution of magnetic bias fields in the Fe_8 system that arise from the weak dipole fields of the clusters themselves. A hole can be "dug" into the distribution by depleting the available spins at a given applied field. Our method is based on the simple idea that after a rapid field change, the resulting short time relaxation of the magnetization is directly related to the number of molecules which are in resonance at the given applied field. Prokof'ev and Stamp have suggested that the short time relaxation should follow a $\sqrt{t}$-relaxation law [Eq. (5.5)]. However, the hole digging method should work with any short time relaxation law—for example, a power law:

$$M(H_z, t) = M_{\text{in}} + (M_{\text{eq}}(H_z) - M_{\text{in}})(\Gamma_{\text{short}}(H_z)t)^{\alpha} \tag{5.7}$$

where Γ_{short} is a characteristic short time relaxation rate that is directly related to the number of molecules which are in resonance at the applied field H_z, and

$0<\alpha<1$ in most cases. $\alpha=0.5$ in the Prokof'ev–Stamp theory [Eq. (5.5)] and Γ_{sqrt} is directly proportional to $P(H_z)$ [Eq. (4.6)]. The *hole digging method* can be divided into three steps (Fig. 43):

1. **Preparing the Initial State.** A well-defined initial magnetization state of the crystal of molecular clusters can be achieved by rapidly cooling the sample from high down to low temperatures in a constant applied field H_z^0. For zero applied field ($H_z=0$) or rather large applied fields ($H_z>1$ T), one yields the demagnetized or saturated magnetization state of the entire crystal, respectively. One can also quench the sample in a small field of

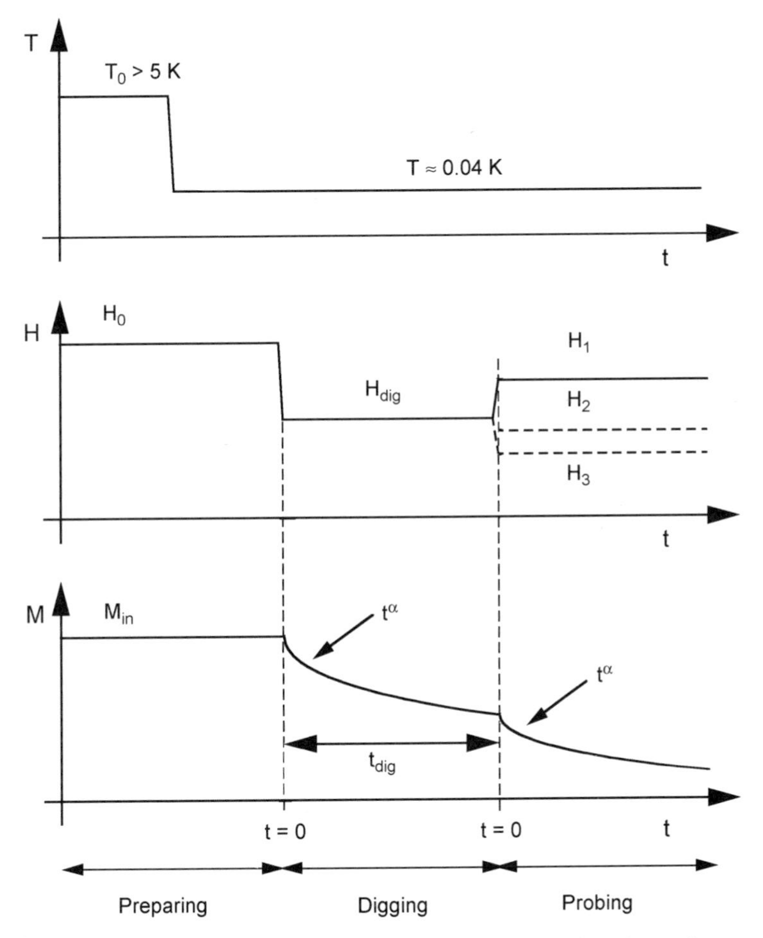

Figure 43. Schema of the hole digging method presenting the time dependence of temperature, applied field, and magnetization of the sample.

few milliteslas yielding any possible initial magnetization M_{in}. When the quench is fast (<1 s), the sample's magnetization does not have time to relax, either by thermal or by quantum transitions. This procedure yields a frozen thermal equilibrium distribution, whereas for slow cooling rates the molecule spin states in the crystal might tend to certain dipolar ordered ground state.

2. **Modifying the Initial State—Hole Digging.** After preparing the initial state, a field H_{dig} is applied during a time t_{dig}, called "digging field and digging time," respectively. During the digging time and depending on H_{dig}, a fraction of the molecular spins tunnel (back and/or forth); that is, they reverse the direction of magnetization.*
3. **Probing the Final State.** Finally, a field H_z^{probe} is applied (Fig. 43) to measure the short time relaxation from which one yields Γ_{short} [Eq. (5.7)] which is related to the number of spins that are still free for tunneling after step 2.

The entire procedure is then repeated many times but at other fields H_z^{probe} yielding $\Gamma_{\mathrm{short}}(H_z, H_{\mathrm{dig}}, t_{\mathrm{dig}})$ which is related to the distribution of spins $P(H_z, H_{\mathrm{dig}}, t_{\mathrm{dig}})$ which are still free for tunneling after the hole digging. For $t_{\mathrm{dig}} = 0$, this method maps out the initial distribution.

3. *Intermolecular Dipole Interaction in Fe$_8$*

We applied the hole digging method to several samples of molecular clusters and quantum spin glasses. The most detailed study has been done on the Fe_8 system. We found the predicted $\sqrt{t}$ relaxation [Eq. (5.5)] in experiments on fully saturated Fe_8 crystals [118, 161] and on nonsaturated samples [129]. Figure 44 displays a detailed study of the dipolar distributions revealing a remarkable structure that is due to next-nearest-neighbor effects [129].† These results are in good agreement with simulations [162, 163].

For a saturated initial state, the Prokof'ev–Stamp theory allows one to estimate the tunnel splitting $\Delta_{\pm S}$. Using Eqs. (3), (9), and (12) of Ref. 143, along with integration, we find $\int \Gamma_{\mathrm{sqrt}} d\xi = c(\xi_0/E_D)(\Delta^2_{\pm S}/4\hbar)$, where c is a constant of the order of unity which depends on the sample shape. With $E_D = 15$ mT, $\xi_0 = 0.8$ mT, $c = 1$, and Γ_{sqrt} [129, 144], we find $\Delta_{\pm 10} = 1.2 \times 10^{-7}$ K which is

*The field sweeping rate to apply H_{dig} should be fast enough to minimize the change of the initial state during the field sweep.

†The peak at 0.04 T as well as the shoulder at 0.02 T and 0.04 T are originated by the clusters which have one nearest-neighbor cluster with reversed magnetization: The peak at 0.04 T corresponds to the reversal of the neighboring cluster along the **a** crystallographic axis, which almost coincides with the easy axis of magnetization, while the shoulder at 0.02 T and 0.04 T are due to the clusters along **b** and **c**.

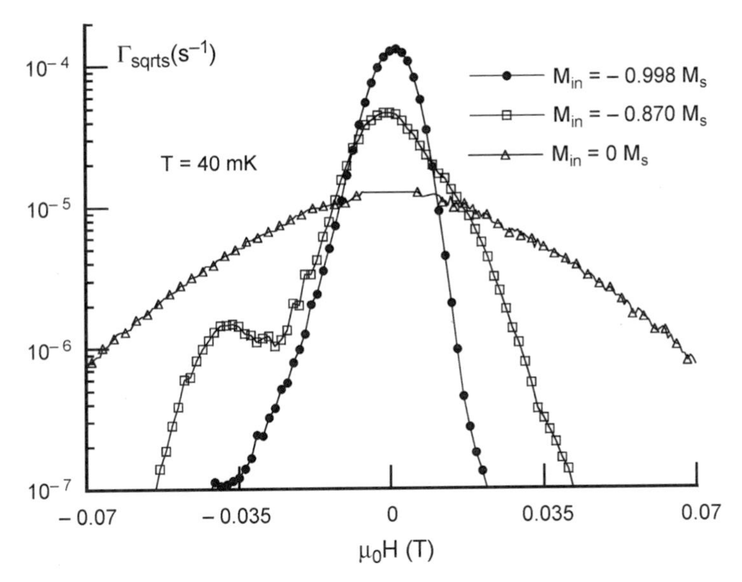

Figure 44. Field dependence of the short time square-root relaxation rates $\Gamma_{sqrt}(H_z)$ for three different values of the initial magnetization M_{in}. According to Eq. (5.6), the curves are proportional to the distribution $P(H_z)$ of magnetic energy bias due to local dipole field distributions in the sample. Note the logarithmic scale for Γ_{sqrt}. The peaked distribution labeled $M_{in} = -0.998\,M_s$ was obtained by saturating the sample, whereas the other distributions were obtained by thermal annealing. $M_{in} = -0.870\,M_s$ is distorted by nearest-neighbor lattice effects.

close to the result of $\Delta_{\pm 10} = 1.0 \times 10^{-7}$ K obtained by using a Landau–Zener method (Section V.A.2) [47]. Whereas the hole digging method probes the longitudinal dipolar distribution (H_z direction), the Landau–Zener method can be used to probe the transverse dipolar distribution by measuring the tunnel splittings Δ around a topological quench. Figure 45 displays such a study for the quantum transition between $m = \pm 10$, and $m = -10$ and 9. Particular efforts were made to align well the transverse field in direction of the hard axis. The initial magnetizations $0 \leq M_{in} \leq M_s$ were prepared by rapidly quenching the sample from 2 K in the present of a longitudinal applied field H_z. The quench takes approximately one second and thus the sample does not have time to relax, either by thermal activation or by quantum transitions, so that the high-temperature "thermal equilibrium" spin distribution is effectively frozen in. For $H_z > 1$ T, one gets an almost saturated magnetization state.

The measurements of $\Delta(M_{in})$ show a strong dependence of the minimal tunnel splittings on the initial magnetization (Fig. 45). They demonstrate that the transverse dipolar interaction between Fe_8 molecular clusters is largest of $M_{in} = 0$ — that is, similar to the longitudinal dipolar interaction (Fig. 44).

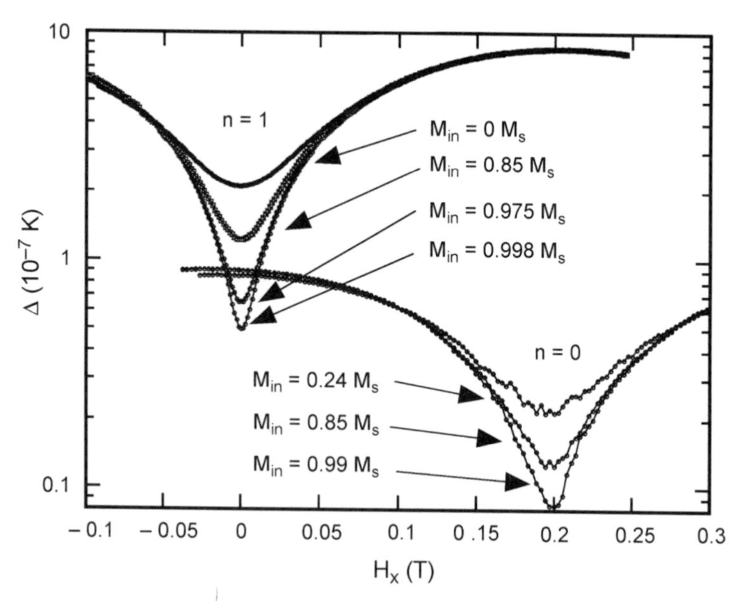

Figure 45. Detailed measurement of the tunnel splitting Δ around a topological quench for the quantum transition between $m = -10$ and $(10-n)$ at $\varphi = 0°$. Note the strong dependence on the initial magnetization M_{in} which demonstrates the transverse dipolar interaction between Fe_8 molecular clusters [129].

4. *Hyperfine Interaction in Fe_8 and Mn_{12}*

The strong influence of nuclear spins on resonant quantum tunneling in the molecular cluster Fe_8 was demonstrated for the first time [144] by comparing the relaxation rate of the standard Fe_8 sample with two isotopic modified samples: (i) ^{56}Fe is replaced by ^{57}Fe, and (ii) a fraction of ^1H is replaced by ^2H. By using the hole digging method, we measured an intrinsic broadening which is driven by the hyperfine fields. Our measurements are in good agreement with numerical hyperfine calculations [136, 144]. For $T > 1.5$ K, the influence of nuclear spins on the relaxation rate is less important, suggesting that spin–phonon coupling dominates the relaxation rate.

Concerning Mn_{12} we did *not* find that the relaxation follows the $\sqrt{t}$-relaxation law at low temperatures [164]. It is well known that the situation in this sample is more complicated due to the fact that there are several coexisting species of Mn_{12} in any crystal, each with different relaxation times. In Ref. 164 we were able to isolate one faster relaxing species. The relaxation could be *approximately* fit to the $\sqrt{t}$-relaxation law, but in fact is better fit to a power law t^{α} with $0.3<\alpha<0.5$ (depending on the applied field). We applied the hole digging method to this species, and we found evidence for intrinsic line broadening below 0.3 K which we suggest comes from nuclear spins in analogy with Fe_8. We also measured the relaxation of Mn_{12} at higher temperature (0.04–5 K) and small fields (< 0.1 T), and we found no evidence for a short time $\sqrt{t}$ relaxation.

5. *Temperature Dependence of the Landau–Zener Tunneling Probability*

In this section we present studies of the temperature dependence of the Landau–Zener tunneling probability P yielding a deeper insight into the spin dynamics of the Fe_8 cluster. By comparing the three isotopic samples (Section V.B.4.) we demonstrate the influence of nuclear spins on the tunneling mechanism and in particular on the lifetime of the first excited states. Our measurements show the need of a generalized Landau–Zener transition rate theory taking into account environmental effects such as hyperfine and spin–phonon coupling.*

All measurement so far were done in the pure quantum regime ($T < 0.36$ K) where transition via excited spin levels can be neglected. We discuss now the temperature region of small thermal activation ($T < 1$ K) where we should consider transition via excited spin levels as well [138, 140].

In order to measure the temperature dependence of the tunneling probability, we used the Landau–Zener method as described in Section V.A.2 with a phenomenological modification of the tunneling probability P (for a negative saturated magnetization):

$$P = n_{-10}P_{-10,10} + P_{th} \tag{5.8}$$

where $P_{-10,10}$ is given by (Eq. 5.2), n_{-10} is the Boltzmann population of the $m = -10$ spin level, and P_{th} is the overall tunneling probability via excited spin levels. $n_{-10} \approx 1$ for the considered temperature $T < 1$ K and a negative saturated magnetization of the sample.

Figure 46 displays the measured tunneling probability P for $^{st}Fe_8$ as a function of a transverse field H_x and for several temperatures. The oscillation of P are seen for all temperatures, but the periods of oscillations decrease for increasing temperature (Fig. 47). This behavior can be explained by the giant spin model [Eq. (5.1)] with fourth-order transverse terms (Section V.A.3.b). Indeed, the tunnel splittings of excited spin levels oscillate as a function of H_x with decreasing periods (Fig. 48).

Figure 49 presents the tunneling probability via excited spin levels $P_{th} = P - n_{-10}P_{-10,10}$. Surprisingly, the periods of P_{th} are temperature-independent in the

*Spin–phonon interactions mainly originate from the perturbation of the crystal field by lattice vibration, which produce both a fluctuating local strain and a fluctuating local rotation [140, 141, 165]. It is sufficient to retain the lowest-order terms which are quadratic with respect to spin operators. The resulting spin–phonon Hamiltonian contains (i) terms that commute with S_z and do not contribute to the relaxation, (ii) terms proportional to S_zS_+ and S_zS_-, and (iii) terms proportional to S_+^2 and S_-^2. Thus, the spin–phonon interaction has matrix elements between states with quantum numbers m and m' if $|m-m'| = 1$ or 2 [140].

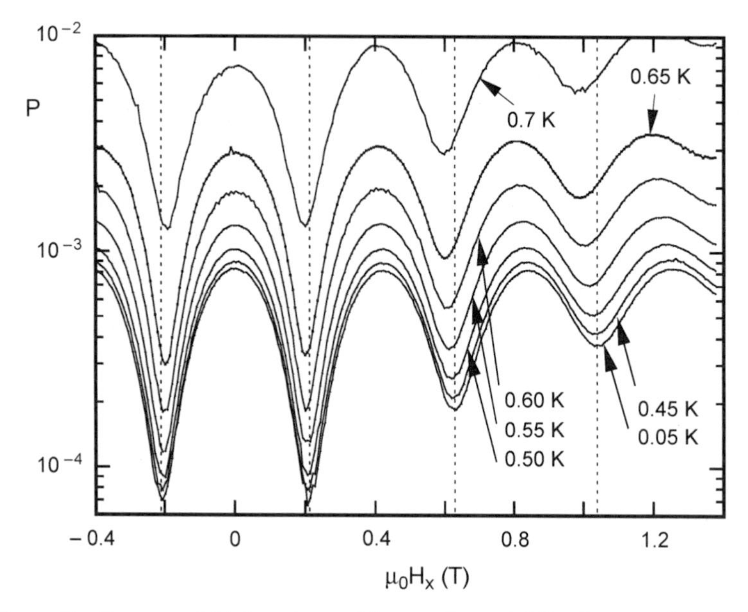

Figure 46. Transverse field dependence of the tunneling probability P at several temperatures, and the ground-state tunneling probability $P_{-10,10}$ measured at $T = 0.05$ K and for $^{st}Fe_8$. The field sweeping rate was 0.14 T/s. The dotted lines indicate the minima of $P_{-10,10}$.

region $T < 0.7$ K. This suggests that only transitions via excited levels $m = \pm 9$ are important in this temperature regime. This statement is confirmed by the following estimation [166].

Using Eq. (5.2), typical field sweeping rates of 0.1 T/s, and tunnel splittings from Fig. 48, one easily finds that the Landau–Zener tunneling probability of excited levels are $P_{-m,m} \approx 1$ for $m < 10$ and $\vec{H} \approx 0$. This means that the relaxation rates via excited levels are mainly governed by the lifetime of the excited levels and the time $\tau_{res,m}$ during which these levels are in resonance. The latter can be estimated by

$$\tau_{res,m} = \frac{\Delta_{-m,m}}{g\mu_B m \mu_0 dH_z/dt} \tag{5.9}$$

The probability for a spin to pass into the excited level m can be estimated by $\tau_m^{-1} e^{-E_{10,m}/k_B T}$, where $E_{10,m}$ is the energy gap between the levels 10 and m, and τ_m is the lifetime of the excited level m. One gets

$$P_{th} \approx \sum_{m=9,8} \frac{\tau_{res,m}}{\tau_m} e^{-E_{10,m}/k_B T} \approx \sum_{m=9,8} \frac{\Delta_{-m,m}}{\tau_m g\mu_B m \mu_0 dH_z/dt} e^{-E_{10,m}/k_B T} \tag{5.10}$$

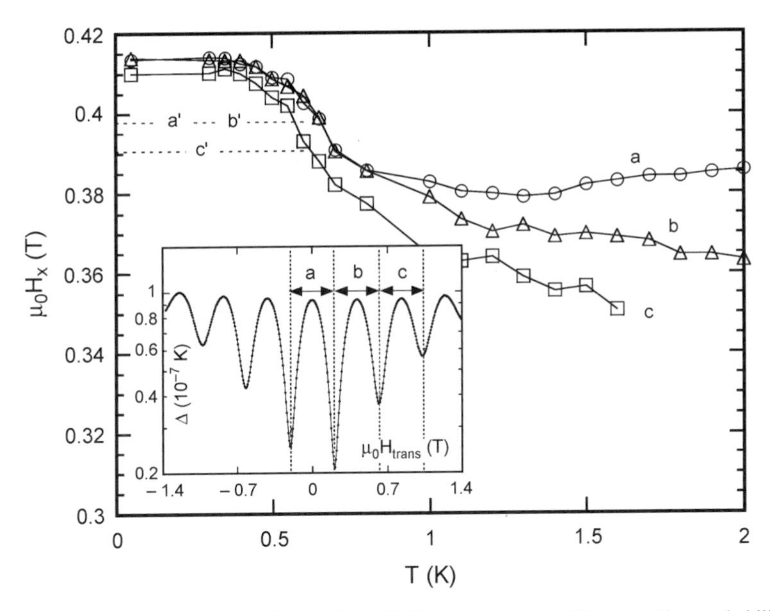

Figure 47. Temperature dependence of the field between minima of the tunneling probability P in Fig. 46. a, b, and c are defined in the inset. The dotted line labeled with a′, b′, and c′ were taken from P_{th} of Fig. 49; see also Ref. 126.

Note that this estimation neglects higher excited levels with $|m|<8$.* Figure 50 displays the measured P_{th} for the three isotopic Fe_8 samples. For $0.4\,K<T<1\,K$ we fitted Eq. (5.10) to the data, leaving *only* the level lifetimes τ_9 and τ_8 as adjustable parameters. All other parameters are calculated using the parameters in Section V.A.3.b. We obtain $\tau_9 = 1.0$, 0.5, and 0.3×10^{-6} s, and $\tau_8 = 0.7$, 0.5, and 0.4×10^{-7} s for ${}^{D}Fe_8$, ${}^{st}Fe_8$, and ${}^{57}Fe_8$, respectively. These results indicate that only the first excited level has to be considered for $0.4\,K<T<0.7\,K$. Indeed, the second term of the summation in Eq. (5.10) is negligible in this temperature interval. It is interesting to note that this finding is in contrast to hysteresis loop measurements on Mn_{12} [120, 167] which were interpreted to have an abrupt transition between thermal assisted and pure quantum tunneling [168]. Furthermore, our result shows clearly the influence of nuclear spins which seem to decrease the level lifetimes τ_m—that is, to increase dissipative effects.

The nuclear magnetic moment and not the mass of the nuclei seems to have the major effect on the dynamics of the magnetization. In fact the mass is increased in both isotopically modified samples whereas the effect on the relaxation rate is opposite. On the other hand, ac-susceptibility measurements

*The probability of phonon induced transitions with $|\Delta m|>2$ are very small [138–140]. Also the Boltzmann factor $e^{-E_{10,m}/k_B T}$ is very small for $m<8$ and $T<1\,K$.

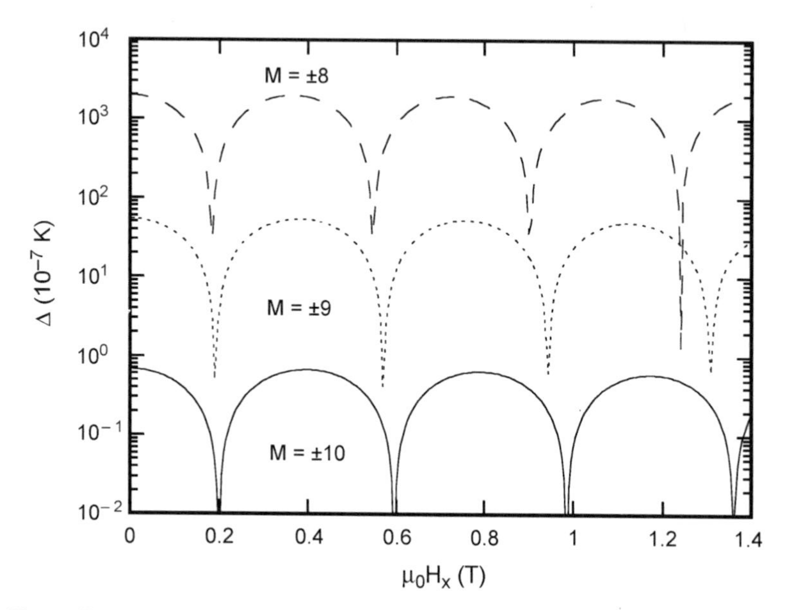

Figure 48. Calculated tunnel splitting $\Delta_{m,m'}$ as a function of the transverse field H_x for quantum transition between $m = \pm 10$, ± 9 and ± 8 (Section V.A.3.b).

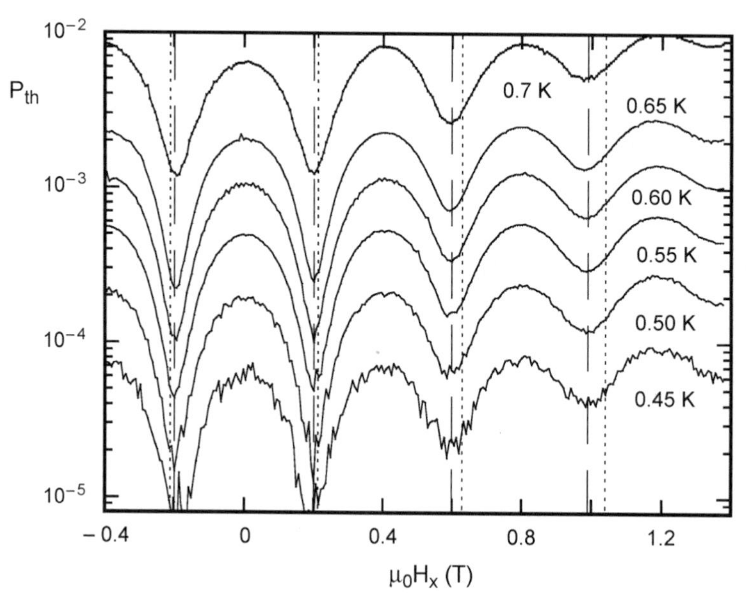

Figure 49. Transverse field dependence of P_{th} which is the difference between the measured tunnel probability P and the ground-state tunnel probability $n_{-10}P_{-10,10}$ measured at $T = 0.05$ K (see Fig. 46). The field sweeping rate was 0.14 T/s. The dotted lines indicate the minima of P_{th}, whereas the dashed lines indicate the minima of $P_{-10,10}$ (see Fig. 47).

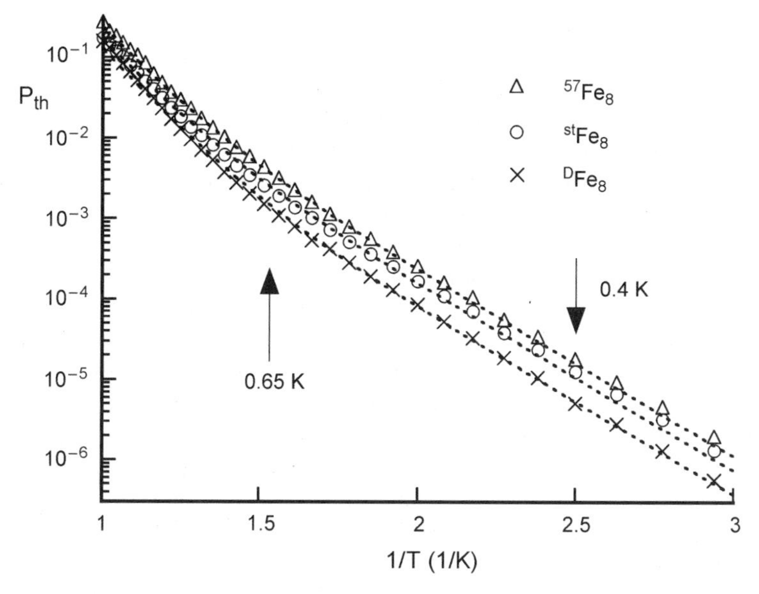

Figure 50. Temperature dependencies of P_{th} for $H_x = 0$ for three Fe_8 samples. The field sweeping rate was 0.14 T/s. The dotted lines are fits of the data using Eq. (5.10).

at $T > 1.5$ K showed no clear difference between the three samples; this suggested that above this temperature, where the relaxation is predominately due to spin–phonon coupling [140, 141], the role of the nuclear spins is less important. Although the increased mass of the isotopes changes the spin–phonon coupling, this effect seems to be small.

We can also exclude that the change of mass for the three isotopic samples has induced a significant change in the magnetic anisotropy of the clusters. In fact the measurements below $T < 0.35$ K, where spin–phonon coupling is negligible, have shown that (i) relative positions of the resonances as a function of the longitudinal field H_z are unchanged,* and (ii) all three samples have the same period of oscillation of Δ as a function of the transverse field H_x [47], a period which is very sensitive to any change of the anisotropy constants.

6. *Conclusion on Molecular Magnets*

In conclusion, we presented detailed measurements which demonstrated that molecular magnets offer a unique opportunity to explore the quantum dynamics

*We observed a small shift of the resonances of the order of magnitude of 1 mT, positive for $^{57}Fe_8$ and negative for $^{D}Fe_8$ ($M_{init} = -M_s$). This can also be attributed to the modified hyperfine fields. However, a quantitative measurement is complicated by the fact that it is impossible to have two crystals with exactly the same shape—that is, the same internal fields.

of a large but finite spin.* We focused our discussion on the Fe_8 molecular magnet because it is the first system where studies in the pure quantum regime were possible. The tunneling in this system is remarkable because it does not show up at the lowest order of perturbation theory.

What remains still debated is the possibility of observing quantum coherence between states of opposite magnetization. Dipole–dipole and hyperfine interactions are sources of decoherence. In other words, when a spin has tunneled through the barrier, it experiences a huge modification of its environment (hyperfine and dipolar) which prohibits the back tunneling. Prokof'ev and Stamp suggested three possible strategies to suppress the decoherence [172]. (i) Choose a system where the NMR frequencies far exceed the tunnel frequencies making any coupling impossible. (ii) Isotopically purify the sample to remove all nuclear spins. (iii) Apply a transverse field to increase the tunnel rate to frequencies much larger than hyperfine field fluctuations. All three strategies are difficult to realize. However, some authors tried to realize the last one by performing EPR experiments in the presence of a large transverse field [173]. Absorption of radio-frequency electromagnetic fields were observed which might be due to induced transitions near the tunnel splitting. However, no experiments showed the oscillatory behavior in the time domain which might be evidenced by a spin-echo type of experiment.

Concerning the perspectives of the field of single molecule magnets, we expect that chemistry is going to play a major role through the synthesis of novel larger spin clusters with strong anisotropy. We want to stress that there are already many other molecular magnets (see, for instance, Refs. 121–124) which are possible model systems. We believe that more sophisticated theories are needed which describe the dephasing effects of the environment onto the quantum system. These investigations are important for studying the quantum character of molecular clusters for applications like "quantum computers." The first implementation of Grover's algorithm with molecular magnets has been proposed [174].

C. Quantum Tunneling of Magnetization in Individual Single-Domain Nanoparticles

The following sections focuses on magnetic quantum tunneling (MQT) studied in individual nanoparticles or nanowires where the complications due to distributions of particle size, shape, and so on, are avoided. The experimental

*Molecules with small spin have also been studied. For example, time-resolved magnetization measurements were performed on a spin 1/2 molecular complex, so-called V_{15} [169]. Despite the absence of a barrier, magnetic hysteresis is observed over a time scale of several seconds. A detailed analysis in terms of a dissipative two-level model has been given, in which fluctuations and splittings are of the same energy. Spin–phonon coupling leads to long relaxation times and to a particular "butterfly" hysteresis loop [170, 171].

evidence for MQT in a single-domain particle or in assemblies of particles is still a controversial subject. We shall therefore concentrate on the necessary experimental conditions for MQT and review some experimental results which suggest that quantum effects might even be important in nanoparticles with $S = 10^5$ or larger. We start by reviewing some important predictions concerning MQT in a single-domain particle.

1. *Magnetic Quantum Tunneling in Nanoparticles*

On the theoretical side, it has been shown that in small magnetic particles, a large number of spins coupled by strong exchange interaction can tunnel through the energy barrier created by magnetic anisotropy. It has been proposed that there is a characteristic crossover temperature T_c below which the escape of the magnetization from a metastable state is dominated by quantum barrier transitions, rather than by thermal over barrier activation. Above T_c the escape rate is given by thermal over barrier activation (Section IV).

In order to compare experiment with theory, predictions of the crossover temperature T_c and the escape rate Γ_{QT} in the quantum regime are relevant. Both variables should be expressed as a function of parameters that can be changed experimentally. Typical parameters are the number of spins S, effective anisotropy constants, applied field strength and direction, coupling to the environment (dissipation), and so on. Many theoretical papers have been published during the last few years [108]. We discuss here a result specially adapted for single-particle measurements, which concerns the field dependence of the crossover temperature T_c.

The crossover temperature T_c can be defined as the temperature where the quantum switching rate equals the thermal one. The case of a magnetic particle, as a function of the applied field direction, has been considered by several authors [175–177]. We have chosen the result for a particle with biaxial anisotropy as the effective anisotropy of most particles can be approximately described by strong uniaxial and weak transverse anisotropy. The result due to Kim can be written in the following form [177]:

$$T_c(\theta) \sim \mu_0 H_{\parallel} \varepsilon^{1/4} \sqrt{1 + a(1 + |\cos\theta|^{2/3})} \frac{|\cos\theta|^{1/6}}{1 + |\cos\theta|^{2/3}} \tag{5.11}$$

where $\mu_0 H_{\parallel} = K_{\parallel}/M_S$ and $\mu_0 H_{\perp} = K_{\perp}/M_s$ are the parallel and transverse anisotropy fields given in Tesla, $K_{\parallel}$ and $K_{\perp}$ are the parallel and transverse anisotropy constants of the biaxial anisotropy, θ is the angle between the easy axis of magnetization and the direction of the applied field, and $\varepsilon = (1 - H/H_{sw}^0)$. Equation (5.11) is valid for any ratio $a = H_{\perp}/H_{\parallel}$. The proportionality coefficient of (5.11) is of the order of unity (T_c is in units of Kelvin) and depends on the approach used for calculation [177]. Equation (5.11) is plotted in Fig. 51 for several values of the ratio a. It is valid in the range $\sqrt{\varepsilon} < \theta < \pi/2 - \sqrt{\varepsilon}$.

The most interesting feature which may be drawn from (5.11) is that the crossover temperature is tunable using the external field strength and direction (Fig. 51) because the tunneling probability is increased by the transverse component of the applied field. Although at high transverse fields, T_c decreases again due to a broadening of the anisotropy barrier. Therefore, quantum tunneling experiments should always include studies of angular dependencies. When the effective magnetic anisotropy of the particle is known, MQT theories give clear predictions with no fitting parameters. MQT could also be studied as a function of the effective magnetic anisotropy. In practice, it is well known for single-particle measurements that each particle is somewhat different. Therefore, the effective magnetic anisotropy has to be determined for each particle (Section III.A.1).

Finally, it is important to note that most of the MQT theories neglect damping mechanisms. In Section IV.A [90] we discussed the case of ohmic damping, which is the simplest form of damping. More complicated damping mechanisms might play an important role. We expect more theoretical work on this in future.

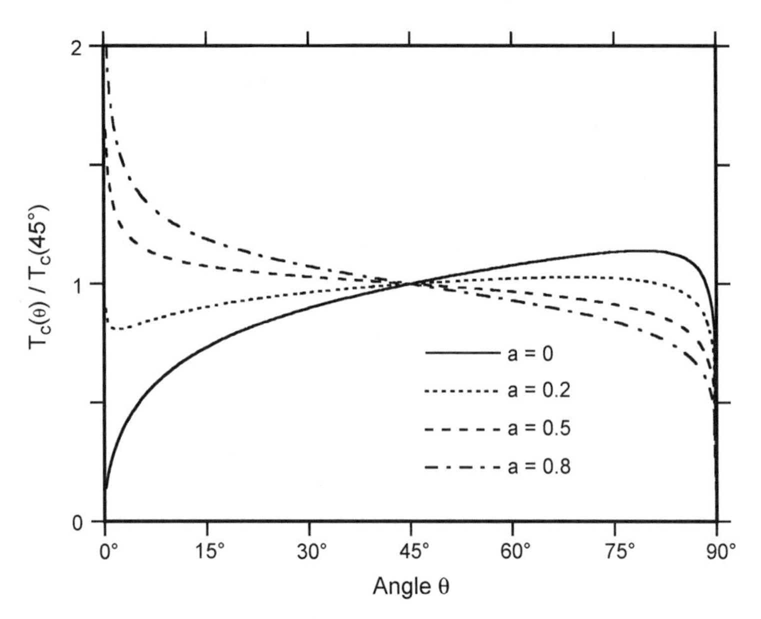

Figure 51. Normalized crossover temperature T_c as given by (5.11) and for several values of the ratio $a = H_{\perp}/H_{\parallel}$.

2. *Magnetization Reversal in Nanoparticles and Wires at Very Low Temperatures*

In order to avoid the complications due to distributions of particle size, shape, and so on, some groups have tried to study the temperature and field dependence of magnetization reversal of individual magnetic particles or wires. Most of the recent studies were done using magnetic force microscopy at room temperature. Low-temperature investigations were mainly performed via resistance measurements (Section II.A).

The first magnetization measurements of individual single-domain nanoparticles at low temperature ($0.1\,\mathrm{K} < T < 6\,\mathrm{K}$) were presented by Wernsdorfer et al. [22]. The detector (a Nb microbridge-DC-SQUID; see Section II.B) and the particles studied (ellipses with axes between 50 and 1000 nm and thickness between 5 and 50 nm) were fabricated using electron beam lithography. Electrodeposited wires (with diameters ranging from 40 to 100 nm and lengths up to 5000 nm) were also studied [35, 74]. Waiting time and switching field measurements (Section IV.B) showed that the magnetization reversal of these particles and wires results from a single thermally activated domain wall nucleation, followed by a fast wall propagation reversing the particle's magnetization. For nanocrystalline Co particles of about 50 nm and below 1 K, a flattening of the temperature dependence of the mean switching field was observed which could not be explained by thermal activation. These results were discussed in the context of MQT. However, the width of the switching field distribution and the probability of switching are in disagreement with such a model because nucleation is very sensitive to factors like surface defects, surface oxidation, and perhaps nuclear spins. The fine structure of pre-reversal magnetization states is then governed by a multivalley energy landscape (in a few cases distinct magnetization reversal paths were effectively observed [59]) and the dynamics of reversal occurs via a complex path in configuration space.

Coppinger et al. [23] used telegraph noise spectroscopy to investigate two-level fluctuations (TLF) observed in the conductance of a sample containing self-assembled ErAs quantum wires and dots in a semi-insulating GaAs matrix. They showed that the TLF could be related to two possible magnetic states of a ErAs cluster and that the energy difference between the two states was a linear function of the magnetic field. They deduced that the ErAs cluster should contain a few tens of Er atoms. At temperatures between 0.35 K and 1 K, the associated switching rates of the TLF were thermally activated, whilst below 0.35 K the switching rate became temperature-independent. Tunneling of the magnetization was proposed in order to explain the observed behavior.

Some open questions remain: What is the object that is really probed by TLF? If this is a single ErAs particle, as assumed by the authors, the switching probability should be an exponential function of time. The preexponential factor

τ_0^{-1} (sometimes called attempt frequency) was found to lie between 10^3 and $10^6\,s^{-1}$ whereas expected values are between 10^9 and $10^{12}\,s^{-1}$. Why must one apply fields of about 2 T in order to measure two-level fluctuations which should be expected near zero field? What is the influence of the measurement technique on the sample?

By measuring the electrical resistance of isolated Ni wires with diameters between 20 and 40 nm, Hong and Giordano studied the motion of magnetic domain walls [24]. Because of surface roughness and oxidation, the domain walls of a single wire are trapped at pinning centers. The pinning barrier decreases with an increase in the magnetic field. When the barrier is sufficiently small, thermally activated escape of the wall occurs. This is a stochastic process that can be characterized by a switching (depinning) field distribution. A flattening of the temperature dependence of the mean switching field and a saturation of the width of the switching field distribution (rms. deviation σ) were observed below about 5 K. The authors proposed that a domain wall escapes from its pinning site by thermal activation at high temperatures and by quantum tunneling below $T_c \sim 5$ K.

These measurements pose several questions: What is the origin of the pinning center which may be related to surface roughness, impurities, oxidation, and so on? The sweeping rate dependence of the depinning field, as well as the depinning probability, could not be measured even in the thermally activated regime. Therefore, it was not possible to check the validity of the Néel–Brown model [9, 10, 86–88] or to compare measured and predicted rms. deviations σ. Finally, a crossover temperature T_c of about 5 K is three orders of magnitude higher than T_c predicted by current theories.

Later, Wernsdorfer et al. published results obtained on nanoparticles synthesized by arc discharge, with dimensions between 10 and 30 nm [36]. These particles were single crystalline, and the surface roughness was about two atomic layers. Their measurements showed for the first time that the magnetization reversal of a ferromagnetic nanoparticle of good quality can be described by thermal activation over a single-energy barrier as proposed by Néel and Brown [9, 10, 86–88] (see Section IV.C). The activation volume, which is the volume of magnetization overcoming the barrier, was very close to the particle volume, predicted for magnetization reversal by uniform rotation. No quantum effects were found down to 0.2 K. This was not surprising because the predicted crossover temperature is $T_c \sim 0.02$ K. The results of Wernsdorfer et al. constitute the preconditions for the experimental observation of MQT of magnetization on a single particle.

Just as the results obtained with Co nanoparticles [36], a quantitative agreement with the Néel–Brown model of magnetization reversal was found on $BaFe_{12-2x}Co_xTi_xO_{19}$ nanoparticles ($0<x<1$) [37], which we will call BaFeO, in the size range of 10–20 nm. However, strong deviations from this model were evidenced for the smallest particles containing about 10^5 μ_B and for temperatures below 0.4 K. These deviations are in good agreement with the theory of macroscopic quantum tunneling of magnetization. Indeed, the measured angular dependence of $T_c(\theta)$ is in excellent agreement with the prediction given by (4.11) (Fig. 52). The normalization value $T_c(45°) = 0.31$ K compares well with the theoretical value of about 0.2 K.

Although the above measurements are in good agreement with MQT theory, we should not forget that MQT is based on several strong assumptions. Among them, there is the assumption of a giant spin; that is, all magnetic moments in the particle are rigidly coupled together by strong exchange interaction. This approximation might be good in the temperature range where thermal activation is dominant, but is it not yet clear if this can be made for very low energy barriers (see, for example, Section IV.C.4). Future measurements might give us the answer.

3. *Quantization of the Magnetization*

In order to give a definite proof that MQT can occur in a magnetic nanoparticle we propose to surge for the energy level quantization of its collective spin state. This was recently evidenced in molecular cluster like Fe_8 having a collective spin state $S = 10$ (Section V.A). In the case of the BaFeO particles with $S \approx 10^5$

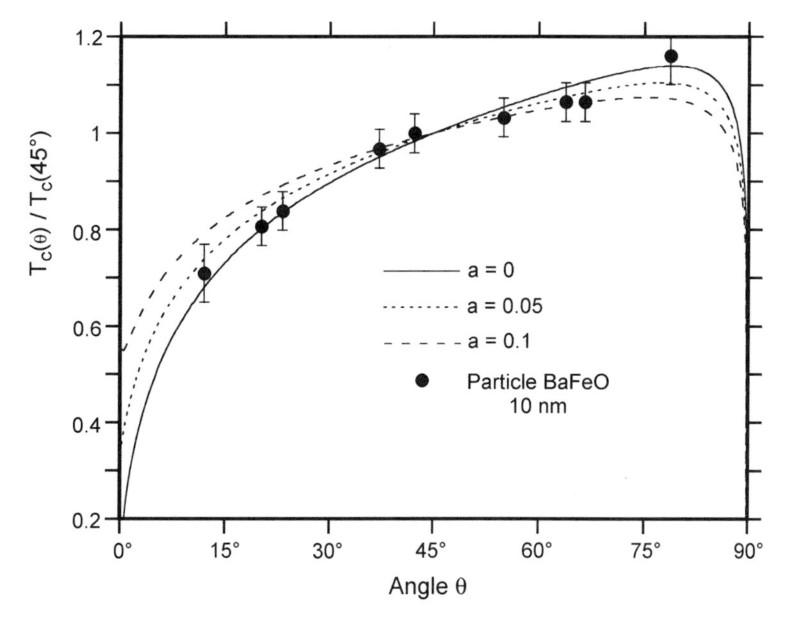

Figure 52. Angular dependence of the crossover temperature T_c for a 10-nm $BaFe_{10.4}Co_{0.8}Ti_{0.8}O_{19}$ particle with $S \approx 10^5$. The lines are given by (5.11) for different values of the ratio $a = H_\perp / H_\parallel$. The experimental data are normalized by $T_c(45°) = 0.31$ K.

[37], the field separation associated with level quantization is rather small: $\Delta H \approx H_a/2S \approx 0.002$ mT where H_a is the anisotropy field. However, for a 3-nm Co cluster with $S \approx 10^3$ the field separation $\Delta H = H_a/2S \sim 0.2$ mT might be large enough to be measurable.

Figure 53 displays schematically the field values of resonances between quantum states of S. When the applied field is ramped in a certain direction, the resonance might occur for fields $H_{res} \approx n \times (H_a/2S)(1/\cos\theta)$, with $n = 1, 2, 3, \ldots$. θ is the angle between the applied field and the easy axis of magnetization. For large spins S, tunneling might be observable only for fields which are close to the classical switching field (Fig. 53). The resonance fields could be evidenced by measuring switching field distributions (inset of Fig. 53) as a function of the angle θ.

Such a study is presented in Fig. 54 for a 3-nm Fe cluster with $S \approx 800$. The estimated field separation $\Delta H = H_a/2S$ is about 0.1 mT whereas the width of the switching field distribution is about ten times larger. We observed sometimes a small periodic fine structure which is close to the expected ΔH. However, this fine structure always disappeared when averaging over more measurements. A possible origin might be hyperfine couplings that broaden the energy levels (Section V.B.4) leading to a complete overlap of adjacent energy levels. It is

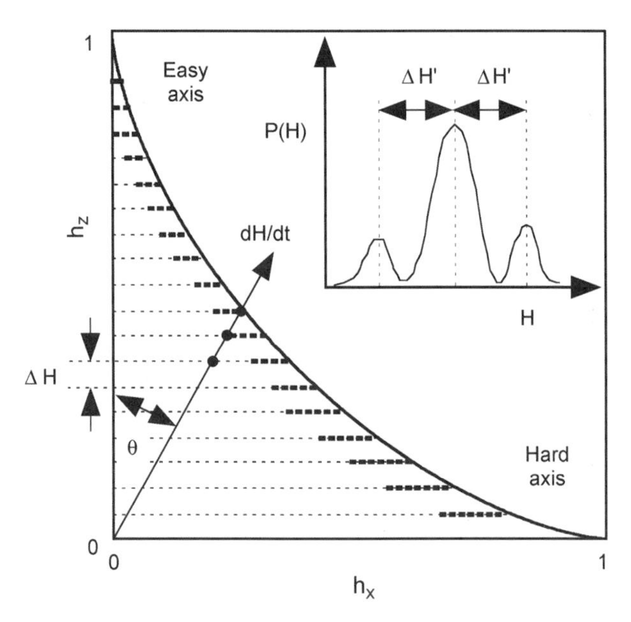

Figure 53. Schematic view of the resonance fields of a giant spin S. The continuous line is the classical switching fields of Stoner–Wohlfarth (Section III.A). The inset presents schematically a switching field histogram with $\Delta H' \approx (H_a/2S)(1/\cos\theta)$.

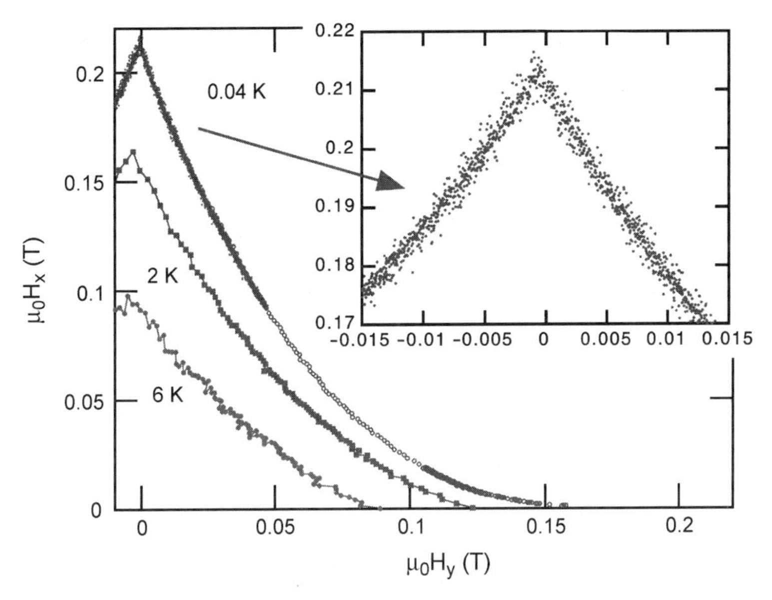

Figure 54. Angular dependence of the switching field of a 3-nm Fe cluster with $S \approx 800$. The inset presents detailed measurements near the easy axis of magnetization.

also important to mention that the switching field distributions were temperature-independent for $0.04\,\text{K} < T < 0.2\,\text{K}$.

In some cases, we observed huge variations of the switching field (Figs. 30 and 31) which might be due to exchange bias of frustrated spin configurations. However, quantum effects are not completely excluded.

Future measurements should focus on the level quantization of collective spin states of $S = 10^2$.

VI. SUMMARY AND CONCLUSION

Nanometer-sized magnetic particles have generated continuous interest as the study of their properties has proved to be scientifically and technologically very challenging. In the last few years, new fabrication techniques have led to the possibility of making small objects with the required structural and chemical qualities. In order to study these objects, new local measuring techniques were developed such as magnetic force microscopy, magnetometry based on micro-Hall probes, or micro-SQUIDs. This led to a new understanding of the magnetic behavior of nanoparticles.

In this chapter we reviewed the most important theories and experimental results concerning the magnetization reversal of single-domain particles and clusters. Special emphasis is laid on single-particle measurements avoiding complications due to distributions of particle size, shape, and so on. Measurements on particle assemblies have been reviewed in Ref. 8. We mainly discuss the low-temperature regime in order to avoid spin excitations.

Section II reviews briefly the commonly used measuring techniques. Among them, electrical transport measurements, Hall probes, and micro-SQUID techniques seem to be the most convenient techniques for low-temperature measurements.

Section III discusses the mechanisms of magnetization reversal in single-domain particles at zero kelvin. For extremely small particles, the magnetization should reverse by uniform rotation of magnetization (Section III.A). For somewhat larger particles, a nonuniform reversal mode occurs like the curling mode (Section III.B). For even larger particles, magnetization reversal occurs via a domain wall nucleation process starting in a rather small volume of the particle (Section III.B.3).

The influence of temperature on the magnetization reversal is reported in Section IV. We discuss in detail the Néel, Brown, and Coffey's theory of magnetization reversal by thermal activation (Section IV).

Finally, Section V shows that for very small systems or very low temperature, magnetization can reverse via quantum tunneling. The boundary between classical and quantum physics has become a very attractive field of research. This section discusses detailed measurements which demonstrated that molecular magnets offer an unique opportunity to explore the quantum dynamics of a large but finite spin. The discussion is focused on the Fe_8 molecular magnet with $S = 10$ because it is the first system where studies in the pure quantum regime were possible. We showed that the understanding of the environmental decoherence is one of the most important issues, in particular for future applications of quantum devices. We then discussed tunneling in nanoparticles and showed how one might give a definite proof of their quantum character at low temperature.

In conclusion, the understanding of the magnetization reversal in nanostructures requires the knowledge of many physical phenomena, and nanostructures are therefore particularly interesting for the development of new fundamental theories of magnetism and in modeling new magnetic materials for permanent magnets or high density recording. Using the quantum character of nanostructures for applications like "quantum computers" will be one of the major concerns of the next decades.

Acknowledgments

The author is indebted to A. Benoit, E. Bonet Orozco, V. Bouchiat, I. Chiorescu, M. Faucher, K. Hasselbach, M. Jamet, D. Mailly, B. Pannetier, and C. Thirion for their experimental contributions and the development of the micro-SQUID technology. The author acknowledges the collaborations with J.-Ph. Ansermet,

B. Barbara, A. Caneschi, A. Cornia, N. Demoncy, B. Doudin, V. Dupuis, O. Fruchart, D. Gatteschi, O. Kubo, J.-P. Nozieres, H. Pascard, C. Paulsen, A. Perez, C. Sangregorio, and R. Sessoli. He is also indebted to W. T. Coffey, A. Garg, J. Miltat, N. Prokof'ev, P. Stamp, A. Thiaville, I. Tupitsyn, and J. Villain for many fruitful and motivating discussions. Finally, O. Fruchart is specially acknowledged for his critical review of the manuscript. This work has been supported by CNRS, DRET, MASSDOTS, and Rhône-Alpes.

References

1. A. Hubert and R. Schäfer, *Magnetic Domains: The Analysis of Magnetic Microstructures*, Springer-Verlag, New York, 1998.
2. Aharoni, *An Introduction to the Theory of Ferromagnetism*, Oxford University Press, London, 1996.
3. A. J. Freemann and R. Wu, *J. Magn. Magn. Mat.* **100**, 497 (1991).
4. G. M. Pastor, J. Dorantes-Dèvila, S. Pick, and H. Dreyssé, *Phys. Rev. Lett.* **75**, 326 (1995).
5. C. Kohl and G. F. Bertsch, *Phys. Rev. B* **60**, 4205 (1999).
6. I. M. L. Billas, A. Châtelain, and W. A. de Heer, *J. Magn. Magn. Mat.* **168**, 64 (1997).
7. S. E. Apsel, J. W. Emmert, J. Deng, and L. A. Bloomfield, *Phys. Rev. Lett.* **76**, 1441 (1996).
8. J. L. Dormann, D. Fiorani, and E. Tronc, *Adv. Chem. Phys.* **98**, 283 (1997).
9. L. Néel, *Ann. Geophys.* **5**, 99 (1949).
10. L. Néel, *C. R. Acad. Sci.* **228**, 664 (1949).
11. A. H. Morrish and S. P. Yu, *Phys. Rev.* **102**, 670 (1956).
12. J. E. Knowles, *IEEE Trans. Mag.* **MAG-14**, 858 (1978).
13. A. Tonomura, T. Matsuda, J. Endo, T. Arii, and K. Mihama, *Phys. Rev. B* **34**, 3397 (1986).
14. H. J. Richter, *J. Appl. Phys.* **65**, 9 (1989).
15. S. J. Hefferman, J. N. Chapman, and S. McVitie, *J. Magn. Magn. Mat.* **95**, 76 (1991).
16. C. Salling, S. Schultz, I. McFadyen, and M. Ozaki, *IEEE Trans. Mag.* **27**, 5185 (1991).
17. N. Bardou, B. Bartenlian, C. Chappert, R. Megy, P. Veillet, J. P. Renard, F. Rousseaux, M. F. Ravet, J. P. Jamet, and P. Meyer, *J. Appl. Phys.* **79**, 5848 (1996).
18. T. Chang and J. G. Chu, *J. Appl. Phys.* **75**, 5553 (1994).
19. M. Ledermann, S. Schultz, and M. Ozaki, *Phys. Rev. Lett.* **73**, 1986 (1994).
20. J. Bansmann, V. Senz, L. Lu, A. Bettac, and K. H. Meiweis-Broer, *J. Electron Spectrosc. Relat. Phenom.* **106**, 221 (2000).
21. K. H. Meiweis-Broer, *Phys. Bl.* **55**, 21 (1999).
22. W. Wernsdorfer, K. Hasselbach, D. Mailly, B. Barbara, A. Benoit, L. Thomas, and G. Suran, *J. Magn. Magn. Mat.* **145**, 33 (1995).
23. F. Coppinger, J. Genoe, D. K. Maude, U. Genner, J. C. Portal, K. E. Singer, P. Rutter, T. Taskin, A. R. Peaker, and A. C. Wright, *Phys. Rev. Lett.* **75**, 3513 (1995).
24. K. Hong and N. Giordano, *J. Magn. Magn. Mat.* **151**, 396 (1995).
25. J. E. Wegrowe, S. E. Gilbert, D. Kelly, B. Doudin, and J.-Ph. Ansermet, *IEEE Trans. Magn.* **34**, 903 (1998).

26. A. D. Kent, S. von Molnar, S. Gider, and D. D. Awschalom, *J. Appl. Phys.* **76**, 6656 (1994).
27. J. G. S. Lok, A. K. Geim, J. C. Maan, S. V. Dubonos, L. Theil Kuhn, and P. E. Lindelof, *Phys. Rev. B* **58**, 12201 (1998).
28. T. Schweinböck, D. Weiss, M. Lipinski, and K. Eberl, *J. Appl. Phys.* **87**, 6496 (2000).
29. V. Gros, Shan-Fab Lee, G. Faini, A. Cornette, A. Hamzic, and A. Fert, *J. Magn. Magn. Mat.* **165**, 512 (1997).
30. W. J. Gallagher, S. S. P. Parkin, Yu Lu, X. P. Bian, A. Marley, K. P. Roche, R. A. Altman, S. A. Rishton, C. Jahnes, T. M. Shaw, and Gang Xiao, *J. Appl. Phys.* **81**, 3741 (1997).
31. J.-E. Wegrowe, D. Kelly, A. Franck, S. E. Gilbert, and J.-Ph. Ansermet, *Phys. Rev. Lett.* **82**, 3681 (1999).
32. L. F. Schelp, A. Fert, F. Fettar, P. Holody, S. F. Lee, J. L. Maurice, F. Petroff, and A. Vaures, *Phys. Rev. B* **56**, R5747 (1997).
33. S. Guéron, M. M. Deshmukh, E. B. Myers, and D. C. Ralph, *Phys. Rev. Lett.* **83**, 4148 (1999).
34. W. Wernsdorfer, K. Hasselbach, A. Benoit, B. Barbara, D. Mailly, J. Tuaillon, J. P. Perez, V. Dupuis, J. P. Dupin, G. Guiraud, and A. Perez, *J. Appl. Phys.* **78**, 7192 (1995).
35. W. Wernsdorfer, B. Doudin, D. Mailly, K. Hasselbach, A. Benoit, J. Meier, J.-Ph. Ansermet, and B. Barbara, *Phys. Rev. Lett.* **77**, 1873 (1996).
36. W. Wernsdorfer, E. Bonet Orozco, K. Hasselbach, A. Benoit, B. Barbara, N. Demoncy, A. Loiseau, D. Boivin, H. Pascard, and D. Mailly, *Phys. Rev. Lett.* **78**, 1791 (1997).
37. W. Wernsdorfer, E. Bonet Orozco, K. Hasselbach, A. Benoit, D. Mailly, O. Kubo, H. Nakano, and B. Barbara, *Phys. Rev. Lett.* **79**, 4014 (1997).
38. E. Bonet, W. Wernsdorfer, B. Barbara, A. Benoit, D. Mailly, and A. Thiaville, *Phys. Rev. Lett.* **83**, 4188 (1999).
39. M. Jamet, W. Wernsdorfer, C. Thirion, D. Mailly, V. Dupuis, P. Mélinon, and A. Pérez, *Phys. Rev. Lett.* **86**, 4586 (2001); cond-mat/0012029.
40. J. Clarke, A. N. Cleland, M. H. Devoret, D. Esteve, and J. M. Martinis, *Science* **239**, 992 (1988).
41. M. Ketchen, D. J. Pearson, K. Stawiaasz, C-H. Hu, A. W. Kleinsasser, T. Brunner, C. Cabral, V. Chandrashekhar, M. Jaso, M. Manny, K. Stein, and M. Bhushan, *IEEE Appl. Supercond.* **3**, 1795 (1993).
42. P. W. Anderson and A. H. Dayem, *Phys. Rev. Lett.* **13**, 195 (1964).
43. C. Chapelier, M. El Khatib, P. Perrier, A. Benoit, and D. Mailly, in *Superconducting Devices and Their Applications, SQUID 91*, H. Koch and H. Lbbig, eds., Springer, Berlin, 1991, p. 286.
44. D. Mailly, C. Chapelier, and A. Benoit, *Phys. Rev. Lett.* **70**, 2020 (1993).
45. W. Wernsdorfer, Ph.D. thesis, Joseph Fourier University, Grenoble, 1996.
46. V. Bouchiat, M. Faucher, C. Thirion, W. Wernsdorfer, T. Fournier, and B. Pannetier, *Appl. Phys. Lett.* **78**, 0 (2001).
47. W. Wernsdorfer and R. Sessoli, *Science* **284**, 133 (1999).
48. S. Mangin, G. Marchal, W. Wernsdorfer, A. Sulpice, K. Hasselbach, D. Mailly, and B. Barbara, *Eur. Phys. Lett.* **39**, 675 (1997).
49. O. Fruchart, J.-P. Nozieres, W. Wernsdorfer, and D. Givord, *Phys. Rev. Lett.* **82**, 1305 (1999).
50. K. Hasselbach, C. Veauvy, and D. Mailly, *Physica C* **332**, 140 (2000).
51. G. Cernicchiaro, K. Hasselbach, D. Mailly, W. Wernsdorfer, and A. Benoit, in *Quantum Transport in Semiconductor Submicron Structures*, Vol. 326 of *NATO ASI Series E: Applied Sciences*, edited by Bernard Kramer, ed., Kluwer Academic Publishers, London, 1996.

52. E. C. Stoner and E. P. Wohlfarth, *Philos. Trans. London Ser. A* **240**, 599 (1948), reprinted in *IEEE Trans. Magn.* **MAG-27**, 3475 (1991).
53. L. Néel, *C. R. Acad. Sci.* **224**, 1550 (1947).
54. A. Thiaville, *J. Magn. Magn. Mat.* **182**, 5 (1998).
55. A. Thiaville, *Phys. Rev. B* **61**, 12221 (2000).
56. R. H. Victora, *Phys. Rev. Lett.* **63**, 457 (1989).
57. H. Pfeiffer, *Phys. Status Solidi* **118**, 295 (1990).
58. H. Pfeiffer, *Phys. Status Solidi* **122**, 377 (1990).
59. W. Wernsdorfer, K. Hasselbach, A. Benoit, G. Cernicchiaro, D. Mailly, B. Barbara, and L. Thomas, *J. Magn. Magn. Mater.* **151**, 38 (1995).
60. A. Perez, P. Melinon, V. Depuis, P. Jensen, B. Prevel, J. Tuaillon, L. Bardotti, C. Martet, M. Treilleux, M. Pellarin, J. L. Vaille, B. Palpant, and J. Lerme, *J. Phys. D* **30**, 709 (1997).
61. M. Jamet, V. Dupuis, P. Mélinon, G. Guiraud, A. Pérez, W. Wernsdorfer, A. Traverse, and B. Baguenard, *Phys. Rev. B* **62**, 493 (2000).
62. C. H. Lee, Hui He, F. J. Lamelas, W. Vavrn, C. Uher, and Roy Clarke, *Phys. Rev. B* **42**, 1066 (1990).
63. D. S. Chuang, C. A. Ballentine, and R. C. O'Handley, *Phys. Rev. B* **49**, 15084 (1994).
64. Ching-Ray Chang, *J. Appl. Phys.* **69**, 2431 (1991).
65. E. Bonet, W. Wernsdorfer, B. Barbara, K. Hasselbach, A. Benoit, and D. Mailly, *IEEE Trans. Mag.* **34**, 979 (1998).
66. H. Frei, S. Shtrikman, and D. Treves, *Phys. Rev.* **106**, 446 (1957).
67. A. Aharoni, *J. Appl. Phys.* **86**, 1042 (1999).
68. A. Aharoni, *J. Appl. Phys.* **82**, 1281 (1997).
69. A. Aharoni, *IEEE Trans. Magn.* **22**, 478 (1986).
70. A. Aharoni, *Phys. Status Solidi* **16**, 3 (1966).
71. Y. Ishii, *J. Appl. Phys.* **70**, 3765 (1991).
72. A. Aharoni, *J. Appl. Phys.* **87**, 5526 (2000).
73. R. Ferré, K. Ounadjela, J. M. George, L. Piraux, and S. Dubois, *Phys. Rev. B* **56**, 14066 (1997).
74. W. Wernsdorfer, K. Hasselbach, A. Benoit, B. Barbara, B. Doudin, J. Meier, J.-Ph. Ansermet, and D. Mailly, *Phys. Rev. B* **55**, 1155 (1997).
75. B. Doudin and J.-Ph. Ansermet, *Nanostructured Mater.* **6**, 521 (1995).
76. J. Meier, B. Doudin, and J.-Ph. Ansermet, *J. Appl. Phys.* **79**, 6010 (1996).
77. Hans-Benjamin Braun, *J. Appl. Phys.* **85**, 6172 (1999).
78. M. Ledermann, R. O'Barr, and S. Schultz, *IEEE Trans. Mag.* **31**, 3793 (1995).
79. W. Wernsdorfer, K. Hasselbach, A. Sulpice, A. Benoit, J.-E. Wegrowe, L. Thomas, B. Barbara, and D. Mailly, *Phys. Rev. B* **53**, 3341 (1996).
80. H. A. M. van den Berg, *J. Appl. Phys.* **61**, 4194 (1987).
81. P. Bryant and H. Suhl, *Appl. Phys. Lett.* **54**, 78 (1989).
82. M. Rührig, W. Bartsch, M. Vieth, and A. Hubert, *IEEE Trans. Mag.* **MAG-26**, 2807 (1990).
83. A. Fernandez, M. R. Gibbons, M. A. Wall, and C. J. Cerjan, *J. Magn. Magn. Mater.* **190**, 71 (1998).
84. C. P. Bean, *J. Appl. Phys.* **26**, 1381 (1955).

85. C. P. Bean and J. D. Livingstone, *J. Appl. Phys.* **30**, 120S (1959).
86. W. F. Brown, *J. Appl. Phys.* **30**, 130S (1959).
87. W. F. Brown, *J. Appl. Phys.* **34**, 1319 (1963).
88. W. F. Brown, *Phys. Rev.* **130**, 1677 (1963).
89. W. T. Coffey, D. S. F. Crothers, J. L. Dormann, Yu. P. Kalmykov, and J. T. Waldron, *Phys. Rev. B* **52**, 15951 (1995).
90. W. T. Coffey, D. S. F. Crothers, J. L. Dormann, Yu. P. Kalmykov, E. C. Kennedy, and W. Wernsdorfer, *Phys. Rev. Lett.* **80**, 5655 (1998).
91. W. T. Coffey, D. S. F. Crothers, J. L. Dormann, Yu. P. Kalmykov, E. C. Kennedy, and W. Wernsdorfer, *J. Phys. Cond. Mater.* **10**, 9093 (1998).
92. I. Klik and L. Gunther, *J. Stat. Phys.* **60**, 473 (1990).
93. I. Klik and L. Gunther, *J. Appl. Phys.* **67**, 4505 (1990).
94. W. T. Coffey, *Adv. Chem. Phys.* **103**, 259 (1998).
95. A. Garg, *Phys. Rev. B* **51**, 15592 (1995).
96. J. Kurkijärvi, *Phys. Rev. B* **6**, 832 (1972).
97. H. L. Richards, S. W. Sides, M. A. Novotny, and P. A. Rikvold, *J. Appl. Phys.* **79**, 5749 (1996).
98. J. M. Gonzalez, R. Ramirez, R. Smirnov-Rueda, and J. Gonzalez, *J. Appl. Phys.* **79**, 6479 (1996).
99. D. Garcia-Pablos, P. Garcia-Mochales, and N. Garcia, *J. Appl. Phys.* **79**, 6021 (1996).
100. D. Hinzke and U. Nowak, *Phys. Rev. B* **58**, 265 (1998).
101. E. D. Boerner and H. Neal Bertram, *IEEE Trans. Mag.* **33**, 3052 (1997).
102. M. Respaud, J. M. Broto, H. Rakoto, A. R. Fert, L. Thomas, and B. Barbara, *Phys. Rev. B* **57**, 2925 (1998).
103. A. E. Berkowitz, J. A. Lahut, I. S. Jacobs, L. M. Levinson, and D. W. Forester, *Phys. Rev. Lett.* **34**, 594 (1975).
104. J. T. Richardson, D. I. Yiagas, B. Turk, J. Forster, and M. V. Twigg, *J. Appl. Phys.* **70**, 6977 (1991).
105. R. H. Kodama, A. E. Berkowitz, E. J. McNiff, Jr., and S. Foner, *Phys. Rev. Lett.* **77**, 394 (1996).
106. R. H. Kodama, *J. Magn. Magn. Mater.* **200**, 359 (1999).
107. A. J. Leggett, S. Chakravarty, A. T. Dorsey, M. P. A. Fisher, A. Garg, and W. Zwerger, *Rev. Mod. Phys.* **59**, 1 (1987).
108. *Quantum Tunneling of Magnetization-QTM'94*, Vol. 301 of *NATO ASI Series E: Applied Sciences*, by L. Gunther and B. Barbara, eds., Kluwer Academic Publishers, London, 1995.
109. Y. Pontillon, A. Caneschi, D. Gatteschi, R. Sessoli, E. Ressouche, J. Schweizer, and E. Lelievte-Berna, *J. Am. Chem. Soc.* **121**, 5342 (1999).
110. A. L. Barra, D. Gatteschi, and R. Sessoli, *Chem. Eur. J.* **6**, 1608 (2000).
111. R. Sessoli, D. Gatteschi, A. Caneschi, and M. A. Novak, *Nature* **365**, 141 (1993).
112. A.-L. Barra, P. Debrunner, D. Gatteschi, Ch. E. Schulz, and R. Sessoli, *Euro. Phys. Lett.* **35**, 133 (1996).
113. M. A. Novak and R. Sessoli, in *Quantum Tunneling of Magnetization-QTM'94*, Vol. 301 of *NATO ASI Series E: Applied Sciences*, L. Gunther and B. Barbara, eds., Kluwer Academic Publishers, London, 1995, pp. 171–188.
114. J. R. Friedman, M. P. Sarachik, J. Tejada, and R. Ziolo, *Phys. Rev. Lett.* **76**, 3830 (1996).

115. L. Thomas, F. Lionti, R. Ballou, D. Gatteschi, R. Sessoli, and B. Barbara, *Nature (London)* **383**, 145 (1996).
116. C. Sangregorio, T. Ohm, C. Paulsen, R. Sessoli, and D. Gatteschi, *Phys. Rev. Lett.* **78**, 4645 (1997).
117. A. Garg, *Phys. Rev. B* **81**, 1513 (1998).
118. T. Ohm, C. Sangregorio, and C. Paulsen, *Eur. Phys. J. B* **6**, 195 (1998).
119. J. A. A. J. Perenboom, J. S. Brooks, S. Hill, T. Hathaway, and N. S. Dalal, *Phys. Rev. B* **58**, 330 (1998).
120. A. D. Kent, Y. Zhong, L. Bokacheva, D. Ruiz, D. N. Hendrickson, and M. P. Sarachik, *Eur. Phys. Lett.* **49**, 521 (2000).
121. A. Caneschi, D. Gatteschi, C. Sangregorio, R. Sessoli, L. Sorace, A. Cornia, M. A. Novak, C. Paulsen, and W. Wernsdorfer, *J. Magn. Magn. Mater.* **200**, 182 (1999).
122. S. M. J. Aubin, N. R. Dilley, M. B. Wemple, G. Christou, and D. N. Hendrickson, *J. Am. Chem. Soc.* **120**, 839 (1998).
123. D. J. Price, F. Lionti, R. Ballou, P. T. Wood, and A. K. Powell, *Phil. Trans. R. Soc. Lond. A* **357**, 3099 (1999).
124. J. Yoo, E. K. Brechin, A. Yamaguchi, M. Nakano, J. C. Huffman, A. L. Maniero, L.-C. Brunel, K. Awaga, H. Ishimoto, G. Christou, and D. N. Hendrickson, *Inorg. Chem.* **39**, 3615 (2000).
125. K. Wieghardt, K. Pohl, I. Jibril, and G. Huttner, *Angew. Chem. Int. Ed. Engl.* **23**, 77 (1984).
126. W. Wernsdorfer, I. Chiorescu, R. Sessoli, D. Gatteschi, and D. Mailly, *Phys. B* **284–288**, 1231 (2000).
127. C. Delfs, D. Gatteschi, L. Pardi, R. Sessoli, K. Wieghardt, and D. Hanke, *Inorg. Chem.* **32**, 3099 (1993).
128. A. Garg, *Eur. Phys. Lett.* **22**, 205 (1993).
129. W. Wernsdorfer, T. Ohm, C. Sangregorio, R. Roberta, D. Mailly, and C. Paulsen, *Phys. Rev. Lett.* **82**, 3903 (1999).
130. L. Landau, *Phys. Z. Sowjetunion* **2**, 46 (1932).
131. C. Zener, *Proc. R. Soc. London, Ser. A* **137**, 696 (1932).
132. E. C. G. Stückelberg, *Helv. Phys. Acta* **5**, 369 (1932).
133. S. Miyashita, *J. Phys. Soc. Jpn.* **64**, 3207 (1995).
134. S. Miyashita, *J. Phys. Soc. Jpn.* **65**, 2734 (1996).
135. G. Rose and P. C. E. Stamp, *Low Temp. Phys.* **113**, 1153 (1998).
136. G. Rose, Ph.D. thesis, The University of British Columbia, Vancouver, 1999.
137. M. Thorwart, M. Grifoni, and P. Hänggi, *Phys. Rev. Lett.* **85**, 860 (2000).
138. M. N. Leuenberger and D. Loss, *Phys. Rev. B* **61**, 12200 (2000).
139. J. Villain, A. Wurger, A. Fort, and A. Rettori, *J. Phys.* I **7**, 1583 (1997).
140. A. Fort, A. Rettori, J. Villain, D. Gatteschi, and R. Sessoli, *Phys. Rev. Lett.* **80**, 612 (1998).
141. M. N. Leuenberger and D. Loss, *Phys. Rev. B* **61**, 1286 (2000).
142. I. Tupitsyn, N. V. Prokof'ev, and P. C. E. Stamp, *Int. J. Mod. Phys. B* **11**, 2901 (1997).
143. N. V. Prokof'ev and P. C. E. Stamp, *Phys. Rev. Lett.* **80**, 5794 (1998).
144. W. Wernsdorfer, A. Caneschi, R. Sessoli, D. Gatteschi, A. Cornia, V. Villar, and C. Paulsen, *Phys. Rev. Lett.* **84**, 2965 (2000).
145. D. Loss, D. P. DiVincenzo, and G. Grinstein, *Phys. Rev. Lett.* **69**, 3232 (1992).

146. J. von Delft and C. L. Hendey, *Phys. Rev. Lett.* **69**, 3236 (1992).
147. R. P. Feynman, R. B. Leighton, and M. Sand, *The Feynman Lectures on Physics*, Vol. 3, Addison-Wesley, London, 1970.
148. A. Garg, *Phys. Rev. Lett.* **83**, 4385 (1999).
149. J. Villain and A. Fort, *Eur. Phys. J. B* **17**, 69 (2000).
150. E. Kececioglu and A. Garg, *Phys. Rev. B* **63**, 064422 (2001).
151. S. E. Barnes, cond-mat/9907257.
152. J.-Q. Liang, H. J. W. Mueller-Kirsten, D. K. Park, and F.-C. Pu, *Phys. Rev. B* **61**, 8856 (2000).
153. S. Yoo and S. Lee, *Phys. Rev. B* **62**, 5713 (2000).
154. R. Lü, H. Hu, J. Zhu, X. Wang, L. Chang, and B. Gu, *Phys. Rev. B* **61**, 14581 (2000).
155. R. Caciuffo, G. Amoretti, A. Murani, R. Sessoli, A. Caneschi, and D. Gatteschi, *Phys. Rev. Lett.* **81**, 4744 (1998).
156. G. Amoretti, R. Caciuffo, J. Combet, A. Murani, and A. Caneschi, *Phys. Rev. B* **62**, 3022 (2000).
157. M. Al-Saqer, V. V. Dobrovitski, B. N. Harmon, and M. I. Katsnelson, *J. Appl. Phys.* **87**, 6268 (2000).
158. R. P. Feynman and F. L. Vernon, *Ann. Phys.* **24**, 118 (1963).
159. N. V. Prokof'ev and P. C. E. Stamp, *J. Low Temp. Phys.* **104**, 143 (1996).
160. N. V. Prokof'ev and P. C. E. Stamp, *Rep. Prog. Phys.* **63**, 669 (2000).
161. T. Ohm, C. Sangregorio, and C. Paulsen, *J. Low Temp. Phys.* **113**, 1141 (1998).
162. T. Ohm, Ph.D. thesis, Joseph Fourier University, Grenoble, 1998.
163. A. Cuccoli, A. Fort, A. Rettori, E. Adam, and J. Villain, *Eur. Phys. J. B* **12**, 39 (1999).
164. W. Wernsdorfer, R. Sessoli, and D. Gatteschi, *Eur. Phys. Lett.* **47**, 254 (1999).
165. A. Abragam and B. Bleaney, *Electron Paramagnetic Resonance of Transition Ions*, Clarendon Press, Oxford, 1970.
166. W. Wernsdorfer, A. Caneschi, R. Sessoli, D. Gatteschi, A. Cornia, V. Villar, and C. Paulsen, *Eur. Phys. Lett.* **50**, 552 (2000).
167. L. Bokacheva, A. D. Kent, and M. A. Walters, *Phys. Rev. Lett.* **85**, 4803 (2000).
168. D. A. Garanin and E. M. Chudnovsky, *Phys. Rev. B* **59**, 3671 (1999).
169. I. Chiorescu, W. Wernsdorfer, B. Barbara, A. Müller, and H. Bögge, *J. Appl. Phys.* **87**, 5496 (2000).
170. W. Wernsdorfer, A. Caneschi, R. Sessoli, D. Gatteschi, A. Cornia, V. Villar, and C. Paulsen, *Phys. Rev. Lett.* **84**, 3454 (2000).
171. V. V. Dobrovitski, M. I. Katsnelson, and B. N. Harmon, *Phys. Rev. Lett.* **84**, 3458 (2000).
172. N. V. Prokof'ev and P. C. E. Stamp, in *Quantum Tunneling of Magnetization—QTM'94*, Vol. 301 of *NATO ASI Series E: Applied Sciences*, L. Gunther and B. Barbara, eds., (Kluwer Academic Publishers, London, 1995), p. 369.
173. E. Del Barco, J. M. Hernandez, J. Tejada, N. Biskup, R. Achey, I. Rutel, N. Dalal, and J. Brooks, *Phys. Rev. B* **62**, 3018 (2000).
174. M. N. Leuenberger and D. Loss, cond-mat/0011415.
175. O. B. Zaslavskii, *Phys. Rev. B* **42**, 992 (1990).
176. M. C. Miguel and E. M. Chudnovsky, *Phys. Rev. B* **54**, 388 (1996).
177. Gwang-Hee Kim and Dae Sung Hwang, *Phys. Rev. B* **55**, 8918 (1997).

DYNAMICAL APPROACH TO VIBRATIONAL RELAXATION

SUSUMU OKAZAKI

Department of Electronic Chemistry, Tokyo Institute of Technology Yokohama, Japan

CONTENTS

Advances in Chemical Physics, Volume 118, Edited by I. Prigogine and Stuart A. Rice.
ISBN 0-471-43816-2

I. INTRODUCTION

Vibrational energy relaxation of solute molecules in the solution, as well as of adsorbate molecules on the solid surface, has been one of the major subjects of current physical chemistry, since energy exchange between the system of interest and its surrounding medium must be closely related to condensed phase chemical reaction dynamics. In this sense, it is of general interest to study what sort of molecular mechanism underlies this energy transfer process. In particular, direct energy dissipation from solute to solvent degrees of freedom without energy redistribution within the solute molecule is very attractive because it must show the essence of condensed phase dynamics. In fact, experimental progress in this field has been accelerated by the development of time-resolved pump-probe spectroscopy, which provides an access to picosecond or faster time-scale processes in the solutions. Using this technique, vibrational energy relaxation has extensively been investigated for various solutions [1–7]. Recently, the experiments have been focused on small polar solutes in polar solvents such as diatomic ions in water [8, 9] since the mechanism of vibrational relaxation is very interesting when the solute vibrational degree of freedom under consideration is strongly coupled solely to the solvent degrees of freedom. In this case, the excited energy dissipates rapidly and directly to the solvent. However, this kind of measurement gives only the value of vibrational relaxation time. From the experiment alone, it is very difficult to obtain molecular information about the relaxation mechanism—for example, the solvent motion that causes the relaxation or the energy transfer pathway to and from the translational, rotational, and vibrational degrees of freedom of the solvent molecules. Detection of the energy transfer to the translation and rotation of the solvent molecules is particularly difficult. Furthermore, experimental analysis for the mechanism of thermal excitation of the solute molecule (i.e., the reverse energy flow to the relaxation) is much more difficult than the above relaxation process. Thus, in addition to the experiment, molecular theory that is able to present essential molecular information about the vibrational energy relaxation is desired.

Now, it is very interesting to investigate the mechanism of energy dissipation from a solute molecule to the solvent by computer simulation. The computer simulation such as molecular dynamics (MD) calculation and Monte Carlo (MC) calculation has history of about a half century as a leading technique for molecular-based science of liquids and solutions [10–13]. Now, the method is widely used for the analysis of various physical and chemical phenomena. In the MD calculation, for example, coupled classical equations of motion are solved numerically for all molecules that constitute the system of interest, assuming certain intermolecular potential functions. Calculated trajectory of the molecules presents structure and dynamics of the condensed matter. Rotational relaxation of the solute is a direct output of the classical MD calculation. Pure dephasing time may also be estimated from the fluctuation of time-dependent instantaneous transition frequency of the solute in the solution. The method of investigating these two factors that cause intensity decrease of the pump-probe spectrum has mostly been established. However, with respect to the vibrational energy relaxation, or more specifically the population relaxation, the calculation method is still unestablished. It should be noted, here, that the molecular vibration is usually quantized. So, in order to pursue the dynamics of vibrational state, we must start with the time-dependent Schrödinger equation. The conventional classical MD calculation cannot be applied to this problem except for the vibration of heavy atoms such as HgI and I_2 where the quantum effect is rather small. Thus, the vibrational population relaxation and its reverse process, the thermal excitation, must be described in terms of nonadiabatic quantum dynamics that follows the time-dependent Schrödinger equation. There is not a concept of state or transition in the classical mechanics.

If the calculation of this kind of quantum dynamics is realized, a microscopic picture for the vibrational relaxation of molecules dissolved in the solvents or adsorbed on the solid surfaces may be obtained. In addition to the relaxation time, molecular mechanism of the relaxation such as pathway of the energy dissipation, or magnitude of coupling between the solute and particular solvent modes, may be clarified. Direct simulation of the system state of interest along real time by solving coupled equations of motion of the system and the solvent molecules may also present an evidence for or against the relaxation models proposed so far. For example, if stepwise probability change of the state is found even in the liquid, an isolated binary collision model may describe the relaxation well after the analogy of the gas phase theory. On the contrary, if it shows continuous relaxation, a description of Langevin equation type may work well for the system dynamics. Search for resonance modes of the solvent or effective collisions is also interesting. However, it is absolutely impossible to solve the fully quantum mechanical equation of motion for the system composed of a few hundred molecules—for example, a few thousand degrees of freedom for the aqueous solution. Then, approximations are inevitable to attain the practical calculation. Hence, the main target of the present theoretical investigation is to solve quantum dynamics of many-body systems by introducing the approximations without losing the essence of the dynamics.

Traditional approaches to this problem have been made widely by many researchers; these approaches may be classified into two types according to the approximations used. The first one is a fully classical approximation, where even a solute oscillator has been described classically. Conventional MD calculation for flexible molecules [10] is a numerical example of this class of the approximation. Generalized Langevin equation [14–16], on the other hand, presents an algebraic way to the solution for this approximation. The method may be applied only to heavy-atom molecule such as HgI and I_2, for which $\hbar\Omega \ll kT$ where $\hbar$ is Planck's constant divided by 2π, Ω is the transition frequency of the system of interest, k is Boltzmann's constant, and T is the temperature. The second one is an application of perturbative quantum mechanical transition rate to Pauli's master equation. This is valid even for high-frequency oscillators (i.e., $\hbar\Omega \gg kT$), which is the case for most molecular systems. Much effort has been made toward the development of this method, and various prescriptions with different level of approximations have been proposed. Among them, Fermi's golden rule formulated by quantum force autocorrelation function has most frequently been adopted [17, 18], taking account of only the first term of the perturbation. Furthermore, the force correlation function has been approximated by the classical one [18] such that it may be evaluated by conventional MD calculation.

However, the former is of limited application as stated above. The number of molecules whose vibrational motion can be described satisfactorily by the classical mechanics is very small. With respect to the latter, the error is still great. Several quantitative investigations of the error caused by the classical approximation of the force has been reported [19–25], where semiempirical treatments for the quantum correction of the classical force autocorrelation function were proposed. It should be noted here that even if the error is corrected by some empirical method giving a good agreement with experiment, Fermi's golden rule does not present microscopic observations for molecular mechanism of the energy transfer between the solute and solvent. Furthermore, influence of the neglect of higher perturbations has been infrequently investigated.

Now, it is required to establish a reliable method that can predict precisely the vibrational relaxation time of the solute molecule in the solution and, at the same time, can present microscopic details of the relaxation process such as relaxation-related molecular motions of the solvent. In the present chapter we review recent development of dynamical approaches to vibrational population relaxation or energy relaxation in the condensed phase—in other words, methods of nonadiabatic quantum dynamics simulation and their applications to the vibrational relaxation. Clearly, one of the most important directions of the study is a development of new computational method that can handle the solvent molecules quantum mechanically.

First, we discuss an application of path integral influence functional theory [26, 27] which is capable of a fully quantum calculation for this problem, based upon harmonic oscillators bath approximation. In this review, major emphasis will be placed on this method, since it is likely to be most powerful among others in the molecular analysis of the vibrational energy relaxation. In particular, a set of new influence functionals will be presented for nonlinear couplings between the solute and solvent [28]. This describes multiphonon processes, which give rise to the relaxation even for the solute vibration with transition frequency where no solvent mode is found. Furthermore, assuming a harmonic oscillator for the solute vibration, too, an exact path integration will be presented for the obtained influence functional. The exact integration corresponds to full inclusion of higher-order perturbations when compared to Fermi's golden rule. Then, the method gives a matrix of contribution from linear and nonlinear couplings and from the first- and higher-order perturbations—that is, the first-order perturbation for the linear coupling, the first-order perturbation for the nonlinear couplings, the higher-order perturbations for the linear coupling, and the higher-order perturbations for the nonlinear couplings. According to this method the multiphonon process, which plays an essential role in the vibrational relaxation for many systems, may be analyzed systematically and quantitatively, presenting a new route to the physics of quantum energy dissipation in the condensed phase.

A discussion is also presented for the path integral centroid molecular dynamics (CMD) method [29, 30]. However, it is numerically hard to evaluate the quantum state of the solute oscillator based upon this CMD approximation. Hence, vibrational energy rather than the vibrational state has been investigated [31]. Comparison between harmonic oscillators bath approximation and CMD approximation is also interesting.

On the other hand, it is very attractive, too, to directly pursue the relaxation process from the vibrational first excited state to the ground state along the time axis, explicitly taking account of the interaction between the solute and solvent. For this kind of analysis, the so-called mixed quantum-classical molecular dynamics calculation [32–34] must be powerful, where time-dependent Schrödinger equation is solved for the quantum degree of freedom of interest while classical equations of motion are solved for the solvent molecules. In the case of a solvated or hydrated electron [33, 35, 36], discrete strong couplings were found, causing great stepwise changes of the probability that the system is found, for example, at an excited state. Furthermore, the potential surface for the solvent degrees of freedom is significantly dependent upon the quantum state occupied by the electron. Then, surface hopping approximation [32, 37–39] may be adequate to describe the dynamics of the system. However, with respect to the present vibrational relaxation, we demonstrate that the molecular vibrational state shows continuous probability change and that the difference in the solute vibrational state little affects the potential energy surface of the solvent molecules. In this rather specific case, mean field approximation [32] must describe the dynamics better than the surface hopping approximation. Here, in order to avoid a logical contradiction, intramolecular degrees of freedom of the solvent molecule are frozen, which should originally be described quantum mechanically. Hence, only the contribution to the relaxation from the solvent translational and rotational motions are taken into account in the present test calculation of this method. Descriptions of the solute and solvent degrees of freedom are summarized in Table I for the methods discussed in this chapter.

Vibrational relaxation of a CN^- ion in the aqueous solution has been attracting much attention as a test system for the various methods stated above. For this diatomic solute, it is not necessary to consider intramolecular energy redistribution because the number of intramolecular degree of freedom is one. Thus, the vibrational relaxation is caused only by the solute–solvent intremolecular energy transfer. Furthermore, transition frequency of CN^- ion in the solution from state $n = 0$ to 1 is about 2080 cm^{-1}, where no solvent mode is found. This means that the multiphonon process is all responsible for the energy exchange between the solute and solvent. For this interesting system, a series of pump-probe experiments have been done by Hochstrasser's group [8, 9], reporting transition frequency dependence of the relaxation time using isotopes of CN^- ion as well as solvent mode dependence of the relaxation time for H_2O and D_2O. Relaxation time as a function of solute concentration can also be read from the data. With this background, computational studies have been done actively for this system [31, 40–43]. Now, it is very interesting to compare the results of these calculations with each other in order to extract the characteristics of the methods.

Population decay curve calculated by the present methods, or measured by the experiment, corresponds to time evolution of quantum-mechanical probability thermally averaged over the initial states of the solvent, i.e. a typical

TABLE I
Description of Solute and Solvent Degrees of Freedom in Various Computational Approaches to Vibrational Energy Relaxation

		Solvent	
		Quantum	Classical
Solute	Quantum	Influence functional theory Centroid MD	Fermi's golden rule Mixed quantum-classical MD
	Classical	—	Classical MD Generalized Langevin equation

many-particle measurement. However, if a measurement for a finite number of solute molecules is done, there must be a certain fluctuation in the observed probability, the magnitude of which depends upon the number of solute molecules. This is the essential behavior of the quantum many-body systems caused by an interference between forward and backward paths in the path integral representation for the time propagation of the wave functions. Star twinkling may be considered as a classical version of this phenomenon. In order to analyze statistics of the fluctuation, higher-order moments of reduced density matrix for the system have been investigated [44]. The higher-order moments together with the first-order moment, which is the ordinary average stated above, gives the distribution via Fourier transformation of the characteristic function. Here, an algebraic description of the influence functional for the higher moments is presented for the system linearly coupled to the harmonic oscillators bath. Then, we can analyze the statistics of the quantum transition probability in the condensed phase. Discussion of the statistics of the diagonal part of density matrix for a simple model system is presented.

After defining Hamiltonians in Section II, we present a brief discussion of traditional approaches to the vibrational relaxation in Section III as an introduction to the review of recent sophisticated methods. In Section IV, fully quantum-mechanical approaches are described based upon path integral techniques—that is, influence functional theory and CMD approximation. A direct simulation of the vibrational state based upon mixed quantum-classical approximation is discussed in Section V, where we demonstrate that mean field approximation is adequate to the case of the vibrational relaxation. In Section VI, we focus our attention on the vibrational relaxation of CN^- ion in the aqueous solution, where the methods are compared with each other. In Section VII, in order to discuss statistics of transition probability fluctuation, a path integral formulation of the higher-order moments of density matrix of the system coupled to the harmonic oscillators bath is presented. A summary for the present status of dynamical approaches to the vibrational relaxation is presented in Section VIII, giving a possible direction of the future development in this field.

II. HAMILTONIAN

In general, the total system is composed of the vibrational degree of freedom of a solute molecule to be studied, its translational and rotational degrees of freedom, and translational, rotational, and vibrational degrees of freedom of the solvent molecules. Distinguished from others, the vibrational mode of the solute molecule under consideration is designated simply as system. All remaining degrees of freedom are called environment or bath. The former is represented by a variable x and the latter by a set of variables $\{q_k\}$. As stated before, x may belong to either a solute molecule in the solution or an adsorbed molecule on the solid. Similarly, $\{q_k\}$ may stand for, in the former case, translational, rotational, and vibrational degrees of freedom of the solvent molecules in the solution and, in the latter case, translational degrees of freedom of ions and atoms which form the matrix. Hereafter, x and $\{q_k\}$ are often simply referred to as solute coordinate and solvent coordinates, respectively.

For example, consider a vibration of solute molecule solved in the solvent. Variable x may be either of a local mode or of a normal mode of the solute molecule. An example of the former is the deviation of intramolecular bond length or bond angle from its equilibrium value. A set of variables $\{q_k\}$ may also be chosen according to the purpose of the study and to the method to be adopted. In one case, $\{q_k\}$ are Cartesian coordinates of atoms which compose the solvent molecules. In another case, they represent Cartesian coordinates of the center of mass of the solvent molecules, some generalized coordinates with respect to the rotation, and normal modes of intramolecular vibration. In contrast to these molecular or local coordinates, solvent motions are often expressed by normal modes delocalized over the solution. A description of the solvent in the harmonic oscillators bath approximation is an example of this type of the representation. In this case, two kinds of definitions are usually adopted. The first one is the normal modes for the quenched structure, and the other is the ones for the instantaneous structure. The former modes all have real number frequencies except for three with respect to the translation of the total system, since the modes are expanded around a local energy minimum. However, for the latter, many modes show imaginary number frequencies indicating that they are not found at the energy minimum. On account of its clear definition, the quenched normal modes are always adopted in the case of glass and solid (i.e., phonon). However, for liquid, the instantaneous normal modes can reproduce the dynamics of the solvent, such as velocity autocorrelation function and force autocorrelation function, better than the quenched normal modes. Thus, although the instantaneous normal modes are not rigorous in a physical sense, they are often used as a set of variables which can describe the solvent motion effectively
[45–47]. This is still the subject of controversy. In any case, here, the term phonon is used, too, for the normal modes of the liquid after the terminology for the solid.

Total Hamiltonian may be written as

$$H(x, \{q_k\}) = H_S(x) + H_B(\{q_k\}) + V_I(x, \{q_k\}) \tag{2.1}$$

where H_S, H_B, and V_I are the Hamiltonian of the system of interest (i.e., solute vibration), the Hamiltonian of the bath (i.e., the solvent), and the interaction between them, respectively. In general, H_S is of the form

$$H_S(x) = -\frac{\hbar^2}{2M}\frac{\partial^2}{\partial x^2} + V_S(x) \tag{2.2}$$

where M is an inertial quantity with respect to the vibration and $V_S(x)$ is the vibrational potential. Two types of the potential function are usually adopted for the vibration. The first one is the harmonic oscillator

$$V_S(x) = \frac{1}{2}M\Omega^2 x^2 \tag{2.3}$$

where Ω represents the transition angular velocity. The second one is the Morse oscillator taking account of the anharmonicity of the system,

$$V_S(x) = D\{\exp(-ax) - 1\}^2 \tag{2.4}$$

where D is the dissociation energy and a is the parameter of curvature of the potential function. For these functions, algebraic expressions for the eigenvalue and the eigenfunction as well as various integrals of the eigenfunction are all presented in mathematical tables.

On the other hand, H_B may be expressed generally as

$$H_B(\{q_k\}) = -\sum_{k=1}^{N} \frac{\hbar^2}{2m_k} \frac{\partial^2}{\partial q_k^2} + V_B(\{q_k\}) \tag{2.5}$$

Mathematical form of the bath potential $V_B(\{q_k\})$ depends upon the definition of coordinates $\{q_k\}$. In many cases, V_B is described by pairwise additive two-body intermolecular interaction such as Lennard-Jones potential between atoms plus Coulombic interaction between atomic partial charges. Here, the number of motional degrees of freedom of the solvent is assumed to be N. If we approximate harmonic oscillators bath for the solvent, we obtain

$$V_B(\{q_k\}) = \sum_{k=1}^{N} \frac{1}{2} m_k \omega_k^2 q_k^2 \tag{2.6}$$

where m_k and ω_k are the inertial quantity and transition angular velocity with respect to q_k, respectively.

With regard to the solute–solvent interaction potential $V_I(x, \{q_k\})$, a few calculation schemes have been proposed. One is a simple summation of atom–atom intermolecular potentials such as LJ and Coulombic potentials which are used in the ordinary classical MD calculations. Since the calculation may be executed very easily and its load on computer is satisfactorily light, this assumption has been employed widely in the theoretical studies of the vibrational relaxation. However, the parameters in the function have not been determined originally to describe the effect of the solvent molecules on the solute vibration. Thus, the potential evaluated by this method may lead to an erroneous result. In order to avoid this, *ab initio* calculations have sometimes been done to determine the vibrational potential in the solution, where instantaneous solvation structure was taken into account to a certain extent [48, 49]. This gives better potential than the primitive calculation above. However, since the calculation requires huge cpu time, the method cannot be used for statistical mechanical calculation of the vibrational relaxation. In a few calculations, inclusion of many-body effect on the vibrational potential has been attempted, too, in more simplified way in order to reproduce transition frequency shift in the solution [50, 51]. However, in most cases, V_I has been evaluated by the method of simple summation of atom–atom interactions stated above at the cost of accuracy of the calculation.

III. TRADITIONAL APPROACHES TO VIBRATIONAL ENERGY RELAXATION RATE

In this section we give a brief discussion about traditional methods of theoretical investigation for vibrational energy relaxation of solute molecules in the solution and adsorbed molecules on the solid. These are classical MD calculation, generalized Langevin equation, and Fermi's golden rule with classical force autocorrelation function. The techniques have been widely applied to the relaxation of various systems. However, limitations of the methods have been pointed out through the application calculations. As an introduction of this review, a description of the methods is presented in order to clarify what is the problem and what kind of new method is required to solve it.

A. Classical Molecular Dynamics

The most primitive procedure to simulate molecular vibration in the solution is an ordinary classical MD calculation based upon flexible model for the molecule. In this rather special case, a translational degree of freedom is provided to each relevant atom (e.g., $\mathbf{r}_i$ for ith atom) instead of an explicit description of the vibration by x. That is, a set of position vectors $\{\mathbf{r}_i\}$ for all atoms are chosen as system variables rather than x and $\{q_k\}$. Then, Newton's equation of motion

$$\mathbf{F}_i = m_i \ddot{\mathbf{r}}_i \tag{3.1}$$

is solved for each atom, where $\mathbf{F}_i$ is the force on atom i and m_i is the atomic mass. The force may be evaluated, for example, by the summation of intramolecular vibrational potential V_S for the oscillator such as harmonic oscillator and Morse oscillator given by Eqs. (2.3) and (2.4), respectively, and intermolecular atom–atom potential V_I such as LJ and Coulombic potentials. As stated before, this is a rough approximation where many-body effects on the intramolecular chemical bond are all neglected. It is this simplification that enables us to execute long-time MD calculations for the vibrational relaxation.

A concept of vibrational state or transition between states has vanished already in the classical approximation. So, large energy is supplied to the solute at $t = 0$ as an initial condition in order to mimic the vibrational excited state classically. One way to do this, for example, for stretching is to set the atom–atom distance far from its equilibrium value, giving large potential energy. Large velocity of the atom with respect to the stretching motion can simulate the excited state, too. The excess energy thus provided to the solute dissipates into solvent degrees of freedom, and the internal energy converges to its equilibrium value according to the equation of motion. This energy relaxation may be monitored by

$$C_E(t) = \frac{\overline{E(t)} - \overline{E(\infty)}}{\overline{E(0)} - \overline{E(\infty)}} \tag{3.2}$$

where $E(t)$ is the summation of potential energy and kinetic energy of the solute atoms with respect to the vibration at $t = t$ averaged over initial conditions of the solvent. Clearly, $\overline{E(\infty)} = kT$. The function starts with $C_E(0) = 1$ and converges to 0 at $t = \infty$, giving a time constant of the relaxation.

Vibrational energy relaxation has been investigated based upon this kind of classical MD calculation for molecules composed of heavy atoms such as I_2 [52] and I_2^- ion [53] and for high-frequency N_2 at very high temperature [54]. For example, the frequency of an I_2^- ion is about 115 cm^{-1}. Considering the thermal energy at 300 K, $kT/\hbar \approx 200$ cm^{-1}, the classical approximation must be good. However, the relaxation time, 0.7 ps, thus estimated does not agree well with the experimental value, 4 ps [55]. It is not clear which causes this discrepancy, the classical approximation or poor potential. In spite of the poor reproducibility of the resultant relaxation time, qualitative discussions may be possible. In fact, for example, the contribution from Coulombic interaction to the relaxation has been investigated comparing the results for the solvents with and without certain partial charges on the atoms.

Classical MD calculations have also been applied to the vibration of molecules such as methylchloride [56, 57] and azulene [58], whose transition energy is much greater than the thermal one. Careful discussion is required for the results of these calculations, since quantum effect is not taken into account.

B. Classical Langevin Equation

Within the classical approximation for the molecular vibration, the generalized Langevin equation with respect to the vibrational degree of freedom x

$$M\ddot{x}(t) = -M\Omega^2 x(t) - \int_0^t d\tau \zeta(t - \tau)\dot{x}(\tau) + F(t) \tag{3.3}$$

must give the same relaxation time as that obtained by the above classical MD calculation. In Eq. (3.3), a harmonic oscillator is assumed for the system [15, 16]. $\zeta(t)$ is the dynamic friction—that is, memory function. According to the fluctuation–dissipation theory of the second kind, it may be related to auto-correlation function of the random force $F(t)$ on x by

$$\zeta(t) = \frac{1}{kT} \langle F(t)F(0) \rangle_{classical} \tag{3.4}$$

where the subscript "classical" is attached in order to stress the difference from the quantum-mechanical correlation function which will appear in the next subsection. Although the equation cannot be solved exactly, an algebraic solution for the energy relaxation function $C_E(t)$ may be obtained by adopting two assumptions; that is, the first one is $\Omega \gg \hat{\zeta}(s)/M$, where $\hat{\zeta}(s)$ is the Laplace transformation of $\zeta(t)$, and the second one is that $\hat{\zeta}(s)$ does not change much around $s = \Omega$ [16]. Then, energy relaxation time τ_E is of the form

$$\tau_E = \frac{1}{2MkT} \int_{-\infty}^{\infty} dt \exp(i\Omega t) \langle F(t)F(0) \rangle_{classical} \tag{3.5}$$

Thus, the value of τ_E may be obtained if we can calculate somehow the auto-correlation function of force with respect to the vibrational degree of freedom x, $\langle F(t)F(0) \rangle_{classical}$. It is interesting to see that essentially the same expression as Eq. (3.5) is obtained from hydrodynamic analysis, too, for the vibrational relaxation [59].

Now, the problem is to evaluate the force autocorrelation function for a given system. The same calculation will appear in the next subsection for the Fermi's golden rule. In general, two methods have been employed for the calculation of the force $F(t)$ with respect to the vibrational degree of freedom of molecule in the condensed phase. According to the first method, an ordinary MD calculation is done for the rigid rotor model, where the vibrational coordinate x is fixed at zero. For example, for stretching mode of a diatomic molecule AB, the bond length r_{AB} is frozen at its equilibrium value throughout the MD calculation. Then, the time evolution of the force $F(t)$ may be obtained by

$$F(t) = \left\{ \frac{m_B}{m_A + m_B} \mathbf{F}_A(t) - \frac{m_A}{m_A + m_B} \mathbf{F}_B(t) \right\} \cdot \frac{\mathbf{r}_{AB}(t)}{r_{AB}} \tag{3.6}$$

along the MD trajectory. An ensemble average of $F(t)F(0)$ is the target auto-correlation function $\langle F(t)F(0) \rangle_{classical}$. The calculations have been done for various systems such as I_2^- [53], iodine molecule [53], methylchloride [56], mercury iodide [60], and bromine [61]. The resultant relaxation times were in good agreement with those obtained by the primitive MD given in Section III.A if the same potential function was adopted. Agreement with the experiment is dependent upon the system or the calculation. In the case of mercury iodide, the calculated relaxation time, $\sim$2 ps, is in satisfactory agreement with the experiment, $\sim$3 ps [62], whereas the agreement is poor for I_2^- as stated above. However, the method presents qualitative analyses for the relaxation. For example, the calculation could reproduce a plateau of the relaxation time for a model oscillator in the supercritical fluid found experimentally near the critical density as a function of the density [63–65].

The other method assumes a harmonic oscillators bath where all the solvent motions are described by a set of normal coordinates [66–68]. The normal mode analysis has been done for both quenched structure and instantaneous structure, the latter being employed in many studies there. The dynamic friction—that is, the Fourier transformation of the force autocorrelation—has been obtained evaluating the couplings between the solute and solvent modes directly. According to this method, contribution of each solvent mode to the relaxation may be analyzed, although exact force autocorrelation function cannot be obtained. In principle, the contribution can be attributed to local modes of the solvent molecules, too, by analyzing the transformation matrix between the local modes and the normal modes.

C. Fermi's Golden Rule With Classical Force Autocorrelation Function

As stated often above, it is impossible to simulate many-body systems fully quantum mechanically. In order to avoid this difficulty, a computationally tractable approximation has been adopted. The method can describe the relaxation time quantitatively, taking account of the quantum effect. In this section, the method based upon Fermi's golden rule with classical force autocorrelation function [18] is briefly discussed, where vibrational degree of freedom is described quantum mechanically while classical approximation is adopted for the solvent degrees of freedom.

Here, interaction potential V_I between the solute and solvent is treated as a perturbation. Then, the first-order time-dependent perturbation theory gives a popular description for the transition rate $k_{i\to f}$ from an initial state i of the system to the final state $f (i \neq f)$,

$$k_{i\to f} = \frac{2\pi}{\hbar}\sum_{\alpha}\sum_{\beta} e^{-\beta E_\alpha}|\langle \alpha, i|V_I|f, \beta\rangle|^2 \delta(E_i - E_f + E_\alpha - E_\beta) \tag{3.7}$$

where α and β represent the initial and final states of the solvent degrees of freedom, respectively, and E_i and E_α, for example, are the energy of the system and the solvent, respectively. Thermal average is taken over the initial states of the bath while the final states are simply summed up, assuming that they are not measured at $t = t$. This is the well-known formula of Fermi's golden rule. Replacing the delta function in Eq. (3.7) by its Fourier transformation, we obtain the following after a certain algebraic rearrangement:

$$\begin{aligned} k_{i\to f} &= \frac{1}{\hbar^2}\int_{-\infty}^{\infty} dt \exp(i\Omega_{if}t)\langle V_{if}^I(t)V_{fi}^I(0)\rangle \\ &= \frac{2}{\hbar^2\{1+\exp(-\beta\hbar\Omega_{if})\}}\int_{-\infty}^{\infty} dt \exp(i\Omega_{if}t)\left\langle \frac{1}{2}[V_{if}^I(t), V_{fi}^I(0)]_+ \right\rangle \end{aligned} \tag{3.8}$$

where $\Omega_{if} = (E_f - E_i)/\hbar$. $V_{if}^I(t) = \exp(iH_B t/\hbar)V_{if}^I \exp(-iH_B t/\hbar)$ is the Heisenberg representation of $V_{if}^I = \langle i|V_I|f\rangle$, $\langle \cdots \rangle$ showing the ensemble average over the initial states of the solvent. $\langle V_{if}^I(t)V_{fi}^I(0)\rangle$ is the quantum-mechanical time correlation function of V_{if}^I in the Heisenberg representation, and its corresponding symmetrized correlation function $\langle \frac{1}{2}[V_{if}^I(t), V_{fi}^I(0)]_+\rangle$ is a real function. Now, the task is to obtain the quantum-mechanical time correlation function of the interaction potential, although the exact solution cannot be obtained even numerically at present. So, one choice to approximate it is to replace the quantum-mechanical symmetrized correlation function by the classical one $\langle V_{if}^I(t)V_{fi}^I(0)\rangle_{classical}$ [18], with the quantum-mechanical description for x being unchanged. Then, the interaction potential may be expanded by the Taylor series with respect to x,

$$V_I = V_I^0 + \left(\frac{\partial V_I}{\partial x}\right)_{x=0} x + \frac{1}{2}\left(\frac{\partial^2 V_I}{\partial x^2}\right)_{x=0} x^2 + \cdots \tag{3.9}$$

If we neglect the higher term than the first, then $V_I = V_I^0 - Fx$, where $F = -(\partial V_I/\partial x)_{x=0}$ is the force at $x = 0$. Thus, we obtain

$$\langle V_{if}^I(t)V_{fi}^I(0)\rangle_{classical} = |\langle i|x|f\rangle|^2 \langle F(t)F(0)\rangle_{classical} \tag{3.10}$$

which implies that the force correlation function $\langle F(t)F(0)\rangle_{classical}$ [e.g., Eq. (3.6)], may be used in place of the interaction correlation function $\langle V_{if}^I(t)V_{fi}^I(0)\rangle_{classical}$. The factor $|\langle i|x|f\rangle|^2$ in the right-hand side of Eq. (3.10) may easily be calculated for a given potential of the system $V_S(x)$. For example, for the harmonic oscillator system, $\langle 1|x|0\rangle = \sqrt{\hbar/2M\Omega}$ for transition fıom the first excited state to the ground state. Thus, the rate may be written finally as

$$k_{1\to 0} = \frac{1}{M\hbar\Omega\{1+\exp(-\beta\hbar\Omega)\}}\int_{-\infty}^{\infty} dt \exp(i\Omega t)\langle F(t)F(0)\rangle_{classical} \tag{3.11}$$

A similar formula can be obtained easily for the Morse oscillator, too. Now, the rate is obtained straightforwardly from the force autocorrelation function which can easily be evaluated by the ordinary MD calculation for the rigid rotor model as shown in the previous section.

Applying this transition rate, together with the assumption of detailed balance such as $k_{1\to 0} = \exp(\beta\hbar\omega)k_{0\to 1}$, to the approximate Bloch–Redfield theory or Pauli's master equation

$$\dot{\rho}_{nn}(t) = \sum_{n' \neq n} \{k_{n'\to n}\rho_{n'n'}(t) - k_{n\to n'}\rho_{nn}(t)\} \tag{3.12}$$

we obtain the time-dependent probability $\rho_{nn}(t)$ that the system is found at state n. This gives experimental relaxation time T_1. Consider a simple case, for example, where the pump pulse excites the solute vibration from $n = 0$ to 1 and the transition energy is much greater than the thermal one; that is, $\hbar\Omega \gg kT$. Then, one choice of the approximation may be a neglect of the transitions except for the one from $n = 1$ to 0. This clearly gives the population decay time $T_1 = 1/k_{1\to 0}$.

Furthermore, if we assume a harmonic oscillator for the solute vibration, the vibrational energy relaxation time τ_E may be formulated [19, 69] as

$$\frac{1}{\tau_E} = \frac{1 - \exp(-\beta\hbar\Omega)}{M\hbar\Omega\{1 + \exp(-\beta\hbar\Omega)\}} \int_{-\infty}^{\infty} dt \exp(i\Omega t)\langle F(t)F(0)\rangle_{classical} \tag{3.13}$$

since only transitions between the neighboring states are permitted. Now, it is interesting to note that the classical limit, $\hbar \to 0$, of this equation coincides with the expression for the fully classical energy relaxation time given by Eq. (3.5) which was derived from the classical Langevin equation.

The calculations have been applied to the molecular vibrations such as methylchloride [69], Si–H stretching on the crystal silicon surface [70], hydrogen bonding model system [71], HDO [72], azide ion [73, 74], and cyanide ion [40], whose transition frequencies are so high, $\hbar\Omega \gg kT$, that the quantum effect cannot be neglected.

However, the method still includes classical approximation for the force autocorrelation function. The approximation is an expedient one to obtain the correlation function indispensable to the calculation. Hence, it is very important to test the validity of the above method to describe the vibrational relaxation of the real system. Although a comparison of the calculated relaxation time with the experimental value might be expected to be an examination for this, it does not work well because the potential function adopted might be invalid. Thus, simple model systems have usually been investigated to test the method.

If we assume a solute harmonic oscillator bilinearly coupled to a solvent harmonic oscillators bath, an algebraic analysis becomes possible for the relaxation time at the Fermi's golden rule level. According to this approximation, Bader and Berne [19] showed a quantum correction for the classical force correlation function which presents the correct vibrational relaxation time. This is clearly the relaxation arising from the single phonon process. Thus, the correction may work for the solute whose transition frequency is covered by the solvent density of states. However, if the transition frequency of the solute is higher or lower than that of the solvent modes, the multiphonon process must become predominant in the energy transfer. In this case, the correction itself causes a large error where even an order of the calculated relaxation time is not reliable [20, 22, 25]. An origin of this error will be shown later by an analysis based upon influence functional theory for the nonlinearly coupled system in the next section. In order to improve this errorneous calculation effectively, various empirical correction coefficients have been tested for particular systems [20–25]. Here, we must be careful that this kind of *ad hoc* correction does not necessarily predict the relaxation time universally for other new systems.

A quantum-mechanical description of the whole system is the straightforward approach to the relaxation by multiphonon processes. One way to do this is to employ a harmonic oscillators bath approximation for the solvent degrees of freedom [see Eq. (2.6)]. An algebraic transition rate at the Fermi's golden rule level was presented not only for the bilinear coupling but also for nonlinear couplings [75–83]. A formal solution of the higher-order terms of the perturbation was also given to describe the transition rate of the system linearly coupled to the bath [83]. However, exact solution which includes the higher-order perturbations for the system nonlinearly coupled to the bath has not yet been given by this technique. Several semiclassical approximations have also been formulated for the vibrational relaxation at the level of Fermi's golden rule such as cummulant expansion of time evolution operator [84], semiclassical expansion of the Pauli's master equation [85], and semiclassical surface hopping [86, 87]. In the next section, however, a route to an exact solution of the vibrational relaxation time for the fully quantum system will be presented based upon influence functional theory.

IV. PATH INTEGRAL APPROACH

In this section, two recently developed methods are discussed. The methods describe the whole system fully quantum mechanically, take account of linear and nonlinear couplings, and consider full perturbation terms. One is the path integral influence functional theory based upon harmonic oscillators bath approximation. The other is CMD calculation which gives an approximate canonical correlation function. In particular, by the former method, quantum states may be investigated explicitly and dissipation of excited energy to the solvent modes can be analysed in detail.

We start with the well-known path integral representation of propagator by Feynman [27]. For one-dimensional system, for example,

$$\begin{aligned} K(x_f, t; x_i, 0) &= \int_{x_i}^{x_f} \mathcal{D}x(t) \exp\left(\frac{i}{\hbar}\int_0^t L dt\right) \\ &= \int_{x_i}^{x_f} \mathcal{D}x(t) \exp\left(\frac{i}{\hbar} S[x(t)]\right) \end{aligned} \tag{4.1}$$

where x is the coordinate of the system of interest, subscripts i and f represent initial and final states, respectively, L is the classical Lagrangian, and $S[x(t)]$ represents the action which is a functional of the path $x(t)$. $\int \mathcal{D}x(t) \cdots$ is the path integral from x_i at $t = 0$ to x_f at $t = t$—that is, summation of all possible paths between x_i and x_f.

In the influence functional theory [26, 27], after time evolution of density matrix for the total system was written using the propagator presented by Eq. (4.1), the solvent degrees of freedom are integrated out to obtain a reduced density matrix for the solute degree of freedom to be studied. Then, effects of the solvent may be expressed by a functional which appears in the integrant of path integral—that is, the influence functional. On the other hand, in the CMD approximation [29, 30], after the discretization of the time integral in the action, the time is replaced by the imaginary time, leading to the classical simulation for necklaces consisting of beads. A particular force is assumed for the center of mass of the necklace, and the trajectory of the centroid is traced by the simulation. Then, the time correlation function with respect to the function of centroid variables is an approximation of the quantum-mechanical canonical correlation function.

A. Influence Functional Theory

First, in this subsection a general theory of influence functional is presented for the conventional system bilinearly coupled to the environment [26, 27, 88–90]. Thereafter, an extension of the theory to nonlinear couplings [28] is given. Using the influence functional obtained there, we show an approximate expression for the survival probability of the excited state, leading to a description for the population decay rate. The exact path integration is presented, too. Finally, a computationally tractable prescription is given to evaluate the time constant of the relaxation, which is a combination of the influence functional theory with conventional MD calculation.

1. *General Theory*

In this theory, a harmonic oscillators bath approximation shown in Eq. (2.6) is usually adopted in order to integrate out the solvent degrees of freedom from the density matrix. Furthermore, interaction potential V_I, which is originally presented by LJ potential and Coulombic interaction between atoms, may be expanded into a power series with respect to the bath coordinates.

$$\begin{aligned} V_I(x, \{q_k\}) &= \sum_{k=1}^{N} f_k(x) q_k + \sum_{k=1}^{N}\sum_{l=1}^{N} g_{kl}(x) q_k q_l \\ &\quad + \sum_{k=1}^{N}\sum_{l=1}^{N}\sum_{m=1}^{N} h_{klm}(x) q_k q_l q_m + \cdots \end{aligned} \tag{4.2}$$

where $f_k(x)$, $g_{kl}(x)$, and $h_{klm}(x)$ are functions of x. Terminating the expansion in Eq. (4.2) at the first order and, further, taking account of only the first term of the Taylor expansion of $f_k(x)$, V_I may be written, as an approximation, by

$$V_I(x, \{q_k\}) = \sum_{k=1}^{N} C_k x q_k \tag{4.3}$$

This corresponds to the bilinear coupling system for which the influence functional theory was first demonstrated by Feynman and Vernon [26].

For simplicity, consider a harmonic oscillator bath of $N=1$. Assuming a factorization of density matrix between the system and bath at $t=0$, the reduced density matrix for the system at $t=t$ may be written as

$$\begin{aligned} \rho(x_f, x'_f; t) &= \langle \psi(x_f; t) \psi^*(x'_f; t) \rangle \\ &= \int_{-\infty}^{\infty}\int_{-\infty}^{\infty} dx_i dx'_i \int_{x_i}^{x_f} \mathcal{D}x(t) \int_{x'_i}^{x'_f} \mathcal{D}x'(t) \exp\left[\frac{i}{\hbar}\{S_S[x(t)] - S_S[x'(t)]\}\right] \\ &\quad \times \int_{-\infty}^{\infty}\int_{-\infty}^{\infty} dq_f dq'_f \sum_n \phi_n^*(q_f) \phi_n(q'_f) \\ &\quad \times \int_{-\infty}^{\infty}\int_{-\infty}^{\infty} dq_i dq'_i \left\{\sum_\ell \exp(-\beta E_\ell)\right\}^{-1} \sum_m \exp(-\beta E_m) \phi_m(q_i) \phi_m^*(q'_i) \\ &\quad \times \int_{q_i}^{q_f} \mathcal{D}q(t) \int_{q'_i}^{q'_f} \mathcal{D}q'(t) \exp\left[\frac{i}{\hbar}\{S_B[q(t)] + S_I[x(t), q(t)] \right. \\ &\qquad \left. - S_B[q'(t)] - S_I[x'(t), q'(t)]\}\right] \\ &\quad \times \rho(x_i, x'_i; 0) \end{aligned} \tag{4.4}$$

where S_S, S_B, and S_I are actions for H_S, H_B, and V_I, respectively, the wave function of the system is represented by ψ, and the eigenfunction for the solvent state of m is represented by ϕ_m. Here, the angle brackets $\langle \cdots \rangle$ stand for a thermal average over the initial bath states, while the final states are all simply summed up assuming that the measurement is not done for the bath at $t = t$. The integrations may be carried out exactly with respect to all bath variables for the bilinearly coupling system, giving

$$\rho(x_f, x_f'; t) = \int_{-\infty}^{\infty} dx_i \int_{-\infty}^{\infty} dx_i' \int_{x_i}^{x_f} \mathcal{D}x(t) \int_{x_i'}^{x_f'} \mathcal{D}x'(t) \times \exp\left[\frac{i}{\hbar}\{S_S[x(t)] - S_S[x'(t)]\}\right] \mathcal{F}[x(t), x'(t)] \rho(x_i, x_i'; 0) \tag{4.5}$$

Here, $\mathcal{F}$ is the influence functional

$$\mathcal{F}[x(t), x'(t)] = \exp\left[-\frac{1}{\hbar}\int_0^t ds_1 \int_0^{s_1} ds_2 \{x(s_1) - x'(s_1)\} \{G(s_1 - s_2)x(s_2) - G^*(s_1 - s_2)x'(s_2)\}\right] \tag{4.6}$$

where

$$G(s) = \int_{-\infty}^{\infty} d\omega \mathcal{J}(\omega)\{z(\omega)\cos\omega s - i\sin\omega s\} \tag{4.7}$$

$$Z(\omega) = \cot\mathrm{h}\left(\frac{\beta\hbar\omega}{2}\right) \tag{4.8}$$

$$\mathcal{J}(\omega) = \frac{C^2}{2m\omega} \tag{4.9}$$

Extension of the analysis to the N oscillators bath is straightforward, resulting in the simple replacement of Eq. (4.9) by

$$\mathcal{J}(\omega) = \sum_{k=1}^{N} \frac{C_k^2}{2m_k\omega_k}\delta(\omega - \omega_k) \tag{4.10}$$

Here, $\mathcal{J}(\omega)$ is the spectral density, representing single-phonon processes.

2. *Perturbative Influence Functional for Nonlinear Couplings*

The purpose of the present subsection is to obtain an expression for the influence functional $\mathcal{F}_{nonlin}[x, x']$ for the system nonlinearly coupled to the harmonic bath as described in Eqs. (2.6) and (4.2), which can describe the relaxation by multiphonon processes such as two-phonon and three-phonon processes [28]. In the present study, it is truncated at the terms higher than the cubic one just for simplicity. Of course, higher-order terms such as the quartic one may also be included in the analysis leading to the similar formula described below. Since exact solution of $\mathcal{F}_{nonlin}[x, x']$ cannot be obtained, we use the perturbation theory. Here, the influence action $A[x, x']$ defined by

$$\mathcal{F}_{nonlin}[x, x'] = \exp\left\{\frac{i}{\hbar} A[x, x']\right\} \tag{4.11}$$

is evaluated instead of $\mathcal{F}_{nonlin}[x, x']$ just for convenience. In order to obtain the perturbative influence action, we make usage of an identity for arbitrary functional $Q[\{q_k\}; \{q'_k\}]$,

$$\begin{aligned}
\Big\langle Q[\{q_k\}; \{q'_k\}] \Big\rangle_0 &\equiv \prod_{k=1}^{N} \mathcal{N}_k \int_{-\infty}^{+\infty} dq_{kf} \int_{-\infty}^{+\infty} dq_{ki} \int_{-\infty}^{+\infty} dq'_{ki} \rho(\{q_{ki}\}; \{q'_{ki}\}) \\
&\quad \times \int_{q_{ki}}^{q_{kf}} \mathcal{D}q_k \int_{q'_{ki}}^{q_{kf}} \mathcal{D}q'_k \exp\left[\frac{i}{\hbar}\left\{S_b[\{q_k\}] - S_b[\{q'_k\}]\right\}\right] \\
&\quad \times Q[\{q_k\}; \{q'_k\}] \\
&= Q\left[\left\{\frac{\hbar}{i}\frac{\delta}{\delta J_k}\right\}; \left\{-\frac{\hbar}{i}\frac{\delta}{\delta J'_k}\right\}\right] \mathcal{G}[\{J_k\}; \{J'_k\}]\Big|_{\{J_k\}=\{J'_k\}=0}
\end{aligned} \tag{4.12}$$

where the generating functional $\mathcal{G}$ is of the form

$$\begin{aligned}
\mathcal{G}[\{J_k\}; \{J'_k\}] &= \left\langle \exp\left[\frac{i}{\hbar}\int_0^t ds \sum_{k=1}^{N} \left\{J_k(s) q_k(s) - J'_k(s) q'_k(s)\right\}\right]\right\rangle_0 \\
&= \exp\left[-\frac{1}{\hbar}\int_0^t ds_1 \int_0^{s_1} ds_2 \sum_{k=1}^{N} \{J_k(s_1) - J'_k(s_1)\}\right. \\
&\quad \left. \times \{\alpha_k(s_1 - s_2) J_k(s_2) - \alpha_k^*(s_1 - s_2) J'_k(s_2)\}\right]
\end{aligned} \tag{4.13}$$

and normalization constant $\mathcal{N}_k$ is chosen such that $\langle 1 \rangle_0 = 1$. If the bath is initially at temperature T, the initial bath density matrix may be expressed by

$$\rho(\{q_{ki}\};\{q'_{ki}\}) = \left\{ \frac{m_k \omega_k}{2\pi\hbar \sinh(\beta\hbar\omega_k)} \right\}^{1/2} \times \exp\left[-\frac{m_k \omega_k}{2\hbar \sinh(\beta\hbar\omega_k)} \times \left\{ (q_{ki}^2 + q'^2_{ki}) \cosh(\beta\hbar\omega_k) - 2q_{ki}\, q'_{ki} \right\} \right] \tag{4.14}$$

and $\alpha_k(\tau)$ may be expressed by

$$\alpha_k(\tau) = \frac{z(\omega_k)}{2m_k\omega_k} \cos(\omega_k \tau) - \frac{i}{2m_k\omega_k} \sin(\omega_k \tau) \tag{4.15}$$

Angular brackets in the above equations denote the average with respect to the bath oscillators uncoupled to the system. We have introduced an external source J_k for each bath coordinate, which is the extension of the technique used in Ref. 91. If $J_k(x) = C_k x(s)$, the functional $\mathcal{G}[\{J_k\};\{J'_k\}]$ is clearly the same as the influence functional for the linear coupling system [26, 27, 88–91]. Now, the second-order perturbative influence phase for the nonlinearly interacting system may be described by the cumulant expansion

$$\begin{aligned} \frac{i}{\hbar} A[x, x'] &= \ln \left\langle \exp\left[\frac{i}{\hbar} \left\{ S_i[x, \{q_k\}] - S_i[x', \{q'_k\}] \right\} \right] \right\rangle_0 \\ &\approx \frac{i}{\hbar} \langle S_i[x, \{q_k\}] - S_i[x', \{q'_k\}] \rangle_0 \\ &\quad + \frac{1}{2} \left(\frac{i}{\hbar} \right)^2 \langle (S_i[x, \{q_k\}] - S_i[x', \{q'_k\}])^2 \rangle_0 \\ &\quad - \frac{1}{2} \left(\frac{i}{\hbar} \right)^2 (\langle S_i[x, \{q_k\}] - S_i[x', \{q'_k\}] \rangle_0)^2 \end{aligned} \tag{4.16}$$

Using Eqs. (4.12) and (4.13), calculation of each element in Eq. (4.16) gives

$$
\begin{aligned}
\frac{i}{\hbar}A[x,x'] = \frac{i}{\hbar}\sum_{k=1}^{N}\int_0^t ds\{-\delta V_k(x(s)) + \delta V_k(x'(s))\} \\
-\frac{1}{\hbar}\sum_{k=1}^{N}\int_0^t ds_1\int_0^{s_1} ds_2\{f_k(x(s_1)) - f_k(x'(s_1))\} \\
\times\{\alpha_k(s_1-s_2)f_k(x(s_2)) - \alpha_k^*(s_1-s_2)f_k(x'(s_2))\} \\
-\frac{1}{\hbar}\sum_{k=1}^{N}\sum_{l=1}^{N}\int_0^t ds_1\int_0^{s_1} ds_2\ \{g_{kl}(x(s_1)) - g_{kl}(x'(s_1))\} \\
\times\{\beta_{kl}(s_1-s_2)g_{kl}(x(s_2)) - \beta_{kl}^*(s_1-s_2)g_{kl}(x'(s_2))\} \\
-\frac{1}{\hbar}\sum_{k=1}^{N}\sum_{l=1}^{N}\sum_{m=1}^{N}\int_0^t ds_1\int_0^{s_1} ds_2\ \{h_{klm}(x(s_1)) - h_{klm}(x'(s_1))\} \\
\times\{\gamma_{klm}(s_1-s_2)h_{klm}(x(s_2)) - \gamma_{klm}^*(s_1-s_2)h_{klm}(x'(s_2))\} \\
-\frac{1}{\hbar}\sum_{k=1}^{N}\sum_{l=1}^{N}\sum_{m=1}^{N}\int_0^t ds_1\int_0^{s_1} ds_2\ \{h_{kll}(x(s_1)) - h_{kll}(x'(s_1))\} \\
\times\{\delta_{klm}(s_1-s_2)h_{kmm}(x(s_2)) - \delta_{klm}^*(s_1-s_2)h_{kmm}(x'(s_2))\} \\
-\frac{1}{\hbar}\sum_{k=1}^{N}\sum_{l=1}^{N}\int_0^t ds_1\int_0^{s_1} ds_2\{f_k(x(s_1)) - f_k(x'(s_1))\} \\
\times\{\varepsilon_{kl}(s_1-s_2)h_{kll}(x(s_2)) - \varepsilon_{kl}^*(s_1-s_2)h_{kll}(x'(s_2))\} \\
-\frac{1}{\hbar}\sum_{k=1}^{N}\sum_{l=1}^{N}\int_0^t ds_1\int_0^{s_1} ds_2\ \{h_{kll}(x(s_1)) - h_{kll}(x'(s_1))\} \\
\times\{\varepsilon_{kl}(s_1-s_2)f_k(x(s_2)) - \varepsilon_{kl}^*(s_1-s_2)f_k(x'(s_2))\}
\end{aligned}
\tag{4.17}
$$

where

$$
\begin{aligned}
\delta V_k(x(\tau)) &= \hbar g_{kk}(x(\tau))\alpha_k(0) \\
\beta_{kl}(\tau) &= 2\hbar\alpha_k(\tau)\alpha_l(\tau) \\
\gamma_{klm}(\tau) &= 6\hbar^2\alpha_k(\tau)\alpha_l(\tau)\alpha_m(\tau) \\
\delta_{klm}(\tau) &= 9\hbar^2\alpha_k(\tau)\alpha_l(0)\alpha_m(0)
\end{aligned}
$$

and

$$
\varepsilon_{kl}(\tau) = 3\hbar\alpha_k(\tau)\alpha_l(0) \tag{4.18}
$$

These are the general form of the perturbative influence phase to be obtained for the nonlinearly coupling system.

In order to understand these results, Feynman diagrams are helpful. In the present case, the vertex functions $f_k(x)$, $g_{kl}(x)$, and $h_{klm}(x)$ are of the form shown in Fig. 1. Each line represents the propagator $\alpha_k(\tau)$ or $\alpha_k^*(\tau)$. Although these propagators should be described by different kinds of lines, they are not distinguished from each other, since the purpose here is just to find out nonzero contributions to the path integral calculation. In Feynman diagram language, the average with respect to the bath coordinates in Eq. (4.12) is represented by sum of diagrams in which every line is terminated by a vertex—that is, "closed diagrams." This method is equivalent to the closed-time path formalism by Schwinger [92] and Keldish [93]. Each term in Eq. (4.17) may be expressed by the closed diagram presented in Fig. 2. In general, the nth cumulant has n vertices. The first-order cumulant has one vertex, and the second-order cumulant has two vertices. Generally, total number of lines from all of the vertices must be an even number to form a closed diagram. Thus, in the case of one-vertex diagram, only even coupling terms (quadratic, quartic, $\cdots$) with respect to the phonon coordinates survive in the potential renormalization term; the first term of Eq. (4.17) comes from the quadratic coupling term in Eq. (4.2). In the second-order cumulant, cross sections between the even (quadratic, quartic, $\cdots$) couplings and odd (linear, cubic, $\cdots$) couplings are not found, since the number of lines from the two vertices is odd ($=$even $+$ odd). Only even–even and odd–odd couplings survive as shown in Fig. 2. Thus, the second term of the right-hand side of Eq. (4.17) is from the linear couplings standing for single-phonon process, the third term is from the quadratic couplings representing two-phonon process, the fourth and fifth terms come from the cubic couplings (i.e., three-phonon process), and both sixth and seventh terms are cross couplings from the linear and cubic couplings.

The second term in Eq. (4.17) may be interpreted as the total influence of N independent oscillators with frequencies ω_k linearly coupled to the system. Now,

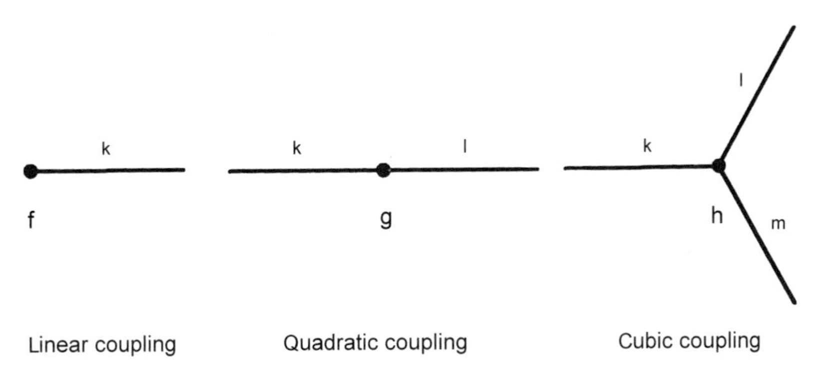

Figure 1. Vertex functions for linear, quadratic, and cubic couplings.

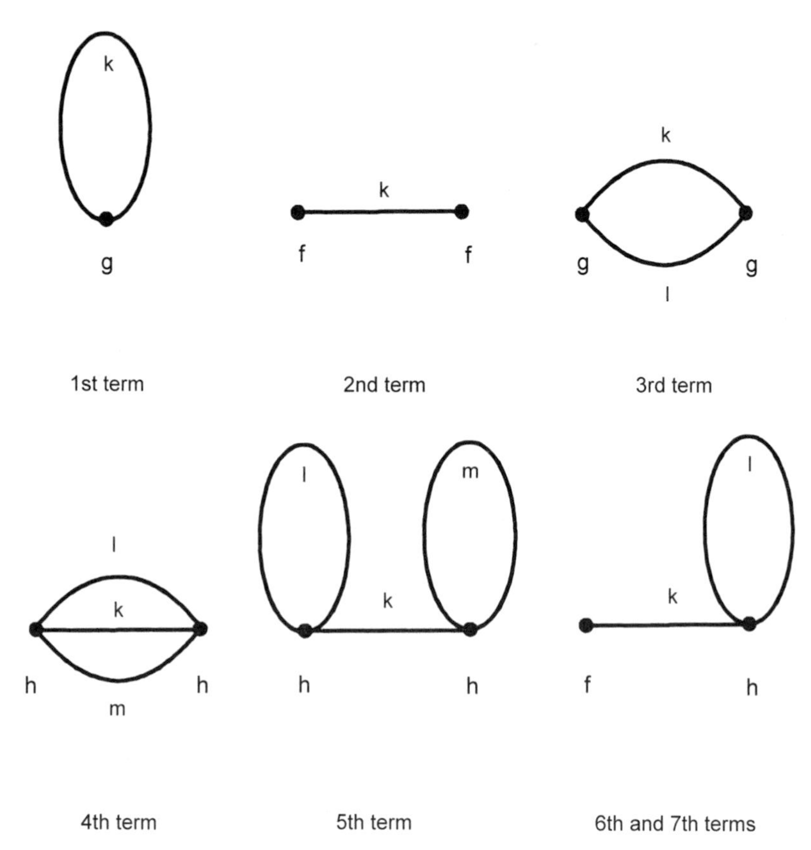

Figure 2. Diagrams representing each term in the second-order cumulant expansion of the influence action. k stands for $\alpha_k(\tau)$ or $\alpha_k^*(\tau)$, where τ denotes the time difference between the line ends.

it is interesting to find that the third term is formally identical to the influence functional for $2N^2$ independent oscillators $\omega_k \pm \omega_l$ linearly coupled to the system. This result is due to the truncation of cumulant expansion of the influence action in Eq. (4.16) at the second order. In the same way, the fourth term may also be regarded as the linear coupling of $4N^3$ independent oscillators with frequencies $\omega_k \pm \omega_l \pm \omega_m$. Similarly, the last three terms in Eq. (4.17) may apparently be assigned to the linear coupling. Thus, in the second-order perturbation theory, the nonlinear couplings formally result in the same expression as the linear coupling influence functional except for the potential renormalization term which may be absorbed into $V_S(x)$.

Now, we rewrite the coupling functions of $f_k(x)$, $g_{kl}(x)$, and $h_{klm}(x)$ in the form of power series,

$$f_k(x) = \sum_w C_k^{(w)} x^w, \qquad g_{kl}(x) = \sum_w C_{kl}^{(w)} x^w, \qquad h_{klm}(x) = \sum_w C_{klm}^{(w)} x^w \tag{4.19}$$

In order to make a physical meaning clearer, let us define the property of the environment, "spectral density", which characterizes the influence on the dynamics of the system, expressed by the strength of the interaction between the system and the bath oscillators. Spectral densities for the linear coupling $\mathcal{J}_{\alpha[1]}^{(wv)}(\omega)$, quadratic coupling $\mathcal{J}_{\beta[1]}^{(wv)}(\omega)$, and $\mathcal{J}_{\beta[2]}^{(wv)}(\omega)$, $\cdots$, are defined by

$$\mathcal{J}_{\alpha[1]}^{(wv)}(\omega) = \sum_{k=1}^{N} \frac{C_k^{(w)} C_k^{(v)}}{2m_k\omega_k} \delta(\omega - \omega_k)$$

$$\mathcal{J}_{\beta[1]}^{(wv)}(\omega) = \sum_{k=1}^{N}\sum_{l=1}^{N} \frac{\hbar C_{kl}^{(w)} C_{kl}^{(v)} (z(\omega_k) + z(\omega_l))}{(2m_k\omega_k)(2m_l\omega_l)} \delta(\omega - \omega_k - \omega_l)$$

$$\mathcal{J}_{\beta[2]}^{(wv)}(\omega) = \sum_{k=1}^{N}\sum_{l=1}^{N} \frac{\hbar C_{kl}^{(w)} C_{kl}^{(v)} (-z(\omega_k) + z(\omega_l))}{(2m_k\omega_k)(2m_l\omega_l)} \delta(\omega - \omega_k + \omega_l)$$

$$\mathcal{J}_{\gamma[1]}^{(wv)}(\omega) = \sum_{k=1}^{N}\sum_{l=1}^{N}\sum_{m=1}^{N} \frac{3\hbar^2 C_{klm}^{(w)} C_{klm}^{(v)} (z(\omega_k)z(\omega_l) + z(\omega_l)z(\omega_m) + z(\omega_m)z(\omega_k) + 1)}{2(2m_k\omega_k)(2m_l\omega_l)(2m_m\omega_m)} \times \delta(\omega - \omega_k - \omega_l - \omega_m)$$

$$\mathcal{J}_{\gamma[2]}^{(wv)}(\omega) = \sum_{k=1}^{N}\sum_{l=1}^{N}\sum_{m=1}^{N} \frac{3\hbar^2 C_{klm}^{(w)} C_{klm}^{(v)} (-z(\omega_k)z(\omega_l) + z(\omega_l)z(\omega_m) + z(\omega_m)z(\omega_k) - 1)}{(2m_k\omega_k)(2m_l\omega_l)(2m_m\omega_m)} \times \delta(\omega - \omega_k - \omega_l + \omega_m)$$

$$\mathcal{J}_{\gamma[3]}^{(wv)}(\omega) = \sum_{k=1}^{N}\sum_{l=1}^{N}\sum_{m=1}^{N} \frac{3\hbar^2 C_{klm}^{(w)} C_{klm}^{(v)} (-z(\omega_k)z(\omega_l) + z(\omega_l)z(\omega_m) - z(\omega_m)z(\omega_k) + 1)}{2(2m_k\omega_k)(2m_l\omega_l)(2m_m\omega_m)} \times \delta(\omega - \omega_k + \omega_l + \omega_m)$$

$$\mathcal{J}_{\delta[1]}^{(wv)}(\omega) = \sum_{k=1}^{N} \frac{1}{2m_k\omega_k} \delta(\omega - \omega_k) \sum_{l=1}^{N}\sum_{m=1}^{N} \frac{9\hbar^2 C_{kll}^{(w)} C_{kmm}^{(v)} z(\omega_l) z(\omega_m)}{(2m_l\omega_l)(2m_m\omega_m)}$$

and

$$\mathcal{J}_{\varepsilon[1]}^{(wv)}(\omega) = \sum_{k=1}^{N} \frac{C_k^{(w)}}{2m_k\omega_k} \delta(\omega - \omega_k) \sum_{l=1}^{N} \frac{3\hbar C_{kll}^{(v)} z(\omega_l)}{2m_l\omega_l} \tag{4.20}$$

It is convenient, too, to introduce

$$\alpha^{(wv)}(\tau) = \int_{-\infty}^{\infty} d\omega \sum_{\kappa=1}^{1} \mathcal{J}_{\alpha[\kappa]}^{(wv)}(\omega)\left[z(\omega)\cos(\omega\tau) - i\sin(\omega\tau)\right]$$

$$\beta^{(wv)}(\tau) = \int_{-\infty}^{\infty} d\omega \sum_{\kappa=1}^{2} \mathcal{J}_{\beta[\kappa]}^{(wv)}(\omega)\left[z(\omega)\cos(\omega\tau) - i\sin(\omega\tau)\right]$$

$$\gamma^{(wv)}(\tau) = \int_{-\infty}^{\infty} d\omega \sum_{\kappa=1}^{3} \mathcal{J}_{\gamma[\kappa]}^{(wv)}(\omega)\left[z(\omega)\cos(\omega\tau) - i\sin(\omega\tau)\right]$$

$$\delta^{(wv)}(\tau) = \int_{-\infty}^{\infty} d\omega \sum_{\kappa=1}^{1} \mathcal{J}_{\delta[\kappa]}^{(wv)}(\omega)\left[z(\omega)\cos(\omega\tau) - i\sin(\omega\tau)\right]$$

and

$$\varepsilon^{(wv)}(\tau) = \int_{-\infty}^{\infty} d\omega \sum_{\kappa=1}^{1} \mathcal{J}_{\varepsilon[\kappa]}^{(wv)}(\omega)\left[z(\omega)\cos(\omega\tau) - i\sin(\omega\tau)\right] \tag{4.21}$$

Using these abbreviations, we obtain

$$\begin{aligned}
\frac{i}{\hbar}A[x,x'] = &-\frac{i}{\hbar}\sum_{w}\int_0^t ds\; G_1^{(w)}\left\{x(s)^w - x'(s)^w\right\} \\
&-\frac{1}{\hbar}\sum_{w,v}\int_0^t ds_1 \int_0^{s_1} ds_2\; \left\{x(s_1)^w - x'(s_1)^w\right\} \\
&\times\left\{G_2^{(wv)}(s_1-s_2)x(s_2)^v - G_2^{(wv)*}(s_1-s_2)x'(s_2)^v\right\}
\end{aligned} \tag{4.22}$$

where

$$G_1^{(w)} = \sum_{k=1}^{N} \frac{\hbar C_{kk}^{(w)} z(\omega_k)}{2m_k\omega_k} \tag{4.23}$$

and

$$\begin{aligned}
G_2^{(wv)}(s_1-s_2) = &\ \alpha^{(wv)}(s_1-s_2) + \beta^{(wv)}(s_1-s_2) + \gamma^{(wv)}(s_1-s_2) \\
&+ \delta^{(wv)}(s_1-s_2) + 2\varepsilon^{(wv)}(s_1-s_2)
\end{aligned} \tag{4.24}$$

When $f_k(x)$, $g_{kl}(x)$, and $h_{klm}(x)$ are all linear in x [see also Eq. (4.31) below], $\mathrm{Re}[\hbar G_2^{(11)}(\tau)]$ corresponds to the quantum force–force time correlation function, where $\mathrm{Re}[\hbar\alpha^{(11)}(\tau)]$, $\mathrm{Re}[\hbar\beta^{(11)}(\tau)], \cdots$ are the contributions from the component couplings. In the same manner, $z(\omega) \times \mathcal{J}^{(11)}(\omega)$'s in Eqs. (4.20) and (4.21) are the Fourier transform of the component correlation functions. At high temperature, the linear coupling contribution, $\alpha^{wv}(\tau)$, may be obtained based upon an approximation for $z(\omega)$,

$$z(\omega) \approx (\beta\hbar\omega/2)^{-1} \tag{4.25}$$

that is, the first term in the high-temperature expansion by Caldeira and Leggett [89]. Substituting this into the expression of $\alpha^{(wv)}(\tau)$ in Eq. (4.21), we get a classical limit of the force–force time correlation function for the linear coupling case. The function may also be obtained directly by evaluating the canonical ensemble average of coordinate autocorrelation function of independent oscillators. They clearly show that quantum correction for the classical correlation should be $\beta\hbar\omega/2\coth(\beta\hbar\omega/2)$ for the linear coupling, as proposed by Bader and Berne [19]. The classical approximation of $\mathrm{Re}[\hbar G_2^{(11)}(\tau)]$ may also be evaluated by taking $\hbar \to 0$ limit; it is clear that the quantum correction for the linearly coupling system does not work for the nonlinearly interacting system (i.e., two-phonon and three-phonon processes), since the corrections to be made for the multiphonon processes are different from the one for the single-phonon process. This will be discussed later in detail.

3. *Time-Dependent Transition Probability*

Now, we are ready to evaluate time-dependent transition probability of the system between eigenstates of H_S—for example, ϕ_i and ϕ_f. The system is, at present, arbitrary. The probability that the system was initially found at the state ϕ_i and finally at ϕ_f after time t may be expressed by

$$\begin{aligned} P_{i\to f}(t) = &\int_{-\infty}^{+\infty} dx_f \int_{-\infty}^{+\infty} dx'_f\, \phi_f^*(x_f)\phi_f(x'_f) \\ &\times \int_{-\infty}^{+\infty} dx_i \int_{-\infty}^{+\infty} dx'_i\, J(x_f, x'_f, t; x_i, x'_i, 0)\phi_i(x_i)\phi_i^*(x'_i) \end{aligned} \tag{4.26}$$

where $J(x_f, x'_f, t; x_i, x'_i, 0)$ is the propagating function [26, 27] defined by

$$\begin{aligned} &J(x_f, x'_f, t; x_i, x'_i, 0) \\ &\quad = \int_{x_i}^{x_f} \mathcal{D}x \int_{x'_i}^{x'_f} \mathcal{D}x' \exp\left[\frac{i}{\hbar}\left\{S_S[x] - S_S[x']\right\}\right] \mathcal{F}_{nonlin}[x, x'] \end{aligned} \tag{4.27}$$

If the system is also a harmonic oscillator—that is, the potential $V_S(x)$ is quadratic with respect to x, and if the coupling is linear in x [as in Eq. (4.31) below]—the path integral may be accomplished exactly since the second-order perturbative influence functional is quadratic in x, too. Even if the potential $V_S(x)$ is higher than quadratic, approximation techniques are available for the integration. The first one is the steepest descent method which has been applied to bilinear couplings [88–90]. The second one is the Taylor expansion of the influence functional. The third one is numerical calculation such as the Monte Carlo calculation proposed by Cline and Wolynes [94]. Before showing an exact calculation of the path integral for the harmonic oscillator system, it is interesting to apply the second technique to obtain an approximate expression for the time-dependent probability.

First, Taylor expansion of the perturbative influence functional $\mathcal{F}_{nonlin}[x, x']$ up to the first order leads to the probability, after taking $t \to \infty$ limit, for the nondegenerate system of the form

$$P_{i\to f}(t) = \frac{2t}{\hbar^2}\mathrm{Re}\left[\int_0^\infty ds \sum_{w,v} G_2^{(wv)}(s)\langle\phi_i|x^w|\phi_f\rangle \times \langle\phi_f|x^v|\phi_i\rangle \exp\{i(\Omega_i - \Omega_f)s\}\right] \tag{4.28}$$

for $\phi_f \neq \phi_i$, and

$$P_{i\to i}(t) = 1 - \sum_{f\neq i} P_{i\to f}(t) \tag{4.29}$$

where $\hbar\Omega_n$ is eigenvalue of the state ϕ_n. This is just the same as Fermi's golden rule. Similarly, the first-order cumulant expansion of the influence functional gives the survival probability from the Pauli's master equation based upon the golden rule transition probability. The explicit formulae for these approximations may be found in many textbooks [76–80]. Here, it is interesting to note that the present theory contains Fermi's golden rule and Pauli's formula as an approximation. Systematic calculation such as the Taylor expansion up to the higher-order terms clearly presents higher-order perturbations for the time-dependent transition probability.

Second, we derive the exact formula of $P_{i\to f}(t)$ for the harmonic oscillator system. The system Hamiltonian is, now,

$$H_S = H_0 = -\frac{\hbar^2}{2M}\frac{\partial^2}{\partial x^2} + \frac{1}{2}M\tilde{\Omega}^2 x^2 \tag{4.30}$$

where $\hbar\tilde{\Omega}$ is the energy gap between the system eigenstates in the solution. Further we assume linear interaction with respect to x

$$f_k(x) = C_k^{(1)} x, \qquad g_{kl}(x) = C_{kl}^{(1)} x, \quad \text{and} \quad h_{klm}(x) = C_{klm}^{(1)} x \tag{4.31}$$

The assumption is reasonable because the nonlinear coupling with respect to the system degree of freedom is much smaller than the linear one $\left[\text{e.g., } |C^{(1)}/C^{(2)}| \gg \sqrt{\hbar z(\tilde{\Omega})/2M\tilde{\Omega}}\right]$, for the case of vibrational energy relaxation of molecule in the solution. Furthermore, when the coupling is quadratic in x (i.e., $C^{(2)}x^2$), contribution to the relaxation time is essentially neglegible. In the stochastic theory, this assumption corresponds to the approximation that the force exerted on the system is a Gaussian process [27]. Furthermore, in this case, the transition may be considered as a stationary process because the correlation function is a single time-dependent function as shown in Eq. (4.17). Assuming Eqs. (4.30) and (4.31), the total Hamiltonian is quadratic in x, and we are able to solve the time-dependent transition probability exactly for the influence phase of the form

$$\begin{aligned} A[x, x'] = &-\int_0^t du G_1 x(u)\{x(u) - x'(u)\} \\ &+ i\int_0^t du_1 \int_0^{u_1} du_2 \{x(u_1) - x'(u_1)\} \\ &\times \{G_2(u_1 - u_2)x(u_2) - G_2^*(u_1 - u_2)x'(u_2)\} \end{aligned} \tag{4.32}$$

where $G_1 = G_1^{(1)}$ and $G_2 = G_2^{(11)}$. It is clear from the definition that

$$G_2^*(u_1 - u_2) = G_2(u_2 - u_1) \tag{4.33}$$

Now, the first thing to do here is to calculate the propagating function $J(x_f, x_f', t; x_i, x_i', 0)$ according to Eq. (4.27) for the system action

$$S_S[x] = S_0[x] = \int_0^t du \left\{ \frac{1}{2} M\dot{x}(u)^2 - \frac{1}{2} M\tilde{\Omega}^2 x(u)^2 \right\} \tag{4.34}$$

as well as the influence phase expressed in Eq. (4.32). Consider classical paths x_{cl} and x'_{cl} which satisfy

$$\text{Re}\left[\frac{\delta S_0[x]}{\delta x(s)}\bigg|_{cl} + \frac{\delta A[x, x']}{\delta x(s)}\bigg|_{cl} \right] = \text{Re}\left[-\frac{\delta S_0[x']}{\delta x'(s)}\bigg|_{cl} + \frac{\delta A[x, x']}{\delta x'(s)}\bigg|_{cl} \right] = 0 \tag{4.35}$$

Here, the subscript cl means that the functional derivatives are along the classical paths. Variable transformation such as $z=(x+x')/2$ and $z'=x-x'$ leads to separation of variables for the coupled differential equation of x and x' given by Eq. (4.35), which presents an algebraic solution for the classical paths [88–90]. Since $S_0[x]$, $S_0[x']$, and $A[x,x']$ are all quadratic with respect to the paths x and x', Taylor expansion of the argument of the exponential function in Eq. (4.27) (i.e., $S_0[x]-S_0[x']+A[x,x']$) around the classical paths up to the second order gives the exact description. Thus,

$$
\begin{aligned}
J(x_f, x'_f, t; x_i, x'_i, 0) \\
&= \exp\left\{\frac{i}{\hbar}\left(S_0[x_{cl}] - S_0[x'_{cl}] + A[x_{cl}, x'_{cl}]\right)\right\} \\
&\times \int_0^0 \mathcal{D}y \int_0^0 \mathcal{D}y' \exp\left\{\frac{i}{\hbar}\int_0^t ds\left[\left(\left.\frac{\delta S_0[x]}{\delta x(s)}\right|_{cl} + \left.\frac{\delta A[x,x']}{\delta x(s)}\right|_{cl}\right)y(s)\right.\right. \\
&\qquad \left.\left. + \left(-\left.\frac{\delta S_0[x']}{\delta x'(s)}\right|_{cl} + \left.\frac{\delta A[x,x']}{\delta x'(s)}\right|_{cl}\right)y'(s)\right]\right\} \\
&\times \exp\left\{\frac{i}{2\hbar}\int_0^t ds_1 \int_0^t ds_2 \left[y(s_1)y(s_2)\left(\left.\frac{\delta^2 S_0[x]}{\delta x(s_1)\delta x(s_2)}\right|_{cl} + \left.\frac{\delta^2 A[x,x']}{\delta x(s_1)\delta x(s_2)}\right|_{cl}\right)\right.\right. \\
&\qquad + y(s_1)y'(s_2)\left.\frac{\delta^2 A[x,x']}{\delta x(s_1)\delta x'(s_2)}\right|_{cl} + y'(s_1)y(s_2)\left.\frac{\delta^2 A[x,x']}{\delta x'(s_1)\delta x(s_2)}\right|_{cl} \\
&\qquad \left.\left. + y'(s_1)y'(s_2)\left(-\left.\frac{\delta^2 S_0[x']}{\delta x'(s_1)\delta x'(s_2)}\right|_{cl} + \left.\frac{\delta^2 A[x,x']}{\delta x'(s_1)\delta x'(s_2)}\right|_{cl}\right)\right]\right\}
\end{aligned}
\tag{4.36}
$$

where

$$
y(s) = x(s) - x_{cl}(s) \tag{4.37}
$$

and

$$
y'(s) = x'(s) - x'_{cl}(s) \tag{4.38}
$$

are deviations from the classical paths. Functional derivatives in the equation may easily be obtained. Direct differentiation gives the first-order functional derivatives

$$
\begin{aligned}
\frac{\delta A[x,x']}{\delta x(s)} = -G_1 + i\left\{\int_0^s du\, x(u) G_2(s-u) + \int_s^t du\, x(u) G_2^*(s-u)\right. \\
\left. - \int_0^t du\, x'(u) G_2^*(s-u)\right\}
\end{aligned}
\tag{4.39}
$$

and

$$\frac{\delta A[x,x']}{\delta x'(s)} = G_1 + i\left\{ -\int_0^t du x(u) G_2(s-u) + \int_0^s du x'(u) G_2^*(s-u) + \int_s^t du x'(u) G_2(s-u) \right\} \tag{4.40}$$

and the second-order derivatives of the form

$$\frac{\delta^2 A[x,x']}{\delta x(s_1)\delta x(s_2)} = i\{\theta(s_1-s_2)G_2(s_1-s_2) + \theta(s_2-s_1)G_2^*(s_1-s_2)\} \tag{4.41}$$

$$\frac{\delta^2 A[x,x']}{\delta x'(s_1)\delta x'(s_2)} = i\{\theta(s_1-s_2)G_2^*(s_1-s_2) + \theta(s_2-s_1)G_2(s_1-s_2)\} \tag{4.42}$$

$$\frac{\delta^2 A[x,x']}{\delta x(s_1)\delta x'(s_2)} = -iG_2^*(s_1-s_2) \tag{4.43}$$

$$\frac{\delta^2 A[x,x']}{\delta x'(s_1)\delta x(s_2)} = -iG_2(s_1-s_2) \tag{4.44}$$

where $\theta(\tau)$ is Heaviside step function. Discretization of the fluctuation paths y and y' leads to the final formula of the propagating function:

$$\begin{aligned}
&J(x_f, x_f', t; x_i, x_i', 0) \\
&= \lim_{\mathcal{P}\to\infty} \left(\frac{m_s}{2\pi\hbar\varepsilon}\right)^{\mathcal{P}} \exp\left\{\frac{i}{\hbar}(S_0[x_{cl}] - S_0[x_{cl}'] + A[x_{cl}, x_{cl}'])\right\} \\
&\times \int_{-\infty}^{\infty} dy_1 \cdots dy_{\mathcal{P}-1} \int_{-\infty}^{\infty} dy_1' \cdots dy_{\mathcal{P}-1}' \exp\left\{\sum_{a=1}^{\mathcal{P}-1} (\mu_a y_a + \mu_a' y_a')\right\} \\
&\times \exp\left\{\sum_{a=1}^{\mathcal{P}-1}\sum_{b=1}^{\mathcal{P}-1} \left(y_a[\xi_{ab} + \eta_{ab}]y_b + y_a \zeta_{ab}^* y_b' + y_a' \zeta_{ab} y_b + y_a'[\xi_{ab}^* + \eta_{ab}^*]y_b'\right)\right\} \\
&= \lim_{\mathcal{P}\to\infty} \left(\frac{m_s}{2\pi\hbar\varepsilon}\right)^{\mathcal{P}} \exp\left\{\frac{i}{\hbar}(S_0[x_{cl}] - S_0[x_{cl}'] + A[x_{cl}, x_{cl}'])\right\} \\
&\times \left(\frac{\pi^{\mathcal{P}-1}}{\sqrt{\det \mathbf{Z}}}\right) \exp\left(\frac{1}{4}\mathbf{m}^t\mathbf{Z}^{-1}\mathbf{m}\right)
\end{aligned} \tag{4.45}$$

where $\mathcal{P}$ is the number of discretization of the paths, $\varepsilon = t/\mathcal{P}$ is the discretized time, and μ_a, μ'_a, ξ_{ab}, η_{ab}, and ζ_{ab} are abbreviations for

$$\mu_a = \frac{i\varepsilon}{2\hbar}\left(- M\ddot{x}_a\big|_{cl} - M\tilde{\Omega}^2 x_a\big|_{cl} + \frac{\partial A}{\partial x_a}\bigg|_{cl} - M\ddot{x}_{a+1}\big|_{cl} - M\tilde{\Omega}^2 x_{a+1}\big|_{cl} + \frac{\partial A}{\partial x_{a+1}}\bigg|_{cl}\right) \tag{4.46}$$

$$\mu'_a = \frac{i\varepsilon}{2\hbar}\left(M\ddot{x}'_a\big|_{cl} + M\tilde{\Omega}^2 x'_a\big|_{cl} + \frac{\partial A}{\partial x'_a}\bigg|_{cl} + M\ddot{x}'_{a+1}\big|_{cl} + M\tilde{\Omega}^2 x'_{a+1}\big|_{cl} + \frac{\partial A}{\partial x'_{a+1}}\bigg|_{cl}\right) \tag{4.47}$$

$$\xi_{ab} = \frac{i\varepsilon}{2\hbar}\left\{ \frac{M}{\varepsilon^2}(2\delta_{a,b} - \delta_{a+1,b} - \delta_{a,b+1}) - \frac{M\tilde{\Omega}^2}{4}(2\delta_{a,b} + \delta_{a+1,b} + \delta_{a,b+1}) \right\} \tag{4.48}$$

$$\eta_{ab} = \frac{i\varepsilon^2}{8\hbar}\left(\frac{\partial^2 A}{\partial x_a \partial x_b}\bigg|_{cl} + \frac{\partial^2 A}{\partial x_a \partial x_{b+1}}\bigg|_{cl} + \frac{\partial^2 A}{\partial x_{a+1} \partial x_b}\bigg|_{cl} + \frac{\partial^2 A}{\partial x_{a+1} \partial x_{b+1}}\bigg|_{cl}\right) \tag{4.49}$$

and

$$\zeta_{ab} = \frac{i\varepsilon^2}{8\hbar}\left(\frac{\partial^2 A}{\partial x'_a \partial x_b}\bigg|_{cl} + \frac{\partial^2 A}{\partial x'_a \partial x_{b+1}}\bigg|_{cl} + \frac{\partial^2 A}{\partial x'_{a+1} \partial x_b}\bigg|_{cl} + \frac{\partial^2 A}{\partial x'_{a+1} \partial x_{b+1}}\bigg|_{cl}\right) \tag{4.50}$$

respectively. Here, we used the trapezoidal rule for the numerical integration. **m** and **Z** are matrices

$$\mathbf{m} = \begin{pmatrix} \{\mu_a\} \\ \{\mu'_a\} \end{pmatrix} \tag{4.51}$$

and

$$\mathbf{Z} = -\begin{pmatrix} \{\xi_{ab} + \eta_{ab}\} & \{\zeta^*_{ab}\} \\ \{\zeta_{ab}\} & \{\xi^*_{ab} + \eta^*_{ab}\} \end{pmatrix} \tag{4.52}$$

respectively, where the matrix **Z** is symmetric. Here, we have adopted the mid-point prescription for the discretized paths. Substitution of the propagating function of Eq. (4.45) into Eq. (4.26) gives a very useful expression for the time-dependent transition probability. The calculation is now free from the sign problem, and all the variables in the equation are accessible from classical MD calculation followed by the normal mode analysis.

The present approximation of the harmonic oscillator system must be valid for the vibrational relaxation of molecules, since $V_S(x)$ may be assumed to be intrinsically quadratic with respect to x when the system was initially in a low vibrationally excited state such as the first excited state and is finally found in the ground state. If highly excited states are of interest, anharmonicity in $V_S(x)$ must be taken into account. In this case, numerical path integration may be done based upon the present influence functional using a sophisticated method such as the one proposed by Cline and Wolynes [94].

4. *Discussion*

The approximation used in the previous subsection to evaluate the influence functional is just the cumulant expansion of influence action. It is interesting to note that the first-order expansion of the present influence functional corresponds to Fermi's golden rule for the case of the Taylor expansion and Pauli's formula for the case of the cumulant expansion. The exact path integration for the transition probability from the second-order perturbative influence functional, of course, includes much more physics than Fermi's golden rule, with full perturbations being taken into account. The second-order perturbative influence functional is exact for the system linearly coupled to the harmonic bath coordinates. This is clear from the fact that the vertex functions $f_k(x)$ can form only one kind of closed diagram which has just two vertices (see the second diagram in Fig. 2). However, if the coupling is anharmonic with respect to the bath degrees of freedom, the cumulant expansion must be done to the infinite order in order to obtain an exact description for the influence functional. This is easily understood, too, with the help of the diagram. For example, we can draw a ring diagram connecting an arbitrary number of vertex functions $g_{kl}(x)$. This implies that even when the interaction Hamiltonian includes only quadratic coupling term, the exact influence functional may be described only by infinite-order cumulant expansion. Diagrams from the third-order perturbative influence functional

$$\begin{aligned} A_3[x,x'] = & \frac{1}{6}\left(\frac{i}{\hbar}\right)^3 \left\langle (S_i[x,\{q_k\}] - S_i[x',\{q'_k\}])^3 \right\rangle_0 \\ & -\frac{1}{2}\left(\frac{i}{\hbar}\right)^3 \left\langle (S_i[x,\{q_k\}] - S_i[x',\{q'_k\}])^2 \right\rangle_0 \left\langle S_i[x,\{q_k\}] - S_i[x',\{q'_k\}] \right\rangle_0 \\ & +\frac{1}{3}\left(\frac{i}{\hbar}\right)^3 \left(\langle S_i[x,\{q_k\}] - S_i[x',\{q'_k\}]\rangle_0\right)^3 \end{aligned} \tag{4.53}$$

are shown in Fig. 3. Expression for the third-order functional may easily be obtained in a way similar to that of the case of the second-order functional. The number of vertices corresponds to the number of times, which is the same as the order of cumulant expansion. In this case, this is three. Thus, three-time correlations with respect to the bath coordinates are involved.

Now, a question arises. When does one expect the second-order perturbative influence functional to be valid? As we can see in Eq. (4.32), the force exerted

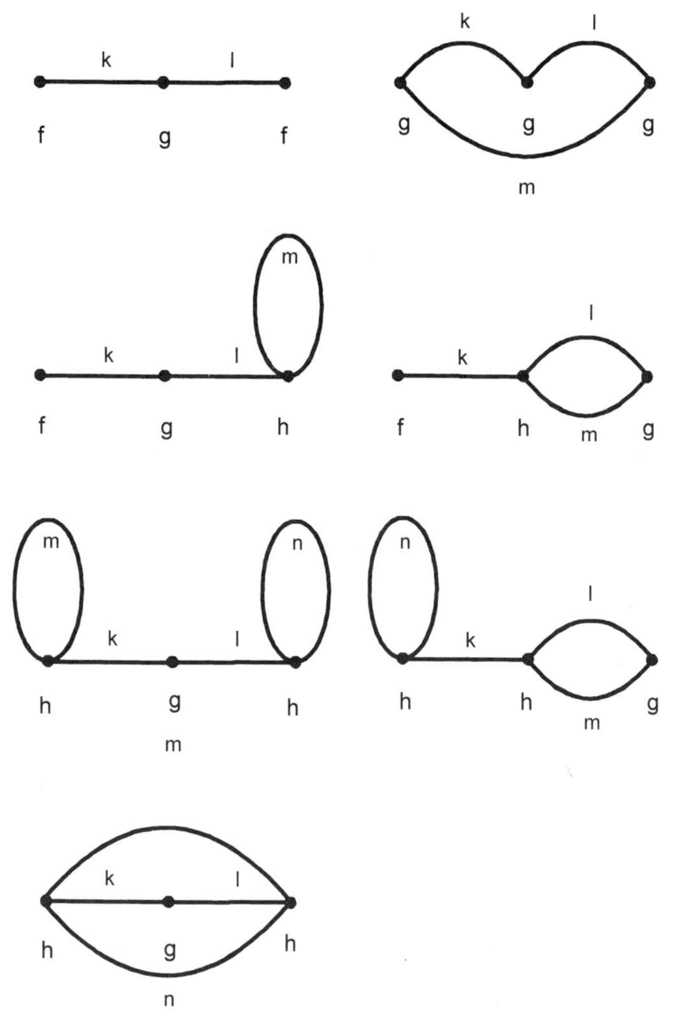

Figure 3. Diagrams representing each term in the third-order cumulant expansion of the influence action.

on the system may be considered as a Gaussian process. Similarly, in more general terms, $v = w = n$ in Eq. (4.22) presents the nth order derivative of the potential energy with respect to x a Gaussian process. Furthermore, the cross term (i.e., $v \neq w$) gives a two-time cross-correlation function. Thus, the second-order cumulant expansion is valid when a Gaussian process is dominant in the interaction between the system and bath or when the many-time correlations greater than two such as three-time correlations is not important in the dynamics.

We showed an exact expression for the path integral for the harmonic oscillator system based upon the second-order perturbative influence functional. For the system with arbitrary potential $V_S(x)$, steepest descent may be applied to do the approximate path integration [90]. As shown in Eqs. (4.17), (4.21), and (4.22), the shape of the influence functional for the nonlinearly coupled system is the same as that for the bilinearly interacting system. This implies that the path $x(t)$ can be considered as a stochastic variable following the generalized Langevin equation in the same way as the case of the conventional bilinear coupling system [88–90]. Thus, the real and imaginary parts of $\alpha_k(\tau)$, $\beta_{kl}(\tau)$, ... may be referred to as the "noise" and "dissipation" kernel, respectively [91]. Of course, both the time-dependent friction $\Gamma(\tau)$ defined by $\mathrm{Im}[\hbar G_2(\tau)] = \partial\Gamma(\tau)/\partial\tau$ and the autocorrelation function of the fluctuating force $R(\tau)$ described by $\mathrm{Re}[\hbar G_2(\tau)] = 1/2\langle R(\tau)R(0) + R(0)R(\tau)\rangle$ are different from those for the bilinear coupling system. These may be described by two-time autocorrelation functions of the bath coordinate. It is also clear that the nth-order perturbative influence functional gives n-time correlation function. By the path integral formulation, we may be able to connect to Fokker–Planck equation and quantum master equation, too, regarding the path $x(t)$ as a stochastic variable [88, 89, 91, 95, 96].

Now it is interesting to discuss about quantum correction by Bader and Bern [19] for the force–force time correlation function which is found in the Fermi's golden rule for the time constant of vibrational energy relaxation. The quantum correction of the form $(\beta\hbar\omega/2)\coth(\beta\hbar\omega/2)$ for the spectral density of force, the Fourier transformation of the force autocorrelation function, has been proposed. In fact, Eq. (4.20) shows that the method is correct for the bilinearly interacting system. In this case, it works well even for the multi-phonon process from the higher-order perturbation than the first—that is, n-times multiplication of the spectral density for the self-n-phonon process. More precisely, this kind of quantum correction is valid so far as the interaction between the system and bath is linear with respect to the bath coordinate. However, it is clear from the equation that the correction is not correct for the multiple coupling system with the interaction such as $\sum_k\sum_l C_{kl}xq_kq_l$. If the coupling constant C_{kl} may be factorized by the contribution from each bath mode as assumed by Egorov and Skinner [83] like $C_{kl} = C_kC_l$, the correction makes a sense, again, even for the multiple coupling system. Although this assumption might be acceptable for their particular problem, it must not be the case for the vibrational energy relaxation of molecules in the solution. The formulae presented in Eqs. (4.17)–(4.24) can be a help to a new correction method.

As stated above, Egorov and Skinner [83] discussed about two pathways of the relaxation. The first one comes from "high-order coupling" analyzed by Fermi's golden rule. The second one is from higher-order perturbations of "linear coupling." The analysis of the present study, of course, takes account of both the pathways as parts of the contributions to the total relaxation. As far as the vibrational relaxation is concerned, we can evaluate, in principle, all the contributions from these kinds of various pathways separately. Now, it is interesting to complete a table of the contributions for real systems in order to obtain an overview of the relaxation mechanism.

Finally, a computationally tractable recipe for the evaluation of the vibrational energy relaxation is presented. The first step is classical MD calculation, with all the intramolecular degrees of freedom being fixed. The second is a quench of the solution giving all the intramolecular degrees of freedom of the whole system including the vibrational degree of freedom of interest. The third is a normal mode analysis and the calculation of the coupling constant which can be obtained from the numerical differentiation of potential energy around the quenched structure. The second step may be skipped if the instantaneous normal mode is preferred. The last step is the calculation of time-dependent probability of the system based upon the above equations such as Eqs. (4.20)–(4.24), (4.26), and (4.45).

Through the present section, we have described a way to connect the liquid state quantum dynamics to quantum field theory based upon the harmonic oscillators bath approximation for the solvent. Along with the present theory, more sophisticated approximations may be promising.

B. Centroid Molecular Dynamics

The second method coming from path integral technique is CMD, developed by Voth et al. [29, 30]. The calculation is dynamical one that is likely to be a quantum version of the MD calculation for the flexible molecules. Since reviews of the CMD calculation have already been given [29, 30], only a rough description of the method is presented here.

According to this method, an atom i is represented by a necklace of P beads—that is, the discretized imaginary time path integral. Trajectory of the center of mass of the necklace is traced according to the classical-like equation of motion,

$$m_i \mathbf{r}_i^C = \mathbf{F}_i^C(\mathbf{r}_i^C) \tag{4.54}$$

where $\mathbf{r}_i^C$ stands for the Cartesian coordinates of the center of mass. The force $\mathbf{F}_i^C$ with respect to the center of mass is assumed to be

$$\mathbf{F}_i^C(\mathbf{r}_i^C) = \frac{\int \cdots \int d\mathbf{r}_{i1} \cdots d\mathbf{r}_{iP} \delta(\mathbf{r}_i^C - \mathbf{r}_i^0)(1/P) \sum_{\alpha=1}^{P} (-dV/d\mathbf{r})|_{\mathbf{r}=\mathbf{r}_{i\alpha}} \exp\{-(S_P[\mathbf{r}_{i1}, \cdots, \mathbf{r}_{iP}]/\hbar)\}}{\int \cdots \int d\mathbf{r}_{i1} \cdots d\mathbf{r}_{iP} \delta(\mathbf{r}_i^C - \mathbf{r}_i^0) \exp\{-(S_P[\mathbf{r}_{i1}, \cdots, \mathbf{r}_{iP}]/\hbar)\}} \tag{4.55}$$

where S_P is the action of P beads and $\boldsymbol{r}_i^0 = \sum_\alpha \boldsymbol{r}_{i\alpha}/P$. The force may be considered as a mean force averaged over all configurations of the beads weighted by $\exp(-S_P/\hbar)$ fixing the center of mass $\boldsymbol{r}_i^0$ of the necklace at $\boldsymbol{r}_i^C$. A few procedures are available to evaluate this force. The most primitive one is to execute a Monte Carlo calculation for an ensemble of the necklaces for all atoms in the solution, fixing their centroids at a given configuration. Then, the force on each centroid is obtained as an average, according to which one MD step can be made. Repetition of this calculation presents a time evolution of all centers of mass of the necklaces. More efficient methods than this are also available [29, 30].

The most important result of this theory is that the time correlation function $C_C(t) = \langle A_C(t) B_C(0) \rangle$ of observables $A_C(t)$ and $B_C(t)$ which are certain functions of centroid variables $\boldsymbol{r}_i^C(t)$ and $\boldsymbol{r}_i^C$ is an approximation for the quantum mechanical canonical correlation function. The function may be transformed to quantum mechanical time correlation function $C(t)$ in the Fourier space by

$$\tilde{C}(\omega) = \frac{\beta\hbar\omega}{2} \left\{ \coth\left(\frac{\beta\hbar\omega}{2} \right) + 1 \right\} \tilde{C}_C(\omega) \tag{4.56}$$

where $\tilde{C}(\omega)$ and $\tilde{C}_C(\omega)$ are Fourier transformations of $C(t)$ and $C_C(t)$, respectively. Inverse transformation of $\tilde{C}(\omega)$ to the time space gives the approximate quantum-mechanical time correlation function.

However, since all CMD calculations must be done in the position representation, it is hard to obtain time evolution of quantum states. So, energy relaxation function $C_E(t)$ given by Eq. (3.2) was monitored for the relaxation [31], which is the same as the case of classical MD calculations. Furthermore, as clearly shown in Eqs. (4.54) and (4.55), the method is different from ordinary real-time simulations. Direct observation of the calculated trajectory of molecules does not give the real motion, although the time correlation function reflects the real dynamics.

Using the CMD, vibrational energy relaxation of CN^- ion in the aqueous solution has been studied very recently [31]. This will be discussed later in detail, comparing the result with those by other techniques.

V. DIRECT SIMULATION BY MIXED QUANTUM-CLASSICAL MOLECULAR DYNAMICS

Dynamical simulations of solute oscillator as well as the solvent molecules have been discussed for both classical and quantum systems. However, the dynamic variable traced there was not the quantum vibrational state but the position coordinate of the atom or the path centroids. In these calculations, transition between states was investigated indirectly through the energy change. On the other hand, the transition rate between the states could be described by Fermi's golden rule and influence functional theory. According to these methods, however, real-time coupling or energy flow between solute and solvent couldn't be observed directly, since only force autocorrelation function and spectral densities were calculated there, respectively.

If we can simulate the relaxation process directly—that is, if time evolution of probability that the system is found at each state can be traced as a function of time and, at the same time, if the trajectory of the solvent molecules can be obtained—a deep understanding of molecular mechanism of the relaxation must be attained. For example, we may catch solvent motions which cause quantum transitions of the solute. Furthermore, it is interesting to find whether the probability change is continuous one or discrete one. Statistics of the coupling is of interest, too.

In this section an application of mixed quantum-classical molecular dynamics method to the vibrational relaxation of a solute molecule in the solution is discussed. The Fermi's golden rule with classical force correlation function might be grouped into the same class as the present approximation in a sense that, in the both methods, the vibrational degree of freedom under consideration is described quantum mechanically while all the remaining solvent degrees of freedom are approximated to follow the classical mechanics. However, the essential difference between the present method and the golden rule is that, in the present calculation, the time-dependent Schrödinger equation for the system is coupled to Newton equations for the solvent degrees of freedom and that these coupled equations of motion are solved numerically in order to obtain real-time trajectory of both solute and solvent at a time. In particular, the time-dependent wave function for the solute is expanded by a set of time-independent eigenfunctions of the isolated solute. This expansion coefficient can be employed as a dynamic variable for the time-dependent Schrödinger equation rather than the vibrational coordinate x itself. Then, time evolution of quantum state of the solute and molecular motions of the solvent may be simulated simultaneously, permitting energy exchange between them by the intermolecular interactions.

However, a logical contradiction of physics inevitably arises from the contact of quantum system with classical degrees of freedom. For example, a certain trajectory of the classical system corresponds to a continuous measurement for the quantum system, which determines the quantum state definitely all the time. In the experiment, however, the quantum state is measured at $t = t$ for the first time. In order to avoid this discrepancy, the time-dependent probability that the quantum system is found at each state must be investigated without determining the state definitely. Then, the motion of the classical system becomes a probabilistic one, resulting in an infinitely branched classical paths. This clearly conflicts with the classical approximation for the solvent. In principle, this kind of contradiction cannot be removed completely as far as the mixed quantum-classical approximation is adopted. Hence, calculation procedure must be determined such that the contradiction is minimized enough to be accepted. For two limiting cases, physically reasonable methods have been proposed. The first one is surface hopping approximation presented for the system whose interaction with the environment can be ideally described by a series of infinitely short-time collisions followed by finite periods of the isolation. A typical example of this may be dynamics of a molecule in the gas phase. The other one is mean field approximation which is valid for the classical system whose potential surface depends little on the quantum state to be studied.

In the present section we demonstrate that the mean field approach may be employed for the investigation of vibrational relaxation of, at least, CN^- ion in the aqueous solution.

A. Mean Field Approximation

With respect to the Hamiltonian of the total system presented by Eq. (2.1), we assume that vibrational degree of freedom of the solute molecule x represents the rapid motion and that motions of other remaining degrees of freedom of the solvent $\{q_k\}$ may be considered to be slow compared with the solute vibration. According to the standard mixed quantum-classical approximation, the time-dependent Schrödinger equation is applied to the vibration

$$\begin{aligned} i\hbar \frac{\partial \psi(x,t)}{\partial t} &= H\psi(x,t) \\ &= [H_S + V_I(x, \{q_k(t)\})]\psi(x,t) \end{aligned} \tag{5.1}$$

where $\psi(x, t)$ is the wave function for the vibrational degree of freedom x at $t = t$. In this expression, interaction potential V_I is a function of time through the motion of solvent molecules $\{q_k(t)\}$. This causes nonadiabatic transition to the solute vibration. On the other hand, classical equation of motion is adopted for the classical degrees of freedom $\{q_k(t)\}$,

$$m_k \ddot{q}_k = F_k^C + F_k^Q \tag{5.2}$$

The first term of the right-hand side of the equation corresponds to the force of ordinary interactions between classical systems. The second term is the one from the quantum system. Here, in the mean field approximation, Hellmann–Feynman force is assumed for the latter

$$F_k^Q = \int_{-\infty}^{\infty} dx\psi^*(x,t)\left(-\frac{\partial V_I}{\partial q_k}\right)\psi(x,t) \tag{5.3}$$

where the force is averaged over the quantum-mechanical degree of freedom. Along the trajectory which follows these equations of motion, total energy

$$E(t) = \int_{-\infty}^{\infty} dx\psi^*(x,t)(H_S + V_I)\psi(x,t) + \sum_{k=1}^{N}\frac{1}{2}m_k\dot{q}_k^2 + V_B(\{q_k\}) \tag{5.4}$$

is conserved, that is, $dE(t)/dt = 0$. In particular, if algebraic solution of eigenfunction $\phi_n(x)$ as well as its eigenvalue E_n can be obtained for stationary state Schrödinger equation with time-independent Hamiltionian H_S

$$H_S\phi_n(x) = E_n\phi_n(x) \tag{5.5}$$

it is convenient to expand $\psi(x, t)$ by $\phi_n(x)$ and E_n

$$\psi(x,t) = \sum_n c_n(t)\phi_n(x)\exp\left(-\frac{iE_nt}{\hbar}\right) \tag{5.6}$$

where $c_n(t)$ is the expansion coefficient. The expansion is usually terminated at a certain excited state $n_{\max}$, neglecting very high excited states. Choice of the parameter $n_{\max}$ depends upon the system to be studied. The coefficient $c_n(t)$ as well as the phase term $\exp(-iE_nt/\hbar)$ represents the time variation of the wave function. Now, inserting Eq. (5.6) into Eq. (5.1), we obtain coupled differential equations for $c_n(t)$:

$$i\hbar\begin{pmatrix}\dot{c}_0(t)\\ \dot{c}_1(t)\\ \dot{c}_2(t)\\ \vdots\end{pmatrix} = \begin{pmatrix} V_{00} & V_{01}\exp(-i\Omega_{01}t) & V_{02}\exp(-i\Omega_{02}t) & \cdots \\ V_{10}\exp(-i\Omega_{10}t) & V_{11} & V_{12}\exp(-i\Omega_{12}t) & \cdots \\ V_{20}\exp(-i\Omega_{20}t) & V_{21}\exp(-i\Omega_{21}t) & V_{22} & \cdots \\ \vdots & \vdots & \vdots & \ddots \end{pmatrix}\begin{pmatrix}c_0(t)\\ c_1(t)\\ c_2(t)\\ \vdots\end{pmatrix} \tag{5.7}$$

where

$$V_{nm} = \int_{-\infty}^{\infty} dx \phi_n^*(x) V_I(x, \{q_k(t)\}) \phi_m(x) \tag{5.8}$$

is time-dependent parameter through the motion of the classical system $\{q_k(t)\}$. The expansion, Eq. (5.6), may be done, for example, by assuming the harmonic oscillator and the Morse oscillator for the system. Then, the coupled equations of motion, Eqs. (5.2) and (5.7), may easily be solved numerically using, for example, the predictor–corrector method. A different set of equations for $c_n(t)$ may be obtained, too. Inserting the expansion Eq. (5.6) into the Lippmann–Schwinger equation, coupled integral equations of Volterra type of the second kind are obtained. Of course, the equations give the same trajectory of $c_n(t)$ and $\{q_k(t)\}$ as that achieved from the above differential equations of motion, Eq. (5.7). However, the integral equation is usually not employed because it requires very long cpu time to be solved.

Solving these coupled equations of motion, we get a trajectory of both quantum and classical degrees of freedom $c_n(t)$ and $\{q_k(t)\}$, respectively, starting from a given initial configuration. With respect to the quantum system, what we want to obtain is an expansion of the wave function by a set of eigenstates of the system in the solution but not in vacuum. The eigenfunction of the oscillator in the solution is obtained from the stationary-state Schrödinger equation with Hamiltonian $H_S + V_I$:

$$H'\phi'_n(x) = (H_S + V_I)\phi'_n(x) = E'_n\phi'_n(x) \tag{5.9}$$

which may be solved, in the numerical calculation, by diagonalizing the matrix

$$H'_{mn} = \int_{-\infty}^{\infty} dx \phi_m^*(x)(H_S + V_I)\phi_n(x) \tag{5.10}$$

Then, the coefficient $c'_n(t)$ in the expansion of $\psi(x, t)$ by the eigenfunction $\phi'_n(x)$ in the solution,

$$\psi(x, t) = \sum_n c'_n(t)\phi'_n(x) \tag{5.11}$$

may be calculated by

$$c'_n(t) = \int_{-\infty}^{\infty} dx \phi_n'^*(x)\psi(x, t) \tag{5.12}$$

Squared absolute value of this expansion coefficient

$$P_n(t) = |c'_n(t)|^2 \tag{5.13}$$

presents the probability that the system is found at state n in the solution. If an initial condition, $c'_1(0) = 1$ and $c'_n(0) = 0$ for $n \neq 1$, is adopted, the survival probability $P_1(t)$ averaged over initial configurations of the solvent corresponds to the population decay of the excited state measured by pump-probe experiment.

B. Application to Vibrational Relaxation

Before starting a demonstration of mean field approximation for the present vibrational relaxation, it is instructive to give a short discussion of the solvated electrons [33–36], focusing our attention on the method. The method assumes two characteristics for the system. The first one is stepwise changes of the wave function. Before and after the change, the electron does not have strong coupling with the solvent, showing little change in the wave function in this period. That is, the quantum system may be considered practically to be isolated between very short but strong couplings with the solvent. During this isolation period, the system may be assumed to occupy one of its quantum states. The second is that potential surface for the solvent depends much on the occupied state of the electron. If we compare a contracted distribution of the electron in the ground state with the expanded one in the first excited state, we can easily understand that this is the case for the solvated electron. Then, trajectory of the classical system depends much upon the quantum state of the electron. Considering the fact that the quantum system loses its coherency with the solvent after the short-time coupling between them, it is physically reasonable to assume that the electron chooses its state following the quantum-mechanical probability. In accordance with the removal of uncertainty with regard to the electronic state, the solvent may be assumed, too, to move on a potential surface determined by the occupied state of the electron. In this case, surface hopping approximation may be employed, where the electronic state is decided periodically or after the strong coupling according to the calculated probability. The method has been powerful to simulate the relaxation of solvated electrons. Several versions of the method have also been proposed in accordance with the individual conditions [33–36].

Now, it is interesting to examine the dynamics of vibrational relaxation in relation to the approximation to be adopted for mixed-quantum classical molecular dynamics method. In this case, atomic distribution of the solute molecule in the solution certainly changes when vibrational state of the solute makes a transition, for example, from the first excited state to the ground state. However, the change is very small or local. So, it is expected that the interaction energy of the solvent molecule with the solute depends little on the solute vibrational state. In order to test this, MD calculation has been performed for the aqueous solution of sodium cyanide [43]. In the calculation, vibrational state of CN^- ion ($\tilde{\Omega} \approx 2080\,\mathrm{cm}^{-1}$ in the solution) was fixed to be the first excited state $n = 1$, and the Hellmann–Feynman force given by Eq. (5.3) was assumed for the solvent. Starting from a certain initial configuration of the solution, a trajectory of the classical system was obtained. Along this trajectory, interaction potential energy between the solute and one of its nearest-neighbor water molecules $\langle 1|V_I^i|1\rangle$ was calculated as a function of time. Here, V_I^i is pair interaction between the solute and the solvent molecule i, where i represents the arbitrarily chosen nearest-neighbor molecule. Along the same trajectory, interaction energy $\langle 0|V_I^i|0\rangle$ may also be calculated just by assuming that the solute vibration is in the ground state $n = 0$. Then, the difference between these two interaction potentials represents the distance between the potential surfaces of the solvent molecule for $n = 0$ and 1.

In Fig. 4, the calculated difference $|\langle 1|V_I^i|1\rangle - \langle 0|V_I^i|0\rangle|$ is plotted along the trajectory. In the figure, the magnitude of the interaction, $\langle 1|V_I^i|1\rangle$, and the total potential energy of the molecule i, $\sum_{j\neq i} V_{ij} + \langle 1|V_I^i|\rangle$, including the interactions with all other solvent molecules, are presented, too, for comparison. As clearly shown, a difference in energy which comes from the difference in the quantum state is only 10^1–$10^2\,\mathrm{J\,mol^{-1}}$. This is about 1/1000 as small as their original value. Furthermore, the value is still more than one order of magnitude smaller than the thermal energy of the solvent, $\frac{1}{2}kT \approx 10^3\,\mathrm{J\,mol^{-1}}$. This is in marked contrast to the case of hydrated electron where the energy difference is comparable to their own values, 10^4–$10^5\,\mathrm{J\,mol^{-1}}$ [97]. Since the energy difference is much smaller than the thermal one in the case of the vibrational relaxation, it is expected that trajectory of the solvent does not depend much on the quantum state.

Trajectories of a water molecule may also be calculated separately assuming, for example, two different states for the quantum system. The calculated trajectories, Cartesian coordinates of center of mass, are presented in Fig. 5 for the same nearest–neighbor molecule as in Fig. 4 assuming the first excited state and the ground state for the solute vibration. Since the quantum state was assumed to be $n = 1$ before $t = 0$, the change of the state to $n = 0$ at $t = 0$ corresponds to a sudden transition of the quantum system to the ground state which must give rise to the maximum difference in the trajectory of the classical system on the relaxation. In spite of chaotic properties of liquid water where even a small difference in force may cause a great difference in its trajectory, the motion of the water molecule is little influenced by the quantum state at least for initial 0.5 ps. Only a small difference is found after about $t = 0.5$ ps. Considering the fact that the structural relaxation time of liquid water is as short as about 1 ps,

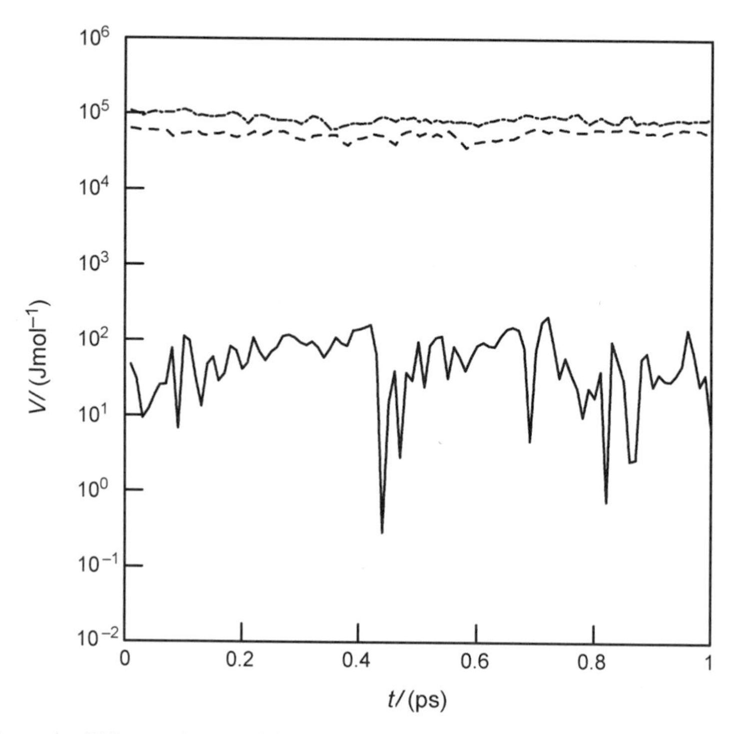

Figure 4. Difference in potential energy surface of a nearest-neighbor solvent molecule i interacting with the solute quantum system. The difference is plotted along the trajectory. Solid line: $|\langle 1|V_l^i|1\rangle - \langle 0|V_l^i|0\rangle|$. Broken line: $|\langle 1|V_l^i|1\rangle|$. Dash–dotted line: $|\sum_{j\neq i} V_{ij} + \langle 1|V_l^i|1\rangle|$.

the classical degrees of freedom may be interpreted as losing their structural memory before the difference in their trajectory caused by full transition of the solute quantum state becomes great. In other words, force on the quantum system, or time-dependent potential for the quantum system, may be evaluated correctly for a period comparable to the time constant of the structural memory by approximating a bundle of classical trajectories by an appropriately determined single path. Time evolution of the quantum system described by the time-dependent potential thus obtained must present dynamics of the system with a satisfactory accuracy. An analogy to generalized Langevin equation, where random force with a certain memory successfully describes various dynamics of liquids, may help the understanding for the validity of this kind of dynamics. One choice for the classical path is to assume Hellmann–Feynman force on the classical degrees of freedom—that is, the mean field approximation or the Ehrenfest method. However, it must be noted that although trajectory of the classical system based upon mean field approximation for the present

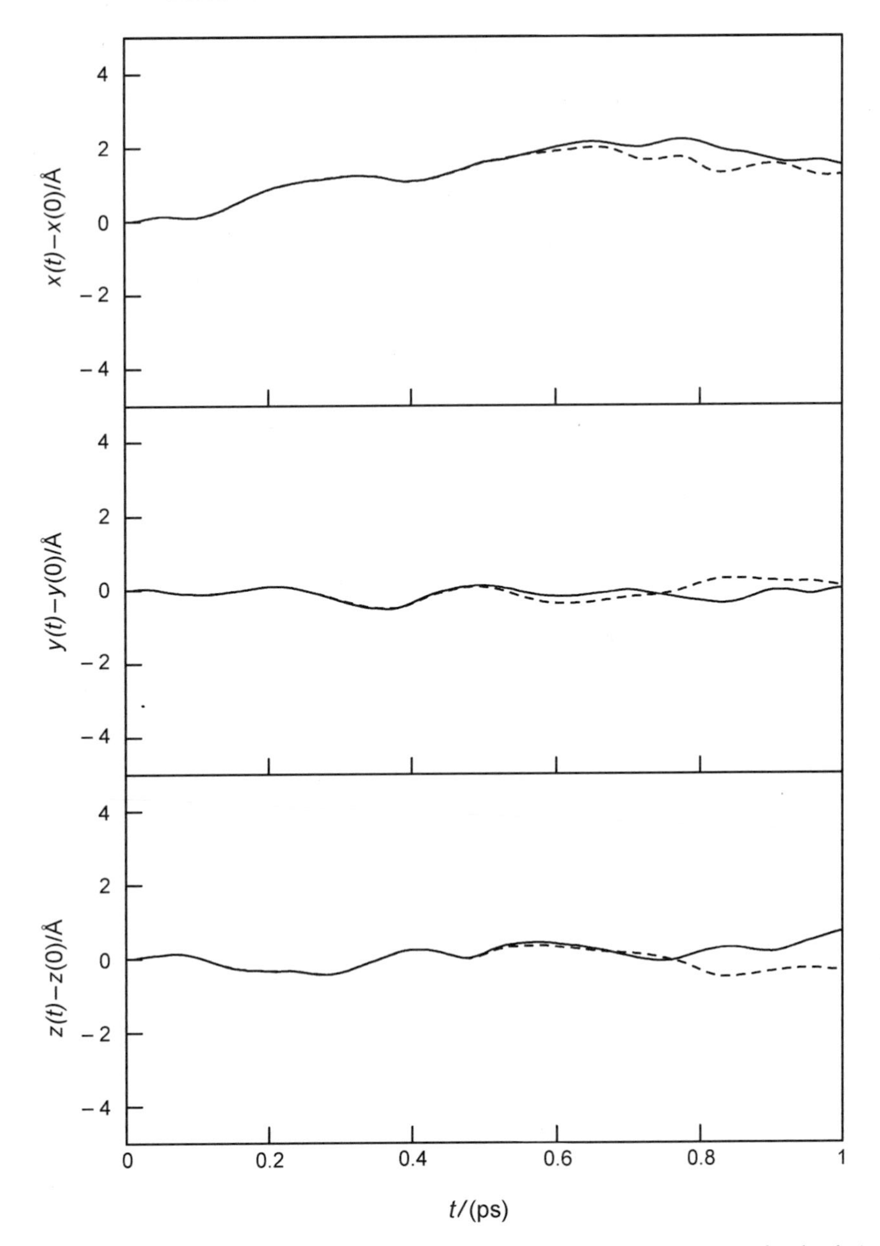

Figure 5. Calculated trajectories of Cartesian coordinates of the center of mass of molecule i assuming the solute quantum state to be $n = 1$ (solid line) and $n = 0$ (broken line).

vibrational relaxation may give reasonable time-dependent potential for the quantum system, long-time dynamics of the classical system itself might have little physical meaning.

In addition to the quantum state dependence of potential surface for the classical system, the time variation pattern of the wave function must be examined. Figure 6 shows the calculated probability $|c'_n(t)|^2$ that the system is found at $n=0$, 1, 2, and 3 as a function of time starting from an arbitrarily chosen equilibrated initial solvent configuration and assuming that the quantum system occupied the first excited state at $t=0$ (i.e., $|c'_1(0)|^2=1$) in accordance with time-resolved pump-probe spectroscopy. As clearly shown in the figure, the probability changes continuously. Great stepwise change is not found. Although very small stepwise changes are observed, they are superposed on the continuous change. This indicates that coherence between solute and solvent and between the first excited state and, for example, the ground state cannot be neglected throughout the time. Since the decoherent state, which is assumed in the surface hopping approximation, is not found for the present system, the quantum state may not be determined definitely before measurement. Removal

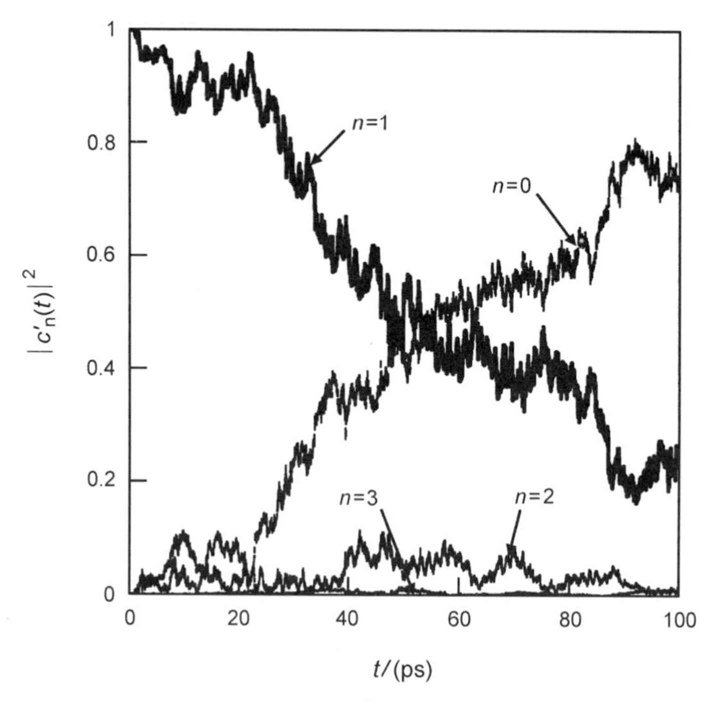

Figure 6. Time-dependent probability $|c'_n(t)|^2$ that the vibrational state of the solute is found at n.

of the uncertainty in the state must cause an artificial disturbance to the system. Together with the small difference between potential surfaces for the classical degrees of freedom stated above, this finding leads to a conclusion that the mean field approximation is better than the surface hopping approximation at least for the vibrational relaxation of CN^- ion in the aqueous solution. It is interesting, too, to examine a solute whose transition frequency is so high that the coupling with the solute becomes a rare event. In this case, the system dynamics might be described by stepwise change of the probability followed by a period of decoherent state, leading to the surface hopping approximation.

Results of the mean field approximation for the vibrational relaxation of CN^- ion in the aqueous solution will be presented in the next section.

VI. CN^- ION IN THE AQUEOUS SOLUTION

The number of intramolecular degrees of freedom of a CN^- ion is just one stretching mode. So, in this case, excited vibrational energy dissipates directly to the solvent degrees of freedom without intramolecular energy redistribution which is often found for molecules composed of more than two atoms. For the present CN^- ion, we can investigate energy transfer from the particular solute mode to the solvent, but we don't need to follow the complicated energy relaxation where a number of intramolecular modes are coupled to the solvent at a time and the excess energy distributed among the intramolecular modes dissipates to the solvent in a competitive manner. The relaxation process of the present system is thus very simple and clear.

Transition frequency of the solute CN^- between the first excited state and the ground state is about $2080\,cm^{-1}$ in the aqueous solution. On the other hand, the transition frequency of translational and rotational librations of liquid water is lower than about $1000\,cm^{-1}$ [98]. Spectra of bending and stretching modes are found at 1600–$1700\,cm^{-1}$ and 3500–$3700\,cm^{-1}$, respectively. Thus, each solvent mode cannot contribute solely to the energy relaxation of the solute due to the mismatch of the frequency between them. In the present system, it is clear that only multiphonon processes are responsible for the relaxation. This presents a place of demonstration of the contribution from nonlinear couplings between solute and solvent and from higher-order perturbations.

Owing to this lack of solvent spectrum at the solute transition frequency, pump or probe light is not disturbed by the solvent absorption, although a very small one caused by the anharmonicity in the modes in real water and/or by the nonlinear response of the dipole moment and polarizability to the molecular motions is found. In this sense, a measurement of the relaxation time of CN^- in the aqueous solution is rather easy to perform. In fact, time-resolved spectroscopic studies have been done for this system extensively by Hochstrasser's group [8, 9]. In their study, a short relaxation time $T_1 = 28$ ps was reported for the solution at 0.22 M. Solute transition frequency dependence of the relaxation time was studied based upon the measurements for various isotopes of the solute. Solvent frequency dependence was also investigated from the experiment for H_2O and D_2O. However, in spite of these studies, molecular picture of the vibrational relaxation has not yet been presented at all.

Thus, the CN^- ion in the aqueous solution is one of the most suitable systems for the test of computational methods and quantum statistical mechanical methods presented here. In fact, computational studies for this system have already been attempted based upon Fermi's golden rule with classical force autocorrelation function [40] in Section III.C, path integral influence functional theory [41, 42, 99] in Section IV.A, CMD [31] in Section IV.B, and mean field approximation for mixed quantum-classical system [43] in Section V.A. Comparison of the results of these calculations must help our understanding of the methods.

In this section, calculated relaxation times from the above four methods are discussed focusing our attention on the approximations on which the methods are based. Efficacy and limitation of the methods are also discussed. A detailed analysis is presented based upon the path integral influence functional theory, which leads to a deep understanding of molecular mechanism of the vibrational relaxation process.

A. Relaxation Time

Four calculations for the vibrational relaxation of CN^- ion in the aqueous solution are summarized in Table II. As clearly read from the table, these calculations are classified into two groups (see also Table I). The first one is full calculation where intramolecular vibrational degrees of freedom of the solvent are explicitly dealt with and the all-solvent degrees of freedom are described quantum mechanically. This corresponds to our calculation based upon path integral influence functional theory [41, 42, 99] and CMD calculation by Voth's group [31]. In the second group, classical approximation is adopted for the solvent degrees of freedom. Consistent with this approximation, the intramolecular degrees of freedom of the solvent (i.e., symmetric and antisymmetric stretchings and bending of water molecule) were frozen, taking account of only translational and rotational degrees of freedom. This includes a calculation based upon Fermi's golden rule with classical force correlation function [40] and the mixed quantum-classical simulation in the mean field approximation [43]. In these calculations, intramolecular vibrations of water cannot take part in the relaxation. Potential function adopted for the CN^- ion is the same among these calculations—that is, a model proposed by Ferarrio et al. [100] (FMK)—except for the calculation by Hynes's group [40]. They treated the value of partial charges on CN^- ion as adjustable parameter to reproduce the experimental

TABLE II
Molecular Dynamics Studies of Vibrational Relaxation of CN^- Ion in the Aqueous Solution

	SO [41, 42, 99]	JPV [31]	RH [40]	TSO [43]	HLH [9]
Method	Influence functional	CMD	Golden rule	MQC MD	Expl.
Further approximation	Cumulant expansion	—	Classical force	Mean field	
System	Quantum	Quantum	Quantum	Quantum	
Bath	Quantum	Quantum	Classical	Classical	
Higher perturbation	Not included	Included	Not included	Included	
Vibration of H_2O	Included	Included	Not included	Not included	
Potential model					
CN^-	FMK	FMK	Adjusted FMK	FMK	
H_2O	TIP4P/HO	SPC/F_2	TIP4P	TIP4P	
T_1 (ps)	7	15	58 138 [101] For original FMK	110	28

CMD, centroid molecular dynamics.
MQC MD, mixed quantum-classical molecular dynamics.
FMK, potential model by Ferrario et al. [73].
TIP4P/HO, TIP4P + harmonic oscillator.

relaxation time. However, the introduction of *ad hoc* parameters make a discussion of quality of the method obscure. So, an MD calculation has been newly executed [101] using the original potential parameters. The table includes the result of this calculation. The common potential function enables us to compare the four methods with each other. However, a short discussion is needed concerning the performance of the potential model. Parameters in the potential function for CN^- were such determined that the structure of NaCN crystal is successfully reproduced. This implies that behavior of the ion in water has not been considered at all. In particular, a primitive extension of the potential function to the mixture based upon the conventional Lorentz–Berthelot rule for LJ terms as well as Coulombic interaction between partial charges on CN^- assumed in FMK model and those on H_2O assumed in TIP4P [102] and SPC/F_2 [103] might give rise to some artifact. Hence, at the present stage we are not very nervous about the agreement of the resultant relaxation time with the experimental one. Agreement in the order of magnitude is important.

In Table II, calculated relaxation time is presented for each method. It is 7 ps for the influence functional theory and 15 ps for the CMD calculation, which are both in satisfactory agreement with the experimental value of 28 ps. In contrast to these, Fermi's golden rule with classical force correlation function gave the relaxation time as long as 138 ps, and the averaged population decay curve obtained by the mixed quantum-classical calculation showed the relaxation time of about 110 ps. The short relaxation times in the first group, where the solvent intramolecular degrees of freedom were explicitly considered and all the solvent degrees of freedom were dealt with quantum mechanically, and the long relaxation times in the second group, where the solvent intramolecular degrees of freedom were frozen and the remaining solvent degrees of freedom were described classically, are both reasonably understood by the difference in the framework of the method as well as the difference in the molecular model for the solvent. In the next section, it will be shown that the classical solvent will result in the relaxation time about eight times longer than the quantum one in multi-phonon processes and that the bending mode of solvent water contributes much to the relaxation. Furthermore, the difference in the calculated relaxation time found between the influence functional calculation and the CMD calculation might be caused by the difference in the potential function adopted for water (i.e., TIP4P/HO for the former and SPC/F_2 for the latter) as well as the different observation for the relaxation time (i.e., population relaxation time for the former and energy relaxation time for the latter).

Potential functions used for the mixed quantum-classical calculation in the mean field approximation and Fermi's golden rule with the classical force auto-correlation function are entirely the same with each other. Thus, these calculations in a similar class of the approximations present the comparable relaxation time 110 ps and 138 ps for the former and the latter, respectively. In more detail, the direct trace of the system dynamics based upon equations of motion essentially includes contribution to the relaxation time from all higher-order perturbations, in the perturbation language, which is in contrast to the first-order perturbation description of Fermi's golden rule. This reasonably explains the slightly shorter relaxation time by the former calculation than the latter, although the difference is too small to be distinguished from the statistical error. Now, it is very interesting to compare these relaxation time with the one based upon influence functional theory. As stated before, in the influence functional theory, $\hbar \to 0$ limit with respect to the bath degrees of freedom gives the relaxation time for the classical solvent. Furthermore, if we neglect the spectral density for ω higher than about $1000\,\mathrm{cm}^{-1}$, we obtain the relaxation time for the solvent water those intramolecular degrees of freedom are all frozen. Then, the relaxation time thus obtained can be compared directly with the one based upon the above mixed quantum-classical molecular dynamics calculation and Fermi's golden rule with classical force correlation function. According to our preliminary calculation [99], it was about 200 ps, where multiphonon processes up to the three phonon coupling were taken into account which can make a contribution to the relaxation even when the bending mode of water is frozen. Thus, adopting the same molecular and potential model, the three methods present the relaxation time of the same order. This guarantees a reliability of the methods although the molecular model adopted there is slightly different from the real one.

In view of the reproducibility of the relaxation time, the first group is clearly superior to the second group. In the former calculations, the relaxation rate was still overestimated a little. However, at present, the good agreement in the order of magnitude of the relaxation time may be stressed, although the potential model should be improved, taking account of the many-body effect in the interaction. This does not disturb much the qualitative analysis of the molecular mechanism of the vibrational relaxation.

B. Relaxation Mechanism

In this section a microscopic picture for the energy transfer in the vibrational relaxation of CN^- ion in the aqueous solution is presented according to the recent analysis based upon path integral influence functional theory [41, 42, 99].

1. Density of States

Density of states of solvent normal modes is presented in Fig. 7. The figure represents the normal modes for the quenched and instantaneous structures of H_2O solution of CN^- ion. From the unitary matrix elements that transfer the normal mode coordinate to the molecular coordinates, we can classify each normal mode into several types of molecular motion [98]. In Fig. 7, the quenched normal modes below $400\,cm^{-1}$ and those between 400 and $1000\,cm^{-1}$ may be assigned to molecular translations (T) and rotational librations (R), respectively [98]. A band found around $1600\,cm^{-1}$ represents intramolecular bending (B) modes of H_2O molecules, and doublet bands around $3700\,cm^{-1}$ come from the intramolecular stretching (S) modes—that is, symmetric and antisymmetric stretchings of H_2O. In some versions of the normal mode analysis, a mode is described as a mixture of T, R, B, and S. However, in the present study, we follow a primitive naming of the mode [98]. Here, an analysis for the instantaneous normal mode is presented because this mode describes dynamics of liquids effectively better than the quenched normal mode. Imaginary frequency modes at $\mathrm{Im}\,\omega_k < 500\,cm^{-1}$ found for the instantaneous normal modes were excluded from the analysis. Although the band gap between T and R modes is not clear from the analysis for the instantaneous normal modes, the modes below $400\,cm^{-1}$ and those between 400 and $1000\,cm^{-1}$ may be referred to as T and R modes, respectively.

The translational and rotational modes of CN^- itself are found around $330\,cm^{-1}$ and $120\,cm^{-1}$, respectively. We note that these modes do not represent "pure" translation or rotation of CN^-. In the diagonalization of the Hessian matrix, the modes for CN^- are mixed with the T modes of H_2O. The C–N vibrational frequency $\tilde{\Omega}$ showed blue shift in the aqueous solution compared with that of the isolated CN^- ion, Ω. The distribution of transition frequency was found at $2071\,cm^{-1} < \tilde{\Omega} < 2099\,cm^{-1}$ for 30 instantaneous structures in the solution.

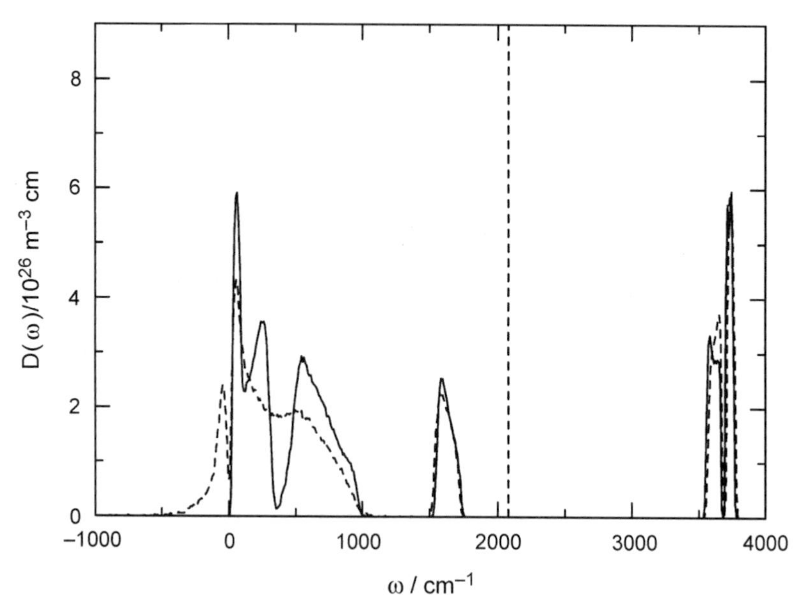

Figure 7. Density of states $D(\omega)$ of the solvent normal modes calculated for the quenched (solid line) and instantaneous (broken line) structures of the aqueous solution of CN^-. CN^- vibrational frequency $\tilde{\Omega}$ is about $2080\,cm^{-1}$.

2. *Spectral Densities*

Calculated single-phonon spectral density $\mathcal{J}_{\alpha[1]}(\omega)$ for the instantaneous normal modes of H_2O solution is presented in Fig. 8. The function $\mathcal{J}_{\alpha[1]}(\omega)$ averaged over 30 configurations is plotted. Within the first-order cumulant expansion of the influence functional or Fermi's golden rule, the relaxation is considered to occur as a result of the resonance between the solute mode and the solvent modes. From this viewpoint, the solute mode frequency $\tilde{\Omega}$ must coincide with the bath mode frequency ω. However, no solvent mode is found near the CN^- vibrational frequency region around $2080\,cm^{-1}$. This clearly indicates that the energy transfer does not occur within the single-phonon process by the linear coupling $C_k^{(1)} x q_k$. The multiphonon process plays a dominant role in the relaxation.

Two-phonon spectral densities $\mathcal{J}_{\beta[1]}(\omega)$ and $\mathcal{J}_{\beta[2]}(\omega)$ are shown in Fig. 9 for the instantaneous normal mode of H_2O solution. In principle, the golden rule for the nonlinear coupling $C_{kl}^{(1)} x q_k q_l$ gives the resonance between the solute mode and the sum- and difference-frequency bath modes, that is, $\omega = \omega_k \pm \omega_l$. However, it is clear from Fig. 9 that sum-frequency spectrum $\mathcal{J}_{\beta[1]}(\omega)$ has large value around the CN^- vibrational frequency, while practically no intensity is found

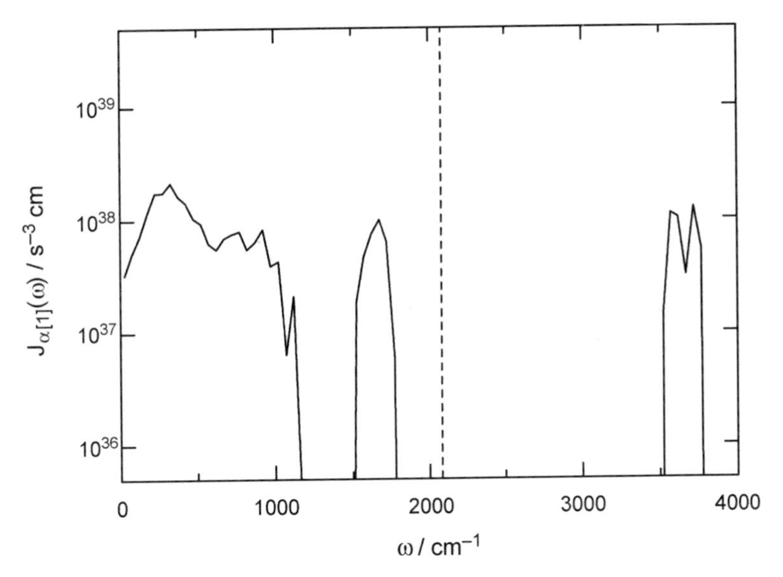

Figure 8. Single-phonon spectral density $\mathcal{J}_{\alpha[1]}(\omega)$ calculated for the instantaneous normal mode. CN^- vibrational frequency $\tilde{\Omega}$ is about 2080 cm^{-1}.

there for the case of difference-frequency spectrum $\mathcal{J}_{\beta[2]}(\omega)$. This implies that the sum-frequency process is dominant in the present solution in the golden rule approximation.

3. *Survival Probabilities*

The time-dependent survival probability averaged over 30 different instantaneous structures is shown in Fig. 10:

$$P(t) = \frac{1}{30}\sum_{c=1}^{30} P_c(t) \tag{6.1}$$

where $P_c(t)$ is the survival probability calculated from Eqs. (4.26) and (4.29), taking account of the first-order term in the cumulant expansion of the influence functional in Eq. (4.27) for the cth solvation structure. In the present calculation, two-phonon processes are considered. The function was integrated with respect to t in order to evaluate the relaxation time T_1. The calculated relaxation time, 7 ps, is of the same order of magnitude as that measured by the pump-probe experiment, $T_1 = 28$ ps [9]. The difference may be caused by the potential functions and/or the harmonic oscillators bath approximation.

Now, it must be reasonable to expect that the value of $P_c(t)$ is not necessarily the same among the solvation structures. To see this, the distribution of the relaxation times for 30 different solvation structures is presented in Fig. 11.

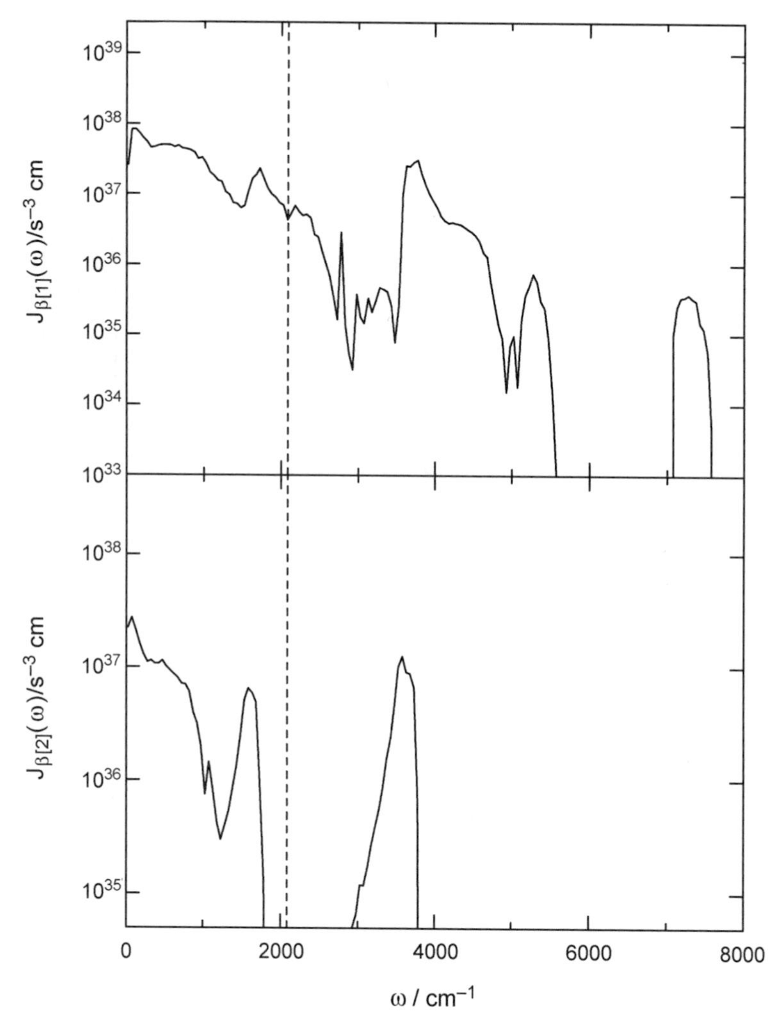

Figure 9. Two-phonon spectral densities $\mathcal{J}_{\beta[1]}(\omega)$ (upper panel) and $\mathcal{J}_{\beta[2]}(\omega)$ (lower panel) of sum- and difference-frequency modes, respectively. CN^- vibrational frequency $\tilde{\Omega}$ is about $2080\,cm^{-1}$.

From the figure, a large variance is found in the relaxation time among the different structures ranging from 0 to 20 ps. The variance reflects the fact that the relaxation rate is fast if the solvation structure has a lot of two-phonon combinations of strongly interacting solvent modes with C–N vibrational mode, and slow if there are few two-phonon combinations. Thus, the relaxation time depends much upon the solvation structure.

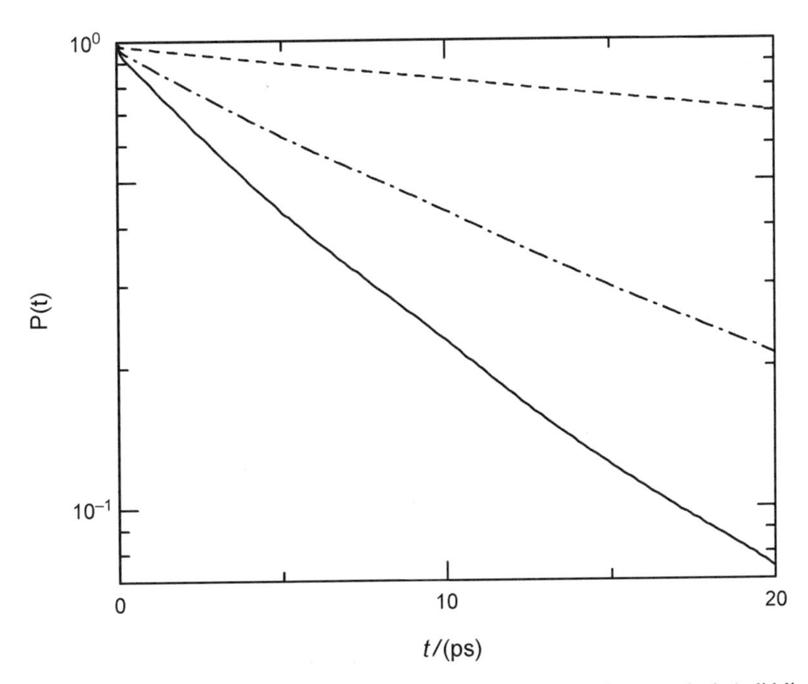

Figure 10. Survival probability for the system of quantum system/quantum bath (solid line), quantum system/classical bath (broken line), and quantum system/classical bath with quantum correction (dash–dotted line) for CN^- ion in the aqueous solution.

Figure 10 also shows the quantum effect of the solvent modes on the survival probability for the instantaneous H_2O solution. In the figure, full quantum survival probability is compared with the quantum-system/classical-bath survival probability which is calculated by approximating $z(\omega_k) \approx z_c(\omega_k)$, $z(\omega_l) \approx z_c(\omega_l)$, and $z(\omega_k \pm \omega_l) \approx [\pm z_c(\omega_k) z_c(\omega_l)]/[z_c(\omega_k) \pm z_c(\omega_l)]$ for Eqs. (4.20), (4.21), and (4.23), where $z_c(\omega) = (\beta\hbar\omega/2)^{-1}$ is the classical approximation of $z(\omega)$. This approximation is essentially the same as the one which has been widely made on the evaluation of the relaxation rate using the classical force–force autocorrelation function. The function might be obtained easily from the time evolution of the classical force exerted on C–N axis by conventional classical MD calculations. However, the classical-bath relaxation time was found to be 59 ps, which is about eight times longer than the quantum-bath relaxation time. Thus, the classical approximation largely underestimate the vibrational energy relaxation rate approximately by a factor of eight.

Recently, a quantum correction has been made as proposed by Bader and Berne [19] by multiplying the spectrum of the correlation function by $z(\omega)/z_c(\omega) = (\beta\hbar\omega/2)\coth(\beta\hbar\omega/2)$. This must be valid for the case of single-phonon

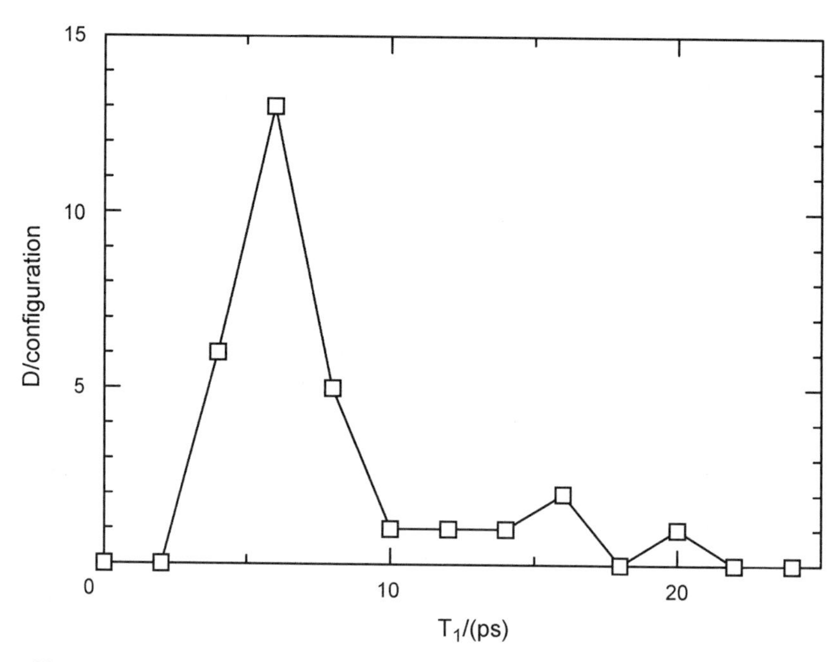

Figure 11. Distribution of the calculated relaxation time for 30 different instantaneous solution structures.

process. Rigorously speaking, however, it is not correct for the multiphonon processes [22, 25, 28, 83]. Now, it is interesting to test this quantum corrections based upon the present spectra. The survival probability of the quantum system in the classical bath with the quantum correction was evaluated by approximating $z(\omega_k) \approx z_c(\omega_k)$ and $z(\omega_l) \approx z_c(\omega_l)$ in Eq. (4.20), leaving $z(\omega_k \pm \omega_l)$ unchanged in Eq. (4.21). The result is given in Fig. 10 where the approximated survival probabilities $P(t)$ are compared with the fully quantum one. It is clear from the figure that although the quantum correction for the classical bath improve the survival probability or the relaxation time for the present system, it is still unsatisfactory. Compare the calculated relaxation time, 13 ps, for the corrected spectrum with that, 7 ps, for the fully quantum system.

It is clear from Eq. (4.20) that the quantum correction to be made depends definitely on the process—that is, number of phonons and sum- or difference-frequency process. However, the relaxation in a real system must be a mixture of these single- and multiphonon processes, and the weight of each process in the relaxation is specific to the system. Thus, it seems to be very difficult to find an effective quantum correction factor that works universally for various systems.

TABLE III
Contribution from Various Pairs of Instantaneous Bath Modes to the Vibrational Relaxation of CN^-

	T	R	B	S
T	0%	0%	24%	0%
R		1%	75%	0%
B			0%	0%
S				0%

4. *Analysis of Bath Modes*

As clearly shown above, the two-phonon coupling plays the dominant role in the vibrational relaxation process of the present system. Now, it is interesting to investigate the bath mode pairs that have large contributions to the relaxation. In order to clarify what sorts of the solvent modes are responsible for the relaxation, we define the "partial two-phonon spectral density" for the pair chosen from T, R, B, and S modes. For example, the sum-frequency RB spectral density may be defined by

$$\mathcal{J}^{RB}_{\beta[1]}(\omega) = \sum_{k\in R}\sum_{l\in B} \frac{\hbar\{C^{(1)}_{kl}\}^2(z(\omega_k)+z(\omega_l))}{(2m_k\omega_k)(2m_l\omega_l)}\delta(\omega-\omega_k-\omega_l) \tag{6.2}$$

According to this definition, the total relaxation rate can be divided into contributions from the pair of the phonon modes. The calculated contribution of each pair is listed in Table III. It is clear from the table that the bending mode of the solvent water is significant in the relaxation process. Neglect of this contribution must lead to much longer relaxation time as obtained in the Fermi's golden rule calculation and the mixed quantum-classical one where the bending of water molecule was frozen. According to our preliminary calculation for the cubic coupling terms, three-phonon processes of various types may be responsible for the relaxation in this rigid solvent model.

VII. QUANTUM PROBABILITY FLUCTUATION

In this section we generalize the path integral influence functional technique to the study of a quantum system with interferences between different histories of paths, in contact with a quantum bath. To examine high moments of squared wave functions, we find that "quantum twinkling," or the existence of the nontrivial moments of the wave function, depends strongly on the interferences between the paths. We also present a particular example of a one-quantum-harmonic-oscillator system coupled with a one-quantum-harmonic-oscillator bath. The low- and high-temperature behavior and the classical bath limit are discussed. We then generalize the results to a many-oscillator bath. One striking result is that the wave function or intensity statistics depend on the number of harmonic oscillators in the bath.

A. Moments of Density Matrix

Consider a system linearly coupled with a single harmonic oscillator bath described by Eq. (4.3) with $N = 1$. Then, the total Hamiltonian is

$$H = H_S + \frac{m}{2}\dot{q}^2 + \frac{m\omega^2}{2}q^2 + Cqx \tag{7.1}$$

where x is the position operator of the system, q, m, and ω are the position operator, mass, and frequency of the bath oscillator, respectively, and C is the coupling constant. H_S is the system Hamiltonian, which is arbitrary at present. Notice that before $t = 0$ there is assumed no interaction between system and bath, so that the initial wave function is factorized.

If the bath is initially in thermal equilibrium and the system can be described by the wave function $\psi_i(x_i)$, the first moment of a conjugated pair of wave functions at time $t = t_f$, which is just a matrix element of the ordinary reduced density operator $\rho(x_f, x'_f; t_f)$ at system coordinates x_f and x'_f, can be written according to Eqs. (4.5)–(4.9) as

$$\begin{aligned} I^{(1)}(x_f, x'_f) &= \rho(x_f, x'_f; t_f) \\ &= \int_{-\infty}^{\infty}\int_{-\infty}^{\infty} dx_i dx'_i J^{(1)}(x_f, x'_f, t_f; x_i, x'_i, 0)\rho^{(1)}(x_i, x'_i; 0) \\ &= \int_{-\infty}^{\infty}\int_{-\infty}^{\infty} dx_i dx'_i \int_{x_i}^{x_f} \mathcal{D}x(t) \int_{x'_i}^{x'_f} \mathcal{D}x'(t) \\ &\quad \times \exp\left[\frac{i}{\hbar}\{S_S[x(t)] - S_S[x'(t)]\}\right] \\ &\quad \times \mathcal{F}^{(1)}[x(t), x'(t)]\psi_i(x_i)\psi_i^*(x'_i) \end{aligned} \tag{7.2}$$

Note that we use $I^{(1)}$ here for convenience to represent the first-moment conjugated pair of wave functions, and $I^{(n)}$ will naturally represent the nth-order moment of conjugated pairs of wave functions. The superscript (1) represents the first moment, averaging over initial bath states. The subscripts i and f are the abbreviations of initial and final, respectively. $J^{(1)}$ is the propagating function for the first moment. The influence functional for the first moment $\mathcal{F}^{(1)}$ may be written as

$$\mathcal{F}^{(1)}[x(t),x'(t)] = \exp\left\{ -\frac{C^2\coth(\beta\hbar\omega/2)}{2m\omega\hbar}A[x(t),x'(t)] + \frac{iC^2}{2m\omega\hbar}B[x(t),x'(t)]\right\} \tag{7.3}$$

$A[x(t),x'(t)]$ and $B[x(t),x'(t)]$ are double path functionals, defined by

$$A[x(t),x'(t)] = \int_0^{t_f}\int_0^{t}\{x(t)-x'(t)\}\cos\omega(t-s)\{x(s)-x'(s)\}dsdt \tag{7.4}$$

$$B[x(t),x'(t)] = \int_0^{t_f}\int_0^{t}\{x(t)-x'(t)\}\sin\omega(t-s)\{x(s)+x'(s)\}dsdt \tag{7.5}$$

These are the same as Eqs. (4.5)–(4.9), although the form is slightly different.

Now, we generalize this influence functional to describe moments higher than the first. The second moment, for example, may be derived in the same way as the first moment as

$$\begin{aligned}
I^{(2)}&(x_f,x'_f,x''_f,x'''_f) \\
&= \langle\psi(x_f)\psi^*(x'_f)\psi(x''_f)\psi^*(x'''_f)\rangle \\
&= \int_{-\infty}^{\infty}\int_{-\infty}^{\infty}\int_{-\infty}^{\infty}\int_{-\infty}^{\infty} dx_i dx'_i dx''_i dx'''_i \\
&\times 2!\int_{x_i}^{x_f}\mathcal{D}x(t)\int_{x'_i}^{x'_f}\mathcal{D}x'(t)\int_{x''_i}^{x''_f}\mathcal{D}x''(t)\int_{x'''_i}^{x'''_f}\mathcal{D}x'''(t) \\
&\times \exp\left[\frac{i}{\hbar}\{S_S[x(t)]-S_S[x'(t)]+S_S[x''(t)]-S_S[x'''(t)]\}\right] \\
&\times \int_{-\infty}^{\infty}\int_{-\infty}^{\infty}\int_{-\infty}^{\infty}\int_{-\infty}^{\infty} dq_f dq'_f dq''_f dq'''_f \sum_k \phi_k^*(q_f)\phi_k(q'_f) \\
&\times \sum_{k'}\phi_{k'}^*(q''_f)\phi_{k'}(q'''_f) \\
&\times \int_{-\infty}^{\infty}\int_{-\infty}^{\infty}\int_{-\infty}^{\infty}\int_{-\infty}^{\infty} dq_i dq'_i dq''_i dq'''_i \left\{\sum_\ell \exp(-\beta E_\ell)\right\}^{-1}
\end{aligned}$$

$$
\begin{aligned}
&\times \sum_m \exp(-\beta E_m)\phi_m(q_i)\phi_m^*(q_i')\phi_m(q_i'')\phi_m^*(q_i''') \\
&\times \int_{q_i}^{q_f} \mathcal{D}q(t) \int_{q_i'}^{q_f'} \mathcal{D}q'(t) \int_{q_i''}^{q_f''} \mathcal{D}q''(t) \int_{q_i'''}^{q_f'''} \mathcal{D}q'''(t) \\
&\times \exp\left[\frac{i}{\hbar}\{S_B[q(t)] + S_I[x(t), q(t)] - S_B[q'(t)] - S_I[x'(t), q'(t)]\right. \\
&\qquad \left. + S_B[q''(t)] + S_I[x''(t), q''(t)] - S_B[q'''(t)] - S_I[x'''(t), q'''(t)]\}\right] \\
&\times \psi(x_i)\psi^*(x_i')\psi(x_i'')\psi^*(x_i''') \\
= &\int_{-\infty}^{\infty}\int_{-\infty}^{\infty}\int_{-\infty}^{\infty}\int_{-\infty}^{\infty} dx_i dx_i' dx_i'' dx_i''' \\
&\times J^{(2)}(x_f, x_f', x_f'', x_f''', t_f; x_i, x_i', x_i'', x_i''', 0)\rho^{(1)}(x_i, x_i'; 0)\rho^{(1)}(x_i'', x_i'''; 0) \\
= &\int_{x_i}^{x_f} \mathcal{D}x(t) \int_{x_i'}^{x_f'} \mathcal{D}x'(t) \int_{x_i''}^{x_f''} \mathcal{D}x''(t) \int_{x_i'''}^{x_f'''} \mathcal{D}x'''(t) \\
&\times \exp\left[\frac{i}{\hbar}\{S_S[x(t)] - S_S[x'(t)] + S_S[x''(t)] - S_S[x'''(t)]\}\right] \\
&\times \mathcal{F}^{(2)}[x(t), x'(t), x''(t), x'''(t)] \\
&\times \psi(x_i)\psi^*(x_i')\psi(x_i'')\psi^*(x_i''')
\end{aligned} \tag{7.6}
$$

where $J^{(2)}$ is the propagating function for the second moment. The factor 2! comes from the permutation of pairs of paths. The influence functional to be calculated is

$$
\begin{aligned}
&\mathcal{F}^{(2)}[x(t), x'(t), x''(t), x'''(t)] \\
&\quad = 2! \int_{-\infty}^{\infty}\int_{-\infty}^{\infty}\int_{-\infty}^{\infty}\int_{-\infty}^{\infty} dq_f dq_f' dq_f'' dq_f''' \sum_k \phi_k^*(q_f)\phi_k(q_f') \\
&\quad \times \sum_{k'} \phi_{k'}^*(q_f'')\phi_{k'}(q_f''') \\
&\quad \times \int_{-\infty}^{\infty}\int_{-\infty}^{\infty}\int_{-\infty}^{\infty}\int_{-\infty}^{\infty} dq_i dq_i' dq_i'' dq_i''' \left(\sum_\ell e^{-\beta E_\ell}\right)^{-1} \\
&\quad \times \sum_m \exp(-\beta E_m)\phi_m(q_i)\phi_m^*(q_i')\phi_m(q_i'')\phi_m^*(q_i''') \\
&\quad \times \int_{q_i}^{q_f} \mathcal{D}q(t) \int_{q_i'}^{q_f'} \mathcal{D}q'(t) \int_{q_i''}^{q_f''} \mathcal{D}q''(t) \int_{q_i'''}^{q_f'''} \mathcal{D}q'''(t)
\end{aligned}
$$

$$
\begin{aligned}
\times \exp \Bigg[\frac{i}{\hbar} \{ & S_B[x(t)] + S_I[x(t), q(t)] - S_B[x'(t)] - S_I[x'(t), q'(t)] \\
& + S_B[x''(t)] + S_I[x''(t), q''(t)] - S_B[x'''(t)] \\
& - S_I[x'''(t), q'''(t)] \} \Bigg]
\end{aligned} \tag{7.7}
$$

Comparing this formula with Eq. (4.4), it is clear that interference between paths comes from the term $\sum_m e^{-\beta E_m} \phi_m \phi_m^* \phi_m \phi_m^*$.

The integration can be performed [44] and we obtain the influence functional for the second moment,

$$
\begin{aligned}
& \mathcal{F}^{(2)}[x(t), x'(t), x''(t), x'''(t)] \\
& \quad = 2! I_0 \left\{ \frac{C^2 \sqrt{A[x(t), x'(t)] A[x''(t), x'''(t)]}}{m\omega\hbar \sinh(\beta\hbar\omega/2)} \right\} \\
& \quad \times F^{(1)}[x(t), x'(t)] F^{(1)}[x''(t), x'''(t)]
\end{aligned} \tag{7.8}
$$

where $I_0(x)$ is the modified Bessel function of zeroth order.

The nth order moment may be described in terms of $2n$-fold path integrals. Extension of the influence functional to the nth order is straightforward using the same technique as adopted for the second moment [44].

$$
\begin{aligned}
F^{(n)} = {} & n! \exp \left\{ -\coth\left(\frac{\beta\hbar\omega}{2}\right) \sum_m A_m + i \sum_m B_m \right\} \\
& \times \prod_{j=2}^{n} \Bigg[\sum_{k_j=0}^{\alpha_j} \sum_{k_j'=0}^{\alpha_j'} \sum_{n_j=0}^{\infty} \left\{ \prod_{j'=1}^{n-2} \sum_{\ell_j^{j'}=0}^{\beta_{j'}} \right\} \\
& \times \frac{\alpha_j! \alpha_j'!}{k_j! k_j'! (\alpha_j - k_j)! (\alpha_j' - k_j')! (k_j + k_j' + n_j)! \{\prod_{j'=1}^{n-2} (\beta_{j'} - \ell_j^{j'})!\} \ell_j^{n-2}!} \Bigg] \\
& \times \left\{ 1 + \coth\left(\frac{\beta\hbar\omega}{2}\right) \right\}^{\alpha_n + n_n + k_n'} \left\{ 1 - \coth\left(\frac{\beta\hbar\omega}{2}\right) \right\}^{\alpha_n' + n_n + k_n} \\
& \times \prod_{j=1}^{n-2} A_j^{k_{n-j+1} + k'_{n-j+1} + n_{n-j+1} + n_{n-j+2} - \ell^1_{n-j+2} + \sum_{j'=1}^{n-4} (\ell^{j'}_{n-j+2+j'} - \ell^{j'+1}_{n-j+2+j'})} \\
& \times A_{n-1}^{n_2 + n_3 - \ell_3^1 + \sum_{j'=1}^{n-3} (\ell^{j'}_{j'+3} - \ell^{j'+1}_{j'+3})} \\
& \times A_n^{n_2 + \sum_{j'=1}^{n-2} \ell^{j'}_{j'+2}}
\end{aligned} \tag{7.9}
$$

where

$$A_j = \frac{C^2}{2m\omega\hbar} A[x^{(2j-1)}(t), x^{(2j)}(t)] \tag{7.10}$$

$$B_j = \frac{C^2}{2m\omega\hbar} B[x^{(2j-1)}(t), x^{(2j)}(t)] \tag{7.11}$$

and

$$\alpha_j = \sum_{j'''=2}^{j-1} n_{j'''} + \sum_{j'''=3}^{j-1} k'_{j'''} \tag{7.12}$$

$$\alpha'_j = \sum_{j'''=2}^{j-1} n_{j'''} + \sum_{j'''=3}^{j-1} k_{j'''} \tag{7.13}$$

$$\beta_{j'} = \begin{cases} n_j & \text{for } j' = 1 \\ \ell_j^{j'-1} & \text{for } j' > 1 \end{cases} \tag{7.14}$$

$$k_\gamma, n_\gamma, \ell_\gamma = 0 \qquad \text{for } \gamma > n \tag{7.15}$$

It is easily found that the equation is of the form

$$F^{(n)} = n! \left[1 + const. \times \sum_{pair} A_j A_{j'} + const. \times \sum_{triplet} A_j A_{j'} A_{j''} + \cdots \right] \times F^{(1)} F^{(1)} \cdots F^{(1)} \tag{7.16}$$

with j ranging from 1 to n.

B. Molecular Vibration Coupled to a One-Harmonic-Oscillator Bath

A specific example of moments with equal final positions x_f for a single harmonic oscillator system coupled with a one-harmonic-bath oscillator has been calculated algebraically. This corresponds to, for example, vibrational dynamics of one molecule in a two-molecule cluster. However, it will be discussed here in detail mainly to obtain an overview of physics of the fluctuations in a quantum bath. Since a single oscillator bath does not behave as a dissipative environment, the system dynamics must be recurrent. However, the statistics of the fluctuations of the system illustrate general characteristics of quantum fluctuations in more realistic baths. For simplicity, the initial system state is assumed to be the ground state. The extension to the case of excited states is straightforward.

1. *Exact Solution*

It is possible to obtain an algebraic solution for the moments by applying steepest descent to the path integral in, for example, Eq. (7.6). However, it seems better to derive the moments directly via propagator $K(x_f, q_f, t_f; x_i, q_i, 0)$ to avoid tedious calculations accompanied by the path integral. The propagator is easily obtained for the normal modes and the inverse transformation to the actual coordinates gives $K(x_f, q_f, t_f; x_i, q_i, 0)$. After some routine algebra, one obtains the first moment [44]

$$\begin{aligned}\langle I\rangle &= \rho^{(1)}(x_f, x_f; t_f)\\ &= \sqrt{\frac{1}{\pi\hbar\{L + R\coth(\beta\hbar\omega/2)\}}}\exp\left[-\frac{x_f^2}{\hbar\{L + R\coth(\beta\hbar\omega/2)\}}\right]\end{aligned} \tag{7.17}$$

In this equation, the symbols

$$\begin{aligned}L = \frac{1}{M\Omega(\omega_1^2 - \omega_2^2)^2}\Bigg[&\frac{\Omega^2}{\omega_1^2\omega_2^2}\{\omega_1(\omega_1^2 - \Omega^2)\sin\omega_2 t_f - \omega_2(\omega_2^2 - \Omega^2)\sin\omega_1 t_f\}^2\\ &+ \{(\omega_1^2 - \Omega^2)\cos\omega_2 t_f - (\omega_2^2 - \Omega^2)\cos\omega_1 t_f\}^2\Bigg]\end{aligned} \tag{7.18}$$

$$\begin{aligned}R = \frac{C^2}{M^2 m\omega(\omega_1^2 - \omega_2^2)^2}\Bigg\{&\frac{\omega^2}{\omega_1^2\omega_2^2}(\omega_1\sin\omega_2 t_f - \omega_2\sin\omega_1 t_f)^2\\ &+ (\cos\omega_2 t_f - \cos\omega_1 t_f)^2\Bigg\}\end{aligned} \tag{7.19}$$

are functions of time as well as potential parameters, where ω_1 and ω_2 represent normal mode frequencies

$$\omega_1^2 = \frac{\Omega^2 + \omega^2 + \sqrt{(\Omega^2 - \omega^2)^2 + (4C^2/Mm)}}{2} \tag{7.20}$$

$$\omega_2^2 = \frac{\Omega^2 + \omega^2 - \sqrt{(\Omega^2 - \omega^2)^2 + (4C^2/Mm)}}{2} \tag{7.21}$$

and M and Ω are, again, mass and frequency of the system, respectively. L and R are greater than or equal to 0. The system that had Gaussian distribution at $t=0$ is Gaussian at $t=t_f$, too. The width of the distribution varies periodically according to Eqs. (7.17), (7.18), and (7.19). The distribution is sharpest when the bath temperature is 0 K and becomes infinitely wide (i.e., constant with respect to x_f) at infinitely high temperature.

The formula for the second moment is

$$\begin{aligned}\langle I^2 \rangle &= \rho^{(2)}(x_f, x_f, x_f, x_f; t_f) \\ &= 2! \frac{1}{\pi\hbar\{L + R\coth(\beta\hbar\omega/2)\}} \exp\left[-\frac{2x_f^2}{\hbar\{L + R\coth(\beta\hbar\omega/2)\}} \right] \\ &\times \sum_{n_2=0}^{\infty} \left[\frac{1}{n_2! 2^{2n_2}} \left\{ \frac{R}{L\sinh(\beta\hbar\omega/2) + R\cosh(\beta\hbar\omega/2)} \right\}^{n_2} \right. \\ &\qquad \left. \times H_{2n_2}\left\{ \frac{x_f}{\sqrt{\hbar\{L + R\coth(\beta\hbar\omega/2)\}}} \right\} \right]^2 \end{aligned} \tag{7.22}$$

where H_n is Hermite polynomial of the nth order.

Since the summation in Eq. (7.22) is composed of several divergent functions which appear both in the numerator and denominator, a discussion about convergence of this series is desirable. The n_2th term of the series can be written as

$$a_{n_2} = \left\{ \frac{D^{n_2}}{n_2! 2^{2n_2}} H_{2n_2}(E) \right\}^2 \tag{7.23}$$

where D and E are abbreviations for

$$D = \frac{R}{L\sinh(\beta\hbar\omega/2) + R\cosh(\beta\hbar\omega/2)} \leq 1 \tag{7.24}$$

$$E = \frac{x_f}{\sqrt{\hbar\{L + R\coth(\beta\hbar\omega/2)\}}} \tag{7.25}$$

respectively. Then the ratio of the n_2th term to the (n_2+1)th term is given by

$$\frac{a_{n_2}}{a_{n_2+1}} \approx \frac{1}{D^2} + \frac{(1/D^2)}{n_2 + (1/2)} + o\left(\frac{1}{n_2^2}\right) \tag{7.26}$$

Gauss's condition for convergence shows that the series converges for $D<1$ and diverges for $D=1$. This means that $\langle I^2 \rangle$ has a finite value at finite temperature while it becomes infinite when $T \to \infty$. The behavior of $\langle I^2 \rangle$ as a function of temperature is quite reasonable.

The interference term in Eq. (7.22) is clearly a function not only of temperature but also of time, system position, and potential parameters. It can vary from 0 to ∞. Considering the fact that this should be always 1 for a Rayleigh process, the fluctuation in the system interacting with the quantum bath has a wide range of behavior. The fluctuation can be greater or less than the conventional wave propagation in random media [104–107] depending upon the temperature, time, system position, and potential parameters. At finite temperature, for example, the difference $\langle I^2 \rangle - \langle I \rangle^2$ approaches 0 rapidly with increasing x_f, while the ratio $\langle I^2 \rangle / \langle I \rangle^2$ becomes infinity. This means that the intensity—that is, the transition probability—is very small for large x_f but shows an extremely large fluctuation. By contrast, a comparably large intensity is found for small x_f, whereas the relative fluctuation is moderate there.

Generalization of the moments to nth order is straightforward [44]. The nth order moment is given by

$$
\begin{aligned}
\langle I^n \rangle = n! \langle I \rangle^n \prod_{j=2}^{n} \Bigg[\sum_{k_j=0}^{\alpha_j} \sum_{k'_j=0}^{\alpha'_j} \sum_{n_j=0}^{\infty} \Bigg\{ \prod_{j'=1}^{n-2} \sum_{\ell_j^{j'}=0}^{\beta_{j'}} \Bigg\} \\
\times \frac{\alpha_j! \alpha'_j!}{k_j! k'_j! (\alpha_j - k_j)! (\alpha'_j - k'_j)! (k_j + k'_j + n_j)! \{\prod_{j'=1}^{n-2} (\beta_{j'} - \ell_j^{j'})!\} \ell_j^{n-2}!} \Bigg] \\
\times \left(\frac{D}{4} \right)^{\alpha_n + \alpha'_n + k_n + k'_n + 2n_n} \exp\left[\frac{\beta \hbar \omega}{2} \left\{ \sum_{j''=3}^{n} (k_{j''} - k'_{j''}) \right\} \right] \\
\times \prod_{j=1}^{n-2} H_{2\{k_{n-j+1} + k'_{n-j+1} + n_{n-j+1} + n_{n-j+2} - \ell^1_{n-j+2} + \sum_{j'=1}^{n-4} (\ell^{j'}_{n-j+2+j'} - \ell^{j'+1}_{n-j+2+j'})\}}(E) \\
\times H_{2\{n_2 + n_3 - \ell^1_3 + \sum_{j'=1}^{n-3} (\ell^{j'}_{j'+3} - \ell^{j'+1}_{j'+3})\}}(E) \\
\times H_{2\{n_2 + \sum_{j'=1}^{n-2} \ell^{j'}_{j'+2}\}}(E)
\end{aligned}
\tag{7.27}
$$

2. *Classical Limit*

Various "Classical limit for the bath" are interesting to investigate. A comparison of the fluctuation in the classical bath with that in the quantum bath discussed so far illustrates a quantum effect on the fluctuation. A formal "classical" limit of the moments may be obtained by taking $\hbar \to 0$ for $\hbar$ for the bath and keeping $\hbar$ for the system unchanged. Of course $\hbar$ has one value, but this corresponds closely with the mixed quantum-classical computations often performed in chemical physics theory. Applying this limit to Eqs. (7.17), (7.22), and (7.27), we get

$$\langle I \rangle_{cl} = \sqrt{\frac{1}{\pi(\hbar L' + (2/\beta\omega)R)}} \exp\left\{ -\frac{x_f^2}{\hbar L' + (2/\beta\omega)R} \right\} \tag{7.28}$$

$$\begin{aligned} \langle I^2 \rangle_{cl} = 2! \langle I \rangle_{cl}^2 \sum_{n_2=0}^{\infty} \Bigg[& \frac{1}{n_2! 2^{2n_2}} \left\{ \frac{M}{(\beta\hbar\omega/2)L' + R} \right\}^{n_2} H_{2n_2} \\ & \times \left\{ \frac{x_f}{\sqrt{\hbar L' + (2/\beta\omega)R}} \right\} \Bigg]^2 \end{aligned} \tag{7.29}$$

and

$$\begin{aligned} \langle I^n \rangle_{cl} = n! \langle I \rangle_{cl}^n \prod_{j=2}^{n} \Bigg[& \sum_{k_j=0}^{\alpha_j} \sum_{k'_j=0}^{\alpha'_j} \sum_{n_j=0}^{\infty} \left\{ \prod_{j'=1}^{n-2} \sum_{\ell_j^{j'}=0}^{\beta_{j'}} \right\} \\ & \times \frac{\alpha_j! \alpha'_j!}{k_j! k'_j! (\alpha_j - k_j)! (\alpha'_j - k'_j)! (k_j + k'_j + n_j)! \{\prod_{j'=1}^{n-2} (\beta_{j'} - \ell_j^{j'})!\} \ell_j^{n-2}!} \Bigg] \\ & \times \left(\frac{D'}{4} \right)^{\alpha_n + \alpha'_n + k_n + k'_n + 2n_n} \\ & \times \prod_{j=1}^{n-2} H_{2\{k_{n-j+1} + k'_{n-j+1} + n_{n-j+1} + n_{n-j+2} - \ell^1_{n-j+2} + \sum_{j'=1}^{n-4} (\ell^{j'}_{n-j+2+j'} - \ell^{j'+1}_{n-j+2+j'})\}} (E') \\ & \times H_{2\{n_2 + n_3 - \ell_3^1 + \sum_{j'=1}^{n-3} (\ell^{j'}_{j'+3} - \ell^{j'+1}_{j'+3})\}} (E') \\ & \times H_{2\{n_2 + \sum_{j'=1}^{n-2} \ell^{j'}_{j'+2}\}} (E') \end{aligned} \tag{7.30}$$

where

$$L' = \frac{1}{M\Omega(\omega_1^2 - \omega_2^2)^2} \{ (\omega_1^2 - \Omega^2) \cos \omega_2 t_f - (\omega_2^2 - \Omega^2) \cos \omega_1 t_f \}^2 \tag{7.31}$$

and D' and E' are abbreviations for

$$D' = \frac{R}{(\beta\hbar\omega/2)L' + R} \tag{7.32}$$

$$E' = \frac{x_f}{\sqrt{\hbar L' + (2/\beta\omega)R}} \tag{7.33}$$

The structural difference in the formula of the moments between quantum and classical baths is just a term $\exp\{(\beta\hbar\omega/2)\sum_{j''=3}^{n}(k_{j''} - k'_{j''})\}$ in the quantum bath, which disappears in the classical limit. The form of the function for $\langle I \rangle$ and $\langle I^2 \rangle$ is the same for both baths, although the difference is still found in the numerical values of the argument—that is, D' and E' in the classical limit instead of D and E in the quantum bath. The difference in structure looks very simple, but it causes a drastic change for the distribution function of I. It is very interesting to find, as shown later, that a distribution which is composed of exponential functions with different arguments becomes a summation of Γ-distributions with single common exponential function when the moments lose the term $\exp\{(\beta\hbar\omega/2)\sum_{j''=3}^{n}(k_{j''} - k'_{j''})\}$.

Exact formulae for the moments in the quantum and classical bath are too complicated to extract information about the fluctuation. Simplified equations based upon some special conditions may provide us good physical insight. In the following two subsections, high-temperature and low-temperature expansions are presented for the second moment as well as for the nth moment.

3. *High-Temperature Behavior*

At sufficiently high temperatures (i.e., $\beta \ll 1$), an approximate description for $\langle I^2 \rangle$ may be obtained by neglecting those terms whose order in β is higher than one. First, D and E in Eqs. (7.24) and (7.25), respectively, may be expanded and approximated by

$$D = 1 - \frac{\beta\hbar\omega}{2}\frac{L}{R} + o(\beta^2) \approx 1 - D_1\beta \tag{7.34}$$

$$E^2 = \frac{\beta\omega}{2}\frac{x_f^2}{R} + o(\beta^2) \approx E_1\beta \tag{7.35}$$

Then, $\langle I^2 \rangle$ for quantum bath can be approximated, too, by

$$\begin{aligned}\langle I^2 \rangle &= 2!\langle I \rangle^2 \left\{ \frac{2}{\pi} K(1 - D_1\beta) - \frac{E_1}{\pi}\beta K'(1 - D_1\beta) \right\} \\ &= 2!\langle I \rangle^2 \left[F\left\{ \frac{1}{2}, \frac{1}{2}, 1; (1 - D_1\beta)^2 \right\} - \frac{1}{4}E_1\beta F\left\{ \frac{3}{2}, \frac{3}{2}, 2; 1 - D_1\beta \right\} \right]\end{aligned} \tag{7.36}$$

where $K(z)$ and $F(\alpha, \beta, \gamma; z)$ represent complete elliptic integral of the first kind and Gauss's hypergeometric function, respectively, and $K(z)'$ is the first derivative of $K(z)$ with respect to z. $K(z)$ and $F((1/2), (1/2), 1; z)$ converge for $z<1$, although they diverge for $z=1$. On the other hand, similar expansions are obtained for classical baths, too, from Eqs. (7.32) and (7.33):

$$D' = 1 - \frac{\beta\hbar\omega}{2}\frac{L'}{R} + o(\beta^2) \approx 1 - D'_1\beta \tag{7.37}$$

$$E'^2 = \frac{\beta\omega}{2}\frac{x_f^2}{R} + o(\beta^2) \approx E_1\beta \tag{7.38}$$

The expression for $\langle I^2 \rangle$ is just the same as Eq. (7.36). Since $E = E'$, a replacement of D_1 by D'_1 gives $\langle I^2 \rangle$ for the classical bath. Both coefficients D_1 and D'_1 are positive. Furthermore, it is clear from Eqs. (7.18) and (7.31) that $D'_1 < D_1$. Thus, $\langle I^2 \rangle$ for the classical bath approaches infinity faster than that for the quantum bath. In other words,

$$\langle I^2 \rangle_{quantum} < \langle I^2 \rangle_{cl} \tag{7.39}$$

For $n>2$, a further difference in $\langle I^n \rangle$ between the quantum and classical baths is the factor

$$\exp\left[\frac{\beta\hbar\omega}{2}\left\{\sum_{j''=3}^{n}(k_{j''} - k'_{j''})\right\}\right] \approx 1 + \frac{\beta\hbar\omega}{2}\sum_{j''=3}^{n}(k_{j''} - k'_{j''}) \tag{7.40}$$

which appears only for the former. At finite temperature, the difference $\langle I^n \rangle_{quantum} - \langle I^n \rangle_{cl}$ can be negative or positive depending upon the conditions. However, it is interesting that, for $\beta \to 0$, the fluctuations in the quantum bath become smaller than in the classical one. This reveals that quantum uncertainty in the bath works to reduce the fluctuations of the transition probability of the system.

4. *Low-Temperature Behavior*

Direct expansions of Eqs. (7.22), (7.27), (7.29), and (7.30) provide the low-temperature behavior of the moments.

The abbreviations

$$P = \frac{1}{4}D \tag{7.41}$$

$$Q = P\exp\left(\frac{\beta\hbar\omega}{2}\right) \tag{7.42}$$

$$P' = \frac{1}{4}D' \tag{7.43}$$

are used, where P and P' are of the order $\exp(-(\beta\hbar\omega/2))$ and $1/\beta$, respectively, with respect to temperature, while Q is the order of unity. The nth order moment for the quantum bath may be written as

$$\begin{aligned}\langle I^n\rangle_{quantum} &= n!\langle I\rangle^n_{quantum}\Big\{1 + {}_nC_2P^2H_2^2 + {}_nC_3P^2QH_2^3 + {}_nC_4P^2Q^2H_2^4 \\ &\quad + \cdots + \frac{{}_nC_2}{2!2!}P^4H_4^2 + \frac{{}_nC_3}{2!2!2!}P^4Q^2H_4^3 \\ &\quad + \left(\frac{2}{2!}+2\right){}_nC_3P^4H_4H_2^2 + \frac{3}{2!}{}_nC_3P^4QH_4^2H_2 \\ &\quad + \cdots + (3+2\cdot 2!)_nC_4P^4H_2^4 \\ &\quad + (5+2\cdot 2!)_nC_4P^4QH_4H_2^3 + \cdots + o(P^6)\Big\} \\ &= n!\langle I\rangle^n_{quantum}\left[1 + \frac{1}{Q^2}\{(1+QH_2)^n - 1 - nQH_2\} + o(e^{-2\beta\hbar\omega})\right]\end{aligned} \tag{7.44}$$

and for the classical bath

$$\begin{aligned}\langle I^n\rangle_{cl} &= n!\langle I\rangle^n_{cl}\Big\{1 + {}_nC_2P'^2H_2^2 + 2{}_nC_3P'^3H_2^3 + \frac{{}_nC_2}{2!2!}P'^4H_4^2 \\ &\quad + (3+2\cdot 2!+2)_nC_4P'^4H_2^4 \\ &\quad + \left(\frac{2}{2!}+2\right){}_nC_3P'^4H_4H_2^2 + \cdots + o(P'^5)\Big\} \\ &= n!\langle I\rangle^n_{cl}\left\{1 +_n C_2\left(\frac{R}{2\beta\hbar\omega L'}\right)^2 + o\left(\frac{R^3}{\beta^3}\right)\right\}\end{aligned} \tag{7.45}$$

where H_2 and H_4 are $H_2(E)$ and $H_4(E)$ or $H_2(E')$ and $H_4(E')$, respectively. It should be noted that odd powers of P are not found for the quantum bath. In contrast to this, all powers appear in the expansion for the classical bath. In spite of the disappearance of the odd power, the form of the former looks much more complicated than that of the latter. Furthermore, it is interesting that factors such as ${}_nC_0=1$, ${}_nC_2=(1/2)n(n-1)$, ${}_nC_3=(1/6)n(n-1)(n-2)$, and so on, are found in each term in both expansions, although a factor ${}_nC_1=n$ does not appear here.

For the classical bath, the same expansion may be done for small R. This includes an important case where interaction between the system and the bath is weak—that is, small C.

It is interesting that the statistics of the fluctuation of quantum system interacting with quantum bath is different from those for wave propagation in classical random media. It becomes just the same in the classical bath limit at low temperature or with weak interaction.

5. A Numerical Example

In order to demonstrate the fluctuation, the first and the second moments $\langle I\rangle$ and $\langle I^2\rangle$, respectively, have been calculated for a harmonic oscillator coupled with one harmonic oscillator bath. The parameters $\Omega=\omega=1000\,\mathrm{cm}^{-1}$ and $M=m=7\,\mathrm{g\,mol}^{-1}$ were adopted, mimicking a molecular vibration. A slightly-strong interaction $C=(1/2)\sqrt{(M\Omega^2/2)(m\omega^2/2)}$ was assumed in order to illustrate the phenomena clearly. In the calculation, the system was initially in the ground state. The moments were calculated as a function of position x_f, time t_f, and temperature T both for the quantum and classical baths according to Eqs. (7.17) and (7.22) and Eqs. (7.28) and (7.29), respectively.

Figure 12 shows the temperature dependence of $\langle I\rangle$ and $\langle I^2\rangle^{1/2}$ at $x_f=0$ and $t_f=50\,\mathrm{fs}$. First, for the quantum bath, the value of $\langle I\rangle$ decreases with increasing temperature indicating a broadening of the positional distribution. The second moment also decreases with increasing temperature. However, the rate of the decrease for the second moment is slower than that of the first moment. This results in an increase in the ratio $\langle I^2\rangle/\langle I\rangle^2$ and, thus, the fluctuation as shown in Fig. 13. Second, a comparison of the functions in Figs. 12 and 13 between the quantum and classical baths is very interesting, too. At low temperature, the averaged intensity $\langle I\rangle$ is considerably different for these two baths. With increasing temperature, $\langle I\rangle$ for the quantum bath approaches the one for the classical bath. The quantum and classical baths present almost the same intensity at high temperature (e.g., at 2500 K), where the quantum effect is very small. The second moment shows behavior similar to that of the first moment as shown in Fig. 12. However, a quantum effect on the second moment still remains even at high temperature. This presents a stronger fluctuation of the intensity I for the classical bath than for the quantum bath as shown in Fig. 13. The figure also indicates that, in this model, the fluctuation of the system in contact with the quantum bath is almost Rayleigh at temperatures lower than about 200 K,

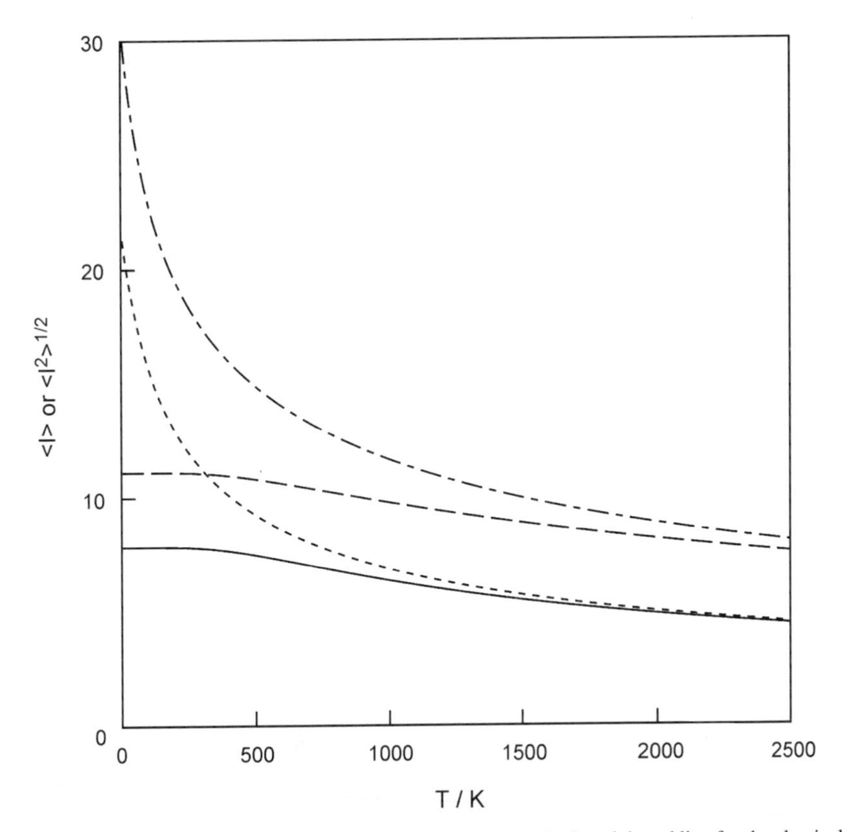

Figure 12. The first moment $\langle I \rangle$ (solid line for the quantum bath and dotted line for the classical bath) and the root mean square of the second moment $\langle I^2 \rangle^{1/2}$ (broken line for the quantum bath and dash–dotted line for the classical bath) as a function of temperature for a harmonic oscillator coupled with one harmonic oscillator bath. $x_f = 0$ and $t_f = 50\,\text{fs}$.

whereas the fluctuation caused by the classical bath shows a very strong deviation from the Rayleigh distribution even at this temperature range.

C. Distribution Function

The statistics of the fluctuations can be easily obtained for some special cases. For a quantum bath at low temperature T [see Eq. (7.44)],

$$\begin{aligned}\langle I^n \rangle &= n!\langle I \rangle^n \{1 - e^{-\beta\hbar\omega}\} + n!\{(1 + const.)\langle I \rangle)\}^n e^{-\beta\hbar\omega} \\ &+ \ const. \times n \times n!\langle I \rangle^n + o(e^{-2\beta\hbar\omega})\end{aligned} \tag{7.46}$$

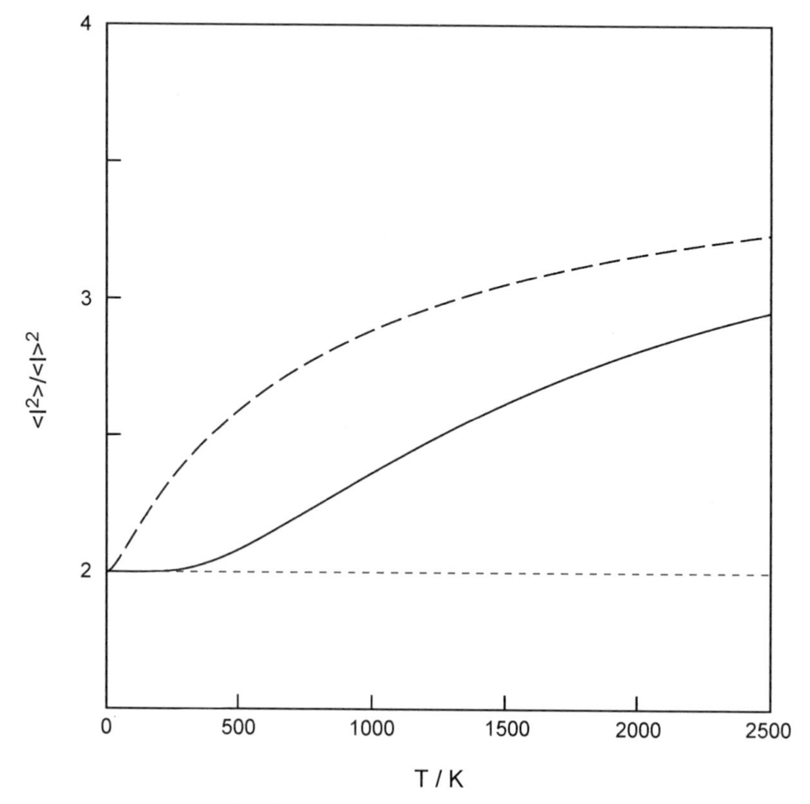

Figure 13. The ratio $\langle I^2\rangle/\langle I\rangle^2$ for a harmonic oscillator coupled with one quantum (solid line) and classical (broken line) harmonic oscillator baths. $x_f = 0$ and $t_f = 50$ fs.

This gives a superposition of exponential distribution functions with different arguments. That is,

$$f_{quant}(I) = \frac{e^{-(I/\langle I\rangle)}}{\langle I\rangle}\left[1 - e^{-\beta\hbar\omega}\left\{1 - QH_2\left(1 - \frac{I}{\langle I\rangle}\right)\right\}\right] + \frac{e^{-(I/(1+QH_2)\langle I\rangle)}}{(1+QH_2)\langle I\rangle}e^{-\beta\hbar\omega} + o(e^{-2\beta\hbar\omega}) \tag{7.47}$$

In the case of the low-temperature expansion up to the first order, two terms come from two possible initial bath states—that is, the ground state and the first excited state. For a classical bath—that is, for $\hbar \to 0$ (with respect to the bath!), at low temperature T or with weak interaction C—we have

$$\langle I^n \rangle = n!\langle I \rangle^n \left\{ 1 + \frac{1}{2}n(n-1) \times \frac{C^4}{\beta^2} \times const. + \frac{1}{6}n(n-1)(n-2) \times \frac{C^6}{\beta^3} \times const. + o\left(\frac{C^8}{\beta^4}\right) \right\} \tag{7.48}$$

This is of the same form as that of wave propagation in classical random media [106], as discussed below. The distribution function is composed of Γ-distributions with one common exponential function $e^{-(I/\langle I \rangle)}$. That is,

$$f_{cl}(I) = \frac{e^{-(I/\langle I \rangle)}}{\langle I \rangle} \left\{ 1 + \frac{C'}{2}\frac{C^4}{\beta^2}\left(2 - 4\frac{I}{\langle I \rangle} + \frac{I^2}{\langle I \rangle^2}\right) \right\} + o\left(\frac{C^6}{\beta^3}\right) \tag{7.49}$$

where C' is a function of masses, frequencies, and time. It is clear that at $T=0$ (initial bath state is certain) or with $C=0$ (no interaction), we have

$$\langle I^n \rangle = n!\langle I \rangle^n \tag{7.50}$$

so that no interference is found except for the permutation. This gives the Rayleigh distribution.

It is worth noting that if the initial system position is certain—that is, $\psi_i(x_i) = \delta(x_i - x_0)$—there is no longer any interference between paths other than the permutation effect, giving

$$\langle I^n \rangle = \langle \psi(x_f)\psi^*(x_f')\psi(x_f'')\psi^*(x_f''')\cdots \rangle = n!\langle I \rangle^n \tag{7.51}$$

This again results in the Rayleigh distribution. Furthermore, if we are interested in the probability that the system is found at a certain position, $x = x_f$, the factor $n!$ also disappears, giving

$$\langle I^n \rangle = \langle \psi(x_f)\psi^*(x_f)\psi(x_f)\psi^*(x_f)\cdots \rangle = \langle I \rangle^n \tag{7.52}$$

In this case, no fluctuations are found. If, instead, the initial system momentum is certain—that is, $\psi_i(x_i) = const.$—the moments are given by Eq. (7.50) and the distribution is, again, Rayleigh.

D. N-Oscillators Bath

Now, we extend our discussion to the N-oscillator bath. The total Hamiltonian is

$$H = H_S + \sum_{k=1}^{N} \left[\frac{m_k}{2}\dot{q}_k^2 + \frac{m_k\omega_k^2}{2}q_k^2 + C_k q_k x \right] \tag{7.53}$$

A formal description of the influence functional is straightforward in terms of the influence functional for a one-oscillator bath, Eq. (7.8) and Eq. (7.16). That is,

$$F_{(N)}^{(n)} = n! \prod_{k=1}^{N} \frac{F_k^{(n)}}{n!} \tag{7.54}$$

The second moment is given by

$$\begin{aligned} F_{(N)}^{(2)} &= 2!\left[\prod_{k=1}^{N} I_0\left\{\frac{C_k^2\sqrt{A[x(t),x'(t)]A[x''(t),x'''(t)]}}{m_k\omega_k\hbar\sinh(\beta\hbar\omega_k/2)}\right\}\right] \\ &\quad\times \exp\left[-\sum_{k=1}^{N}\frac{C_k^2\coth(\beta\hbar\omega_k/2)}{2m_k\omega_k\hbar}\{A[x(t),x'(t)]+A[x''(t),x'''(t)]\}\right. \\ &\qquad\left.+\frac{iC_k^2}{2m_k\omega_k\hbar}\{B[x(t),x'(t)]+B[x''(t),x'''(t)]\}\right] \\ &= 2!\left[\prod_{k=1}^{N} I_0\right] F_{(N)}^{(1)} F_{(N)}^{(1)} \end{aligned} \tag{7.55}$$

The above equation clearly shows that the bath is represented by much more complicated functional $F^{(2)}$ for the second moments than the usual influence functional of the first moment $F^{(1)}$, the latter being often characterized simply by spectral density $J(\omega) = \sum_{k=1}^{N}(\pi C_k^2/2m_k\omega_k)\delta(\omega-\omega_k)$. In other words, when considering moments of conjugated pairs of wave functions beyond first order, one cannot treat the environments as an effectively averaged bath without taking into account the true complexity of the bath degree of freedom—even if exactly harmonic, their number enters explicitly. In general, the fluctuation or non-self-averaging effects of the bath due to quantum interferences of bath paths as we are going to discuss below does play a crucial role in determining the high-order moments of the wave functions. This is a very important conclusion worth noticing. Since real baths are not truely harmonic at the microscopic level, twinkling phenomena may provide a route to characterize their nature more fully.

It is interesting to examine how the number of bath oscillators affects the moments. One simple example, which does not lose the essence of the physics, is obtained by assuming that each oscillator has the same frequency ω, the same mass m, and the same coupling constant $C_k^2 = C_{eff}^2/N$. The first moment influence phase does not depend upon N, keeping the total coupling constant. Substitution of the parameters into Eq. (7.55) gives the same exponential term irrespective of N, but the prefactor becomes, for large N,

$$
\begin{aligned}
2!\prod_{k=1}^{N} I_0 &\left\{ \frac{C_k^2 \sqrt{A[x(t),x'(t)]A[x''(t),x'''(t)]}}{m_k\omega_k\hbar \sinh(\beta\hbar\omega_k/2)} \right\} \\
&= 2! I_0 \left\{ \frac{C_{eff}^2 \sqrt{A[x(t),x'(t)]A[x''(t),x'''(t)]}}{Nm\omega\hbar \sinh(\beta\hbar\omega/2)} \right\}^N \\
&\approx 2!\left[1 + \frac{C_{eff}^4 A[x(t),x'(t)]A[x''(t),x'''(t)]}{4N\{m\omega\hbar \sinh(\beta\hbar\omega/2)\}^2} + o\left(\frac{1}{N^2}\right)\right]
\end{aligned}
\tag{7.56}
$$

This clearly shows a non-Rayleigh character in the distribution, dependent on $1/N$. For $N \to \infty$, the distribution again becomes Rayleigh; the density of interferences between paths decreases with increasing N as the total coupling is kept the same. Substitution of the assumed scaling relation $C_k^2 \approx C_{eff}^2/N$ into Eq. (7.16) for a general harmonic oscillator bath with spectral density $J(\omega) = \sum_{k=1}^{N}(\pi C_k^2/2m_k\omega_k)\delta(\omega - \omega_k)$ presents a similar N dependence of the moments.

It is instructive to make a comparison between quantum bath moments and that of wave propagation in random media. In the latter case, Dashen [106] presented a general formula for the nth moment of the intensity of light as

$$
\langle I^n \rangle = n!\langle I \rangle^n \left\{ 1 + \frac{1}{2}n(n-1) \times const. + \cdots \right\} \tag{7.57}
$$

In the quantum case, summation $\sum_{pair}$ and $\sum_{triplet}$ in Eq. (7.16) clearly give the factors $(1/2)n(n-1)$ and $(1/6)n(n-1)(n-2)$, respectively. Thus, the nth moment is expected to be of the form

$$
\langle I^n \rangle = n!\langle I \rangle^n \left\{ 1 + \frac{1}{2}n(n-1) \times const. + \frac{1}{6}n(n-1)(n-2) \times const. + \cdots \right\} \tag{7.58}
$$

The distribution of I depends upon the actual values of each coefficient. In some conditions, the statistics of the classical and quantum twinkling coincide, while, in other conditions, they differ.

VIII. CONCLUSION

In this chapter, recent progress in the theoretical approach to the dynamics of vibrational population relaxation, or energy relaxation, has been reviewed in addition to the traditional methods conventionally used so far. In particular, path integral influence functional theory and CMD calculation describe the total system quantum mechanically including the intramolecular degrees of freedom of the solvent, predicting the relaxation time which is in the same order as the experimental one based upon rather primitive potential model. On the other hand, time evolution of the vibrational state and trajectory of the solvent molecules can be pursued simultaneously solving coupled time-dependent Schrödinger equation and classical equations of motion, respectively, assuming Hellmann–Feynman force between them. This mixed quantum-classical approach presents real-time analysis of the coupling between solute and solvent in the solution, although the method gave a longer relaxation time compared with the above fully quantum methods because of the classical approximation for the solvent as well as the neglect of the intramolecular vibrational degrees of freedom of the solvent.

In combination with the normal mode analysis for the solution, the influence functional theory describes the contribution of the solvent degrees of freedom to the relaxation in terms of single-phonon process, two-phonon process, and so on, separately; and each normal mode may be assigned easily to translational, rotational, or vibrational mode according to its transition frequency. The method also gives the contribution from the higher-order perturbations as well as the first-order one corresponding to Fermi's golden rule. Furthermore, classical approximations for the solute and/or solvent may be examined only by taking zero limit of the relevant $\hbar$. Thus, this technique is very promising as a dynamical approach to the vibrational relaxation, although there remain a number of things to do such as examination of the harmonic oscillators bath approximation, evaluation of higher-order nonlinear couplings, efficient calculation for the exact path integration, and choice of instantaneous normal mode or quenched mode. Application of short-lived normal modes in the solution to longer-time dynamics of the vibrational relaxation is also an essential approximation to be tested.

Investigation for the validity of the CMD approximation are now still in progress where dynamics is tested for individual systems. It is a promising method, too, although the calculation is not good at analyzing the coupling of the solvent with the solute along real time.

The mixed quantum-classical method should be used for the analysis of molecular mechanism of the relaxation rather than for the prediction of the relaxation time since, as stated above, the method results in longer relaxation time because of the classical approximation of the solvent. However, the direct pursuit of the vibrational state of the solute as well as the motion of the solvent molecules at a time gives a lot of information of the mechanism. For example, magnitude of the coupling between the solute and solvent as a function of time gives a test for relaxation models proposed so far such as isolated binary collision model in the solution. Statistics of the relaxation itself is very interesting, too.

At present, establishment of the potential model which can effectively describe the interaction between vibrational degree of freedom and the solvent degrees of freedom is very important, although the detailed discussion is not presented here. For example, as sometimes mentioned here, the present potential model employed for the CN^- ion are no more than a primitive one, which caused rather great discrepancies between the calculated relaxation times and the experimental value. However, explicit inclusion of many-body interactions leads to unacceptable difficulty in determining the potential function. Thus, in practice, improvements of the potential model will be done first such as the inclusion of molecular and ionic polarizations by the surrounding molecules and ions [108–110], keeping the simple form of potential function amenable to the computation.

A route to the analysis of the quantum fluctuation has also been presented. An interesting statistics of the probability fluctuation generated by the interference between forward and backward paths of the wave function was demonstrated for a simple model of oscillators. Although an experimental technique is not available, at present, for the measurement of this kind of fluctuation, the physical concept is, at least, very attractive.

Acknowledgments

The author thanks Dr. M. Shiga for his great contribution to the study of vibrational relaxation. He is grateful, too, to the co-workers in his group, Dr. S. Miura, T. Terashima, T. Mikami, and M. Satoh, for their contributions to this project. The author also would like to thank Prof. P. G. Wolynes for his distinguished leadership in the study of quantum twinkling. Contributions from Drs. J. Wang and S. A. Schofield are acknowledged, too. The work was supported in part by Grants-in-Aid for Scientific Research on Priority Area "Chemistry of Small Many-Body Systems," "Molecular Physical Chemistry," and "Nano-Mechanics of Atoms and Molecules" from the Ministry of Education, Science, and Culture, Japan.

References

1. A. Laubereau and W. Kaiser, *Rev. Mod. Phys.* **50**, 607 (1978).
2. E. J. Heilweil, M. P. Casassa, R. R. Cavanagh, and J. C. Stephenson, *Annu. Rev. Phys. Chem.* **40**, 143 (1989).
3. T. Elsaesser and W. Kaiser, *Annu. Rev. Phys. Chem.* **42**, 83 (1991).
4. D. L. Abdrews, ed., *Applied Laser Spectroscopy*, VCH, New York, 1992.
5. G. R. Flemming and P. Hänggi, eds., *Activated Barrier Crossing*, World Scientific, River Edge, NJ, 1993.
6. A. Kaiser, ed., *Ultrashort Laser Pulses*, Springer-Verlag, New York, 1993.
7. J. C. Owrutsky, D. Raftery, and R. M. Hochstrasser, *Annu. Rev. Phys. Chem.* **45**, 519 (1994).
8. E. J. Heilweil, F. E. Doany, R. Moore, and R. M. Hochstrasser, *J. Chem. Phys.* **76**, 5632 (1982).
9. P. Hamm, M. Lim, and R. M. Hochstrasser, *J. Chem. Phys.* **107**, 10523 (1997).

10. M. P. Allen and D. J. Tildesley, *Computer Simulation of Liquids*, Clarendon, Oxford, 1987.
11. G. Ciccotti and W. G. Hoover, eds., *Molecular-Dynamics Simulation of Statistical–Mechanical Systems*, North Holland, Amsterdam, 1986.
12. M. P. Allen and D. J. Tildesley, eds., *Computer Simulation in Chemical Physics*, Kluwer, Dordrecht, 1992.
13. K. Binder, ed., *The Monte Carlo Methods in Condensed Matter Physics*, Springer-Verlag, Berlin, 1992.
14. J. P. Hansen and I. R. McDonald, *Theory of Simple Liquids*, Academic Press, London, 1986.
15. S. A. Adelman and R. H. Stote, *J. Chem. Phys.* **88**, 4397 (1988).
16. M. Tuckerman and B. J. Berne, *J. Chem. Phys.* **98**, 7301 (1993).
17. R. Zwanzig, *J. Chem. Phys.* **34**, 1931 (1961).
18. D. W. Oxtoby, *Adv. Chem. Phys.* **47**, 487 (1981).
19. J. S. Bader and B. J. Berne, *J. Chem. Phys.* **100**, 8359 (1994).
20. S. A. Egorov and B. J. Berne, *J. Chem. Phys.* **107**, 6050 (1997).
21. S. A. Egorov and J. L. Skinner, *J. Chem. Phys.* **105**, 7047 (1996).
22. S. A. Egorov and J. L. Skinner, *Chem. Phys. Lett.* **293**, 469 (1998).
23. K. F. Everitt, S. A. Egorov, and J. L. Skinner, *Chem. Phys.* **235**, 115 (1998).
24. K. F. Everitt and J. L. Skinner, *J. Chem. Phys.* **110**, 4467 (1999).
25. S. A. Egorov, K. F. Everitt, and J. L. Skinner, *J. Phys. Chem.* **A103**, 9494 (1999).
26. R. P. Feynman and F. L. Vernon, *Ann. Phys.* (N.Y.) **24**, 118 (1963).
27. R. P. Feynman and A. R. Hibbs, *Quantum Mechanics and Path Integrals*, McGraw-Hill, New York, 1965.
28. M. Shiga and S. Okazaki, *J. Chem. Phys.* **109**, 3542 (1998).
29. G. A. Voth, *Adv. Chem. Phys.* **93**, 135 (1996).
30. G. A. Voth, in *Classical and Quantum Dynamics in Condensed Phase Simulations*, B. J. Berne, G. Ciccotti, and D. F. Coker, eds., World Scientific, Singapore, 1998.
31. S. Jang, Y. Pak, and G. A. Voth, *J. Phys. Chem.* **A103**, 10289 (1999).
32. J. C. Tully, in *Classical and Quantum Dynamics in Condensed Phase Simulations*, B. J. Berne, G. Ciccotti, and D. F. Coker, eds., World Scientific, Singapore, 1998.
33. P. J. Rossky, in *Classical and Quantum Dynamics in Condensed Phase Simulations*, B. J. Berne, G. Ciccotti, and D. F. Coker, eds., World Scientific, Singapore, 1998.
34. D. F. Coker, in *Classical and Quantum Dynamics in Condensed Phase Simulations*, B. J. Berne, G. Ciccotti, and D. F. Coker, eds., World Scientific, Singapore, 1998.
35. F. A. Webster, P. J. Rossky, and R. A. Friesner, *Comp. Phys. Comm.* **63**, 494 (1991).
36. B. Space and D. F. Coker, *J. Chem. Phys.* **94**, 1976 (1991).
37. J. C. Tully and R. K. Preston, *J. Chem. Phys.* **55**, 562 (1971).
38. J. C. Tully, *J. Chem. Phys.* **93**, 1061 (1990).
39. D. S. Sholl and J. C. Tully, *J. Chem. Phys.* **109**, 7702 (1998).
40. R. Rey and J. T. Hynes, *J. Chem. Phys.* **108**, 142 (1998).
41. M. Shiga and S. Okazaki, *Chem. Phys. Lett.* **292**, 431 (1998).
42. M. Shiga and S. Okazaki, *J. Chem. Phys.* **111**, 5390 (1999), **113**, 6451 (2000).
43. T. Terashima, M. Shiga, and S. Okazaki, *J. Chem. Phys.*, in press.

44. S. Okazaki, J. Wang, S. A. Schofield, and P. G. Wolynes, *Chem. Phys.* **222**, 175 (1997).
45. M. Buchner, B. M. Ladanyi, and R. M. Stratt, *J. Chem. Phys.* **97**, 8522 (1992).
46. G. Goodyear and R. M. Stratt, *J. Chem. Phys.* **105**, 10050 (1996).
47. B. M. Ladanyi and R. M. Sratt, *J. Phys. Chem.* **A102**, 1068 (1998).
48. S. Okazaki and I. Okada, *J. Chem. Phys.* **98**, 607 (1993).
49. H. Fukunaga and K. Morokuma, *J. Phys. Chem.* **97**, 59 (1993).
50. J. R. Reimers and R. O. Watts, *Chem. Phys.* **91**, 201 (1984).
51. S. A. J. Frankland and M. Maroncelli, *J. Chem. Phys.* **110**, 1687 (1999).
52. J. K. Brown, C. Harris, and J. C. Tully, *J. Chem. Phys.* **89**, 6687 (1988).
53. I. Benjamin and R. M. Whitnell, *Chem. Phys. Lett.* **204**, 45 (1993).
54. P. B. Visscher and B. L. Holian, *J. Chem. Phys.* **89**, 5128 (1988).
55. D. A. V. Kliner, J. C. Alfano, and P. F. Barbara, *J. Chem. Phys.* **98**, 5375 (1993).
56. R. M. Whitnell, K. R. Wilson, and J. T. Hynes, *J. Phys. Chem.* **94**, 8625 (1990).
57. R. M. Whitnell, K. R. Wilson, and J. T. Hynes, *J. Chem. Phys.* **96**, 5354 (1992).
58. C. Heiderbach, V. S. Vikhrenko, D. Schwarzer, and J. Schroeder, *J. Chem. Phys.* **110**, 5286 (1999).
59. H. Metiu, D. W. Oxtoby, and K. F. Freed, *Phys. Rev.* **A15**, 361 (1977).
60. S. Gnanakaran and R. M. Hochsrasser, *J. Chem. Phys.* **105**, 3486 (1996).
61. B. M. Ladanyi and R. M. Stratt, *J. Chem. Phys.* **111**, 2008 (1999).
62. N. Pugliano, A. Z. Szarka, S. Gnanakaran, M. Triechel, and R. M. Hochstrasser, *J. Chem. Phys.* **103**, 6498 (1995).
63. G. Goodyear and S. C. Tucker, *J. Chem. Phys.* **110**, 3643 (1999).
64. B. J. Cherayil and M. D. Fayer, *J. Chem. Phys.* **107**, 7642 (1997).
65. T. Yamaguchi, Y. Kimura, and N. Hirota, *J. Chem. Phys.* **113**, 2772 (2000), **113**, 4340 (2000).
66. R. E. Larsen, E. F. David, G. Goodyear, and R. M. Stratt, *J. Chem. Phys.* **107**, 524 (1997).
67. G. Goodyear and R. M. Stratt, *J. Chem. Phys.* **107**, 3098 (1997).
68. R. E. Larsen and R. M. Stratt, *J. Chem. Phys.* **110**, 1036 (1999).
69. R. E. Figueirido and R. M. Levy, *J. Chem. Phys.* **97**, 703 (1992).
70. H. Gai and G. A. Voth, *J. Chem. Phys.* **99**, 740 (1993).
71. M. Bruehl and J. T. Hynes, *Chem. Phys.* **175**, 2058 (1993).
72. R. Rey and J. T. Hynes, *J. Chem. Phys.* **104**, 2356 (1996).
73. M. Ferrario, M. L. Klein, and I. R. McDonald, *Chem. Phys. Lett.* **213**, 537 (1993).
74. A. Morita and S. Kato, *J. Chem. Phys.* **109**, 5511 (1998).
75. W. E. Hagston and J. E. Lowther, *Physica.* **70**, 40 (1973).
76. F. Fong, ed., *Radiationless Processes in Molecules and Condensed Phases*, Springer-Verlag, Berlin, 1976.
77. S. H. Lin, ed., *Radiationless Transitions*, Academic Press, New York, 1980.
78. J. Jortner and B. Pullman, eds., *Intramolecular Dynamics*, D. Reidel, Dordrecht, 1982.
79. W. G. Rothschild, *Dynamics of Molecular Liquids*, Wiley, New York, 1984.
80. S. Mukamel, *Principles of Nonlinear Optical Spectroscopy*, Oxford University Press, New York (1995).
81. V. M. Kenkre, A. Tokmakoff, and M. D. Fayer, *J. Chem. Phys.* **101**, 10618 (1994).

82. P. Moore, A. Tokmakoff, T. Keyes, and M. D. Fayer, *J. Chem. Phys.* **103**, 3325 (1995).
83. S. A. Egorov and J. L. Skinner, *J. Chem. Phys.* **103**, 1533 (1995).
84. A. Nitzan and R. J. Silbey, *J. Chem. Phys.* **60**, 4070 (1974).
85. R. Karrlein and H. Grabert, *J. Chem. Phys.* **108**, 4792 (1998).
86. M. F. Herman, *Int. J. Quant. Chem.* **70**, 897 (1998).
87. P. Velev and M. F. Herman, *Chem. Phys.* **240**, 241 (1999).
88. A. Schmidt, *J. Low Temp. Phys.* **49**, 609 (1982).
89. A. O. Caldeira and A. J. Leggett, *Physica.* **A121**, 587 (1983).
90. U. Weiss, *Quantum Dissipative Systems*, World Scientific, Singapore, 1993.
91. B. L. Hu, J. P. Paz, and Y. Zhang, *Phys. Rev.* **D47**, 1576 (1993).
92. J. Schwinger, *J. Math. Phys.* **2**, 407 (1961).
93. L. V. Keldish, *Zh. Eksp. Teor. Fiz.* **47**, 1515 (1964).
94. R. E. Cline, Jr. and P. G. Wolynes, *J. Chem. Phys.* **88**, 4334 (1987).
95. W. H. Louisell, *Quantum Statistical Properties of Radiation*, Wiley, New York, 1973.
96. B. L. Hu, J. P. Paz, and Y. Zhang, *Phys. Rev.* **D45**, 2843 (1992).
97. E. Keszei, S. Nagy, T. H. Murphrey, and P. J. Rossky, *J. Chem. Phys.* **99**, 2004 (1993).
98. A. Pohorille, L. R. Pratt, R. A. Laviolette, M. A. Wilson, and R. D. MacElroy, *J. Chem. Phys.* **87**, 6070 (1987).
99. T. Mikami, M. Shiga, and S. Okazaki, to be published.
100. M. Ferrario, I. R. McDonald, and M. L. Klein, *J. Chem. Phys.* **84**, 3975 (1986).
101. M. Satoh and S. Okazaki, to be published.
102. W. L. Jorgensen, J. Chandrasekhar, J. D. Madura, R. W. Impey, and M. L. Klein, *J. Chem. Phys.* **79**, 926 (1983).
103. J. Lobaugh and G. A. Voth, *J. Chem. Phys.* **106**, 2400 (1997).
104. L. A. Chernov, *Wave Propagation in a Random Medium*, Dover, New York, 1960.
105. J. W. Goodman, *J. Opt. Soc. Am.* **66**, 1145 (1976).
106. R. Dashen, *J. Math. Phys.* **20**, 894 (1979).
107. V. Tartarskii and V. U. Zavorotnyi, *Prog. Opt.* **18**, 204 (1980).
108. L. X. Dang, *J. Chem. Phys.* **97**, 2659 (1992).
109. S. W. Rick, S. J. Stuart, and B. J. Berne, *J. Chem. Phys.* **101**, 6141 (1994).
110. M. C. C. Ribeiro and L. C. J. Almeida, *J. Chem. Phys.* **110**, 11445 (1999).

AUTHOR INDEX

Numbers in parentheses are reference numbers and indicate that the author's work is referred to although his name is not mentioned in the text. Numbers in *italic* show the pages on which the complete references are listed.

SUBJECT INDEX